FOUNDATION SKILLS: PAINTING AND DECORATING, AND MORTAR TRADES

Project Coordinator
Glenn Costin

Contributors
Glenn Costin
Bob Hart
John Martin
Phillip Mitchell
John Rodoreda
Peter Tierney
Rob Young

BUILDING SKILLS SERIES

Foundation Skills: PAINTING and DECORATING, and MORTAR TRADES

TAFE NSW
South Western Sydney Institute

Pearson Australia
(a division of Pearson Australia Group Pty Ltd)
707 Collins Street, Melbourne, Victoria 3008
PO Box 23360, Melbourne, Victoria 8012
www.pearson.com.au
2018 2017 2016
10 9 8 7 6 5

Reproduction and communication for educational purposes
The Australian *Copyright Act 1968* (the Act) allows a maximum of one chapter or 10% of the pages of this work, whichever is the greater, to be reproduced and/or communicated by any educational institution for its educational purposes provided that that educational institution (or the body that administers it) has given a remuneration notice to Copyright Agency Limited (CAL) under the Act. For details of the CAL licence for educational institutions contact Copyright Agency Limited (www.copyright.com.au).

Reproduction and communication for other purposes
Except as permitted under the Act (for example any fair dealing for the purposes of study, research, criticism or review), no part of this book may be reproduced, stored in a retrieval system, communicated or transmitted in any form or by any means without prior written permission. All enquiries should be made to the publisher at the address above.

This book is not to be treated as a blackline master; that is, any photocopying beyond fair dealing requires prior written permission.

Acquisitions Editor: Andrew Brock
Project Editor: Catherine du Peloux Menagé
Content Editor: Helen Doran
Production Controller: Rochelle Deighton
Copy Editor: Julie Ganner
Proofreader: Ron Buck
Copyright and Pictures Editor: Laura Ramsay
Indexer: Frances Paterson
Cover design by Liz Nicholson
Internal design by Pier Vido
Cover photographs from Dreamstime
Typeset by Midland Typesetters, Australia

Printed in Australia by the SOS Print + Media Group

National Library of Australia
Cataloguing-in-Publication Data

National Library of Australia Cataloguing-in-Publication entry
Title:	Foundation skills painting and decorating, and mortar trades. Project coordinator Glenn Costin; contributors: Bob Hart ... [et al.].
ISBN:	9781442527706 (pbk.)
Series:	Building skills series.
Notes:	Other contributors: John Martin, Philip Mitchell, John Rororeda, Peter Tierney, Rob Young. Includes bibliographical references and index.
Subjects:	Interior decoration—Australia—Handbooks, manuals, etc. Decoration and ornament, Architectural—Australia—Handbooks, manuals, etc. Mortar—Handbooks manuals, etc. Construction industry.
Contributors:	Costin, Glenn. Hart, Bob.
Dewey Number:	690.0994

Every effort has been made to trace and acknowledge copyright. However, should any infringement have occurred, the publishers tender their apologies and invite copyright owners to contact them.

Pearson Australia Group Pty Ltd   ABN 40 004 245 943

CONTENTS

Foreword	vii
Acknowledgements	ix
Preface	xi
List of figures	xiii

Chapter 1 Occupational health and safety — 1

OH&S introduction and responsibilities	1
Safe work practices	7
Housekeeping and workplace maintenance	13
Personal protective equipment (PPE) and clothing	16
Guards for tools and equipment	21
Chemical hazards in the construction industry	21
Safety signs and tags	25
Risk assessment	29
Management of site hazards	33
Accident reporting	35
Emergency procedures	40
Worksheets	49
References and further reading	68

Chapter 2 Working effectively in the general construction industry — 70

Historical industry background	71
The role of employers, employees and workplace committees	78
Workplace structure	86
Time management	89
Sequencing of major building activities	92
Basic quality concepts	94
Work in a team	98
Personal development needs	99
Workplace meetings	100
Sustainability—resource efficiency and waste minimisation	102
Sustainable housing	106
Worksheets	109
References and further reading	117

Chapter 3 Planning and organising work — 118

Plan and organise your work	118
Prepare a safe site	121
Determine tools and materials	122
Carry out the task correctly	126
Environmental protection requirements	129
Worksheets	131
References and further reading	139

Chapter 4 Workplace communication — 140

What is communication?	140
The communication process	142
Ensure you get the message	145
Workplace signage	148
The art of clear communication	152
On-site meeting processes	155
Worksheets	159
References and further reading	171

Chapter 5 Basic measuring and calculations — 172

Correct units of measure	172
Basic measuring tools	173
Quantities and costs	174
Measurement, calculations and quantities	175
Calculation of various solid shapes	183
Calculation of percentages	185
Ratios	186
Worksheets	187
References and further reading	196

Chapter 6 Plan interpretation and specifications — 197

Introduction to plan and specification reading	198
Key users of drawings	198
Drawings and their functions	199

Plan and document reading	209	Worksheets	275
Issues for painters, tilers and plasterers	211	References and further reading	287
Residential building structure types	214		
Paint, tile and plaster substrates	220		
Environmental controls	220		
Worksheets	225		
References and further reading	233		

Chapter 8 Hand tools, plant and equipment — 288

Introduction to tools	288
Basic hand tools	289
Trade-specific hand tools	302
Plant and equipment	316
Your hands, eyes and ears: The forgotten tools	331
Worksheets	333
References and further reading	345

Chapter 7 Levelling procedures — 234

Introduction to levelling procedures	235
The tool list	238
Basic levelling procedures	244
Laser line generators	248
Other levelling procedures	251
Checking your levelling equipment	269
Cleaning and storage	274
A closing word	274

Glossary	**346**
Index	**351**
Appendix	**361**

FOREWORD

In Australia, the Building and Construction industry provides employment across a wide range of vocations and services. The industry is divided into three main categories that include residential, commercial and multi-storey construction. It is one of the most diverse single industries and contributes a large percentage to Australia's annual GDP, helping to make this country competitive on the global stage.

This complex industry is made up of many discrete trade areas, which includes those dealt with in this text such as bricklaying, solid plastering, wall and ceiling lining, wall and floor tiling, as well as painting and decorating. This text is designed to meet the needs of the latest national Training Package (CPC08) by providing information and activities that reflect basic vocational and employability skills aligned to painting and decorating and the general mortar trades. The knowledge and skill derived from this text will provide a strong foundation for future learning and will prepare new workers for a long and rewarding career in the industry.

I thank Glenn Costin from Riverina Institute who has contributed his time and expertise to ensure that this text is aligned to the current Training Package and that it provides the essential underpinning information to enhance the knowledge and skill of workers in the painting and decorating and mortar trades.

Rob Young
Assistant Faculty Director–Building
and Construction
TAFE NSW–South Western Sydney Institute

ACKNOWLEDGEMENTS

Acknowledgement is due to the following for permission to reproduce copyright material:

Cover images
Dreamstime
© Aleksander Jovanovic/Dreamstime.com
© Billyfoto/Dreamstime.com
© Arcobaleno/Dreamstime.com

Chapter 1

Figure 1.1	Master Builders Association of Victoria.
Figure 1.2	WorkSafe Victoria.
Figure 1.3	Workplace Health and Safety Queensland.
Figure 1.4	Courtesy SafeWorkSA, Department of the Premier and Cabinet, South Australia.
Figure 1.5	WorkCover ACT.
Figure 1.17	Photo courtesy of Beaver Brands Pty Ltd.
Figures 1.21, 1.23	Product shot provided by Paramount Safety Products.
Figures 1.22, 1.25-1.28	Cigweld Safety Products and Thermadyne.
Figure 1.24	Protector Safety Supply Ltd.
Figure 1.31	Lightning Cleans Pty Ltd.
Figure 1.34	www.chemalert.com.au
Figure 1.35	SAI Global.
Figures 1.36-1.57	SAI Global.
Figure 1.59	Courtesy of WorkCover NSW. Hazard/Incident Report Form (Catalogue No. WCO4921).
Figure 1.60	WorkCover NSW.
Figure 1.61	Hazard/Incident Report Form. Source www.worksafe.vic.gov.au.
Figure 1.63–1.64	Product shots provided by Fast Aid.
Figure 1.75	PSA Products Pty Ltd; LIFESAVER.
Figure 1.76	Australian Competition and Consumer Commission.

Chapter 2

Figure 2.1	Pearson Education Inc, Upper Saddle River, NJ for from Maslow, Abraham I.L., Frager, Robert D., and Daidman James, *Motivation and Personality*, 3rd edition, © 1997. Reprinted with permission.
Figure 2.2	Dreamstime.
Figure 2.9	© Subsurface/Dreamstime.
Figure 2.10	Dreamstime.
Figure 2.11	Dreamstime.
Figure 2.12	Mitchell Library, State Library of NSW Pty Ltd.
Figure 2.13	Elizabeth Farm, A property of the Historic House Trust of New South Wales.
Figure 2.16	Construction Training Australia.

Chapter 3

Figure 3.8	© Eugenesergee/Dreamstime

Chapter 4

Figures 4.16-4.22 Standards Australia International.

Chapter 6

Figures 6.3–6.11	TAFE Construction & Transport ESD, for the use of the graphics.
Figure 6.23	NSW Government Printing Service, for extract from *Standard drawing, symbols, abbreviation, graphical representation*.

Chapter 7

Figure 7.14, 7.15 With kind permission from Apex Tool Group

Figure 7.20, 7.21	With kind permission from Apex Tool Group	Figure 8.87	Kennard's Hire.
Figure 7.23, 7.25	With kind permission from Apex Tool Group	Figure 8.89	Sydney Electrical Appliance Testing, Andrew Hardingham.
Figure 7.29-7.30	With kind permission from Apex Tool Group	Figure 8.97	Department of Employment and Youth Affairs, *Carpentry and Joinery 4* (portable power tools), BTM 14.4, Canberra, AGPS, 1981, Commonwealth of Australia copyright reproduced by permission.
Figure 7.32	With kind permission from Apex Tool Group		

Chapter 8

Figure 8.26	With kind permission from Apex Tool Group.	Figure 8.100	Hanson Construction Materials.
Figure 8.70	Department of Employment and Industrial Relations, *Painting and Decorating: painting equipment, applicators and scaffolding*, BTM 15.2, Canberra, AGPS, 1985, Commonwealth of Australia copyright reproduced by permission.	Figure 8.107	© Eric Inghels/Dreamstime.com.
		Table 2.1	National Training Reform Agenda, ACTU.
Figure 8.71	Dreamstime.		
Figure 8.86	Castle Hill Skills Centre.		

While every effort has been made to trace and acknowledge copyright, in some cases copyright proved untraceable. Should any infringement have occurred, the publishers tender their apologies and invite copyright owners to contact them.

PREFACE

Foundation Skills: Painting and decorating, and mortar trades is an acknowledgment of the need for a more targeted basic skills text for the wet area trades—specifically bricklaying, plastering, tiling, and painting and decorating. The book is adapted from *Basic Building and Construction Skills*, 4th edition. These four trades remain, however, a diverse group. As such, not all elements of the text will have relevance to each and every one of these skill areas. This is unavoidable. What we believe has been achieved, however, through the engaging of various technical experts and advisors, is a work that speaks the language of, and offers insights into, the basic skills of each of these trades. Working on the basis that there is seldom a skill not worth developing, there is then little in this text that will not be invaluable to the aspiring tradesperson, irrespective of their particular field of employment. The text is therefore offered as a significant resource for both teachers and students alike.

In adapting this work for the wet trades, the first four chapters have not significantly changed aside from providing more relevant examples and figures, or altering the text slightly to reflect currency for these particular industries. Chapters 5 and 6, however, have been altered significantly so as to better reflect each trade. In addition, a specific section has been added that deals with issues relevant to painters, tilers and plasterers in particular. Chapters 7 and 8 have been completely rewritten, offering not only more relevance (it is hoped), but more currency.

A work of this nature, covering as it does a raft of skill areas for which no one author could claim expertise, required the input of a number of technical specialists. In some cases this may have been a simple nod that the text was OK, but in others it involved a significant amount of time demonstrating the tools and/or explaining a particular technique. Thus the hands in the photographs are those of tradespeople, and I offer my heartfelt thanks to these individuals, particularly, in acknowledging the following:

Bricklaying	Andrew Tutt (NSW/Vic)
	Peter Gossi (WA)
	Glen Kilpatrick (NSW)
Painting and decorating	Neil Griffiths (NSW)
Plastering	Kevin Hartwig (NSW)
	Darren Johan (NSW/Vic)
Tiling	Llewellyn Biggar (NSW)

Dr Glenn P. Costin
Yackandandah
September 2010

LIST OF FIGURES

1.1	Red card sample as issued in Victoria	6
1.2	Construction induction card sample as issued in Victoria	6
1.3	White card sample as issued in Queensland	6
1.4	White card sample as issued in South Australia	7
1.5	Typical Code of Practice	7
1.6	The human spine	8
1.7	Lifting correctly	9
1.8	Apply push or pull force at around waist level	10
1.9	Correct posture when using a long-handled shovel	10
1.10	Correct posture when using a short-handled shovel	11
1.11	Manhole cover lifter	11
1.12	Multi-wheeled skate	11
1.13	A typical wheelbarrow	11
1.14	Wheelset	11
1.15	Hand trolley	12
1.16	Sheet lifter	12
1.17	Chain block pulley system	12
1.18	Forklift	12
1.19	Suction grip for lifting	12
1.20	Carry grip	13
1.21	Safety cap with detachable accessories	17
1.22	Full welding helmet	17
1.23	Fabric sun brim accessory for a safety cap	17
1.24	Clear wide-vision goggles	18
1.25	Clear-framed spectacles	18
1.26	Face shield	18
1.27	Hearing protection	18
1.28	Folding P1 dust mask with vent	19
1.29	Half-face respirator with P2 dust filters fitted	19
1.30	Gloves	19
1.31	Barrier cream	20
1.32	Foot protection	20
1.33	Circular saw fitted with a guard	21
1.34	Page 1 of a sample MSDS	23
1.35	Examples of dangerous goods labels	24
1.36	Picture signs—smoking prohibited	26
1.37	Word-only messages	26
1.38	Combined picture and word signs	26
1.39	Digging prohibited	27
1.40	No pedestrian access	27
1.41	Water not suitable for drinking	27
1.42	Fire, naked flame and smoking prohibited	27
1.43	Eye protection must be worn	27
1.44	Head protection must be worn	27
1.45	Hearing protection must be worn	27
1.46	Face protection must be worn	27
1.47	Restriction sign	27
1.48	Fire risk	28
1.49	Toxic hazard	28
1.50	Electric shock risk	28
1.51	Forklift hazard	28
1.52	Danger signs	28
1.53	First aid	28
1.54	Emergency (safety) eye wash	28
1.55	Fire alarm	28
1.56	Fire hose reel station	28
1.57	Electrical safety signs and tags	29
1.58	Hazard groups	31
1.59	Sample hazard report form	34
1.60	Typical SWMS format	36
1.61	Sample accident report form	37
1.62	Emergency caused by a partial building collapse	40
1.63	Type B first aid kit	42
1.64	Stored contents of the kit	42
1.65	Use a fire extinguisher to control small outbreaks	42
1.66	The elements necessary for a fire	43
1.67	Class A—ordinary combustible materials: wood, paper and cloth	44
1.68	Class B—flammable and combustible materials: paint, oil and petrol	44
1.69	Class C—flammable gases: LPG	44
1.70	Class E—electrical switchboards	45
1.71	Water extinguisher indicator	45
1.72	Foam extinguisher indicator	45
1.73	Dry chemical powder indicator	46
1.74	Carbon dioxide indicator	46
1.75	A typical fire blanket packet	46
1.76	A fire blanket in use	46
1.77	Hose reel	47
1.78	Various signs	56
2.1	Maslow's hierarchy of human needs: shelter is one of our most basic needs	71
2.2	Primitive post and lintel construction as seen at Stonehenge, Salisbury, England, c. 1800–1400 BC	71
2.3	Roman aqueduct lined with lead and supported by stone arches, built in Segovia, Spain nearly 2000 years ago	72
2.4	The Pantheon, Rome, 120–124 AD, showing dome roof	72
2.5	Byzantine dome on a square base	72
2.6	Barrel or tunnel vault and barrel groin vault	72
2.7	Gothic buttresses similar to those found on Notre Dame, 1163–c. 1250	72
2.8	Reinforced concrete house construction by Le Corbusier, 1930–1965	73
2.9	Steel-framed Eiffel Tower stands 300 m tall	73
2.10	Sydney Opera House designed by Danish architect Jørn Utzon and built 1959–1973	73
2.11	Sydney Tower completed in 1981, stands 304.8 m above street level, including 30 m spire	73
2.12	A wattle-and-daub hut typical of those constructed by early settlers	74
2.13	Elizabeth Farm Parramatta, Australia's oldest surviving building c. 1794	75
2.14	Single-storey semi	75
2.15	Chart to show career path options	86
2.16	Chart showing training and career path	87

2.17	Hierarchy chart showing on-site organisation	88
2.18	Typical Gantt or bar chart for a brick veneer cottage on a concrete slab	91
2.19	A bricklaying team in action	98
3.1	Result of poor preparation and planning	119
3.2	Prepare a checklist	120
3.3	Identifying joint type and characteristics	121
3.4	Identifying waterproofing requirements	121
3.5	Preparing the wall frame ready for the brick veneer	121
3.6	Rough-in the services before fitting linings	121
3.7	Typical chainwire fence type	122
3.8	Barricade tapes	122
3.9	Plan and elevations of an ensuite bathroom	123
3.10	East and west elevations of ensuite	124
3.11	Ensuite location	124
3.12	Determine tools required	125
3.13	Determine equipment required	125
3.14	Double-check details before you start	126
3.15	A building is like a huge jigsaw puzzle	127
3.16	Discuss and solve problems as they occur	127
3.17	Clean work area daily	128
3.18	Cleaning tools using a bucket	128
4.1	Communicating by talking	141
4.2	Communicating = creating understanding	141
4.3	The feedback process	143
4.4	Communicating through plans and drawings	143
4.5	Communicating using hand gestures and facial expression	144
4.6	Communication using pictograms	144
4.7	Communication through warning siren or bell	144
4.8	Communication using rotating amber beacons on machinery	145
4.9	Communication through touch	145
4.10	Communication through smell	145
4.11	Record accurate messages	146
4.12	Communicate using hand signals	147
4.13	Communicate using hand or body gestures	147
4.14	Communication using facial expressions	147
4.15	Communication through body posture	147
4.16	Prohibition (don't do) signs	148
4.17	Mandatory (must do) signs	148
4.18	Hazard warning signs	149
4.19	Danger hazard signs	149
4.20	Emergency information signs	149
4.21	Fire signs	150
4.22	Place signs just above average eye-height	150
4.23	Communicating by sketching a detail	152
4.24	Telling others to stop	153
4.25	Telling others to move forwards	154
4.26	Telling others to move away	154
4.27	Telling others to move left or right	154
4.28	Telling others to cease what they are doing	154
4.29	Typical team meeting	155
4.30	Typical example of a meeting agenda	156
4.31	Typical examples of meeting documents	157
4.32	An example of an informal meeting in progress	158
5.1	Accurate measuring and marking is critical	173
5.2	Select the correct measuring tools	173
5.3	A typical scale rule	173
5.4	The four-fold rule	173
5.5	A typical retractable metal tape	173
5.6	Plan for the process	174
5.7	Which unit to use	174
5.8	Trapezoid	178
5.9	Quadrilateral	178
5.10	Irregular polygon	178
5.11	Polygon	178
5.12	Polygon	179
5.13	Liquid volumes	179
5.14	Wall quantities	180
5.15	Single skin of brickwork	183
5.16	The prism	183
5.17	The cylinder	183
5.18	The cone	184
5.19	The pyramid	184
5.20	Surface area formulae	185
5.21	Wall quantities	193
6.1	Plans and specifications	198
6.2	A typical plan	199
6.3	Perspective view	199
6.4	Pictorial representation	199
6.5	Isometric projection	200
6.6	Typical floor plan	200
6.7	Typical details of elevations	201
6.8	Typical section taken through kitchen and living/dining areas	201
6.9	Slab edge detail	201
6.10	Typical site plan	202
6.11	Common reduction scales	203
6.12	Examples of dimension lines	203
6.13	Typical drawing sheet size and format	204
6.14	Typical title block showing required information	204
6.15	Extract from a standard Department of Housing specification	205
6.16	Symbols for sections in-ground	206
6.17	Symbolic representations of windows for elevations	207
6.18	Graphics for use on site plans	207
6.19	Symbolic representation for floor plans and details	207
6.20	Symbolic representation of fixtures and fittings	208
6.21	Symbolic representation of doors for elevation	209
6.22	Symbolic representation for sections (a)	209
6.23	Symbolic representation for sections (b)	209
6.24	Symbolic representation for floor plans and horizontal sections	209
6.25	Symbolic representation for floor plans and horizontal sections	209
6.26	Cavity brick cottage	210
6.27	Plan of a contemporary ensuite	212
6.28	A contemporary ensuite	212
6.29	A designer's image of the ensuite	212
6.30	Sketch plan of ensuite	213
6.31	Simple fold out of ensuite	213
6.32	Fold out of ensuite ready for client approval and costing	214
6.33	Typical detail of a reinforced concrete strip footing	214
6.34	Slab edge detail in stable soil	215
6.35	Typical waffle-pod slab system	215
6.36	Slab-on-ground for masonry, veneer and clad frames	215
6.37	Typical blob/pad footing with a minimum depth of 200 mm	215

6.38	Vertical section through external walls of brick veneer and timber-frame construction	216	
6.39	Suspended concrete floor with balcony projection at first floor level	217	
6.40	Vertical section through an external timber frame wall and timber awning window	217	
6.41	Vertical section through an external brick veneer wall and horizontally sliding aluminium window	218	
6.42	Vertical section through an external cavity brick wall and timber door jamb	218	
6.43	Vertical section through a conventional roof	219	
6.44	Detail A and B from Figure 6.43	219	
6.45	A simple truss showing all members	219	
6.46	Truss roof layout showing erection sequence	219	
6.47	Rubble drain	221	
6.48	Porous pipe encased in stone rubble	221	
6.49	Typical available agricultural pipes	221	
6.50	Building site describing surface water drainage and soil retention	222	
6.51	Plan of ground floor of building	228	
6.52	Plan	230	
7.1	Square or perpendicular	235	
7.2	Checking for 'square'	235	
7.3	A 'straight and true' corner	235	
7.4	Plumb	236	
7.5	In wind	236	
7.6	Out of wind	236	
7.7	Checking tiles with entrance corner for wind	237	
7.8	Tiles laid 'square' to bath	237	
7.9	Perpendicular to stair stringer	237	
7.10	Straight edge (with spirit level)	238	
7.11	Level bubble	238	
7.12	Digital level	239	
7.13	Line level	239	
7.14	String lines	239	
7.15	Chalk line	240	
7.16	Plumb bobs and frame	240	
7.17	Flat-headed tripod	240	
7.18	Domed tripod	241	
7.19	Four-fold rule	241	
7.20	Retractable tape measure—8 m	241	
7.21	Wind-up tape measure—100 m	241	
7.22	Manual laser	242	
7.23	Pendulum auto-levelling laser	242	
7.24	Self-levelling laser	242	
7.25	A multi-line laser line generator	243	
7.26	An automatic optical level or 'dumpy'	243	
7.27	Survey staff	244	
7.28	The water level in use	244	
7.29	Checking a level by 'end-for-ending'	245	
7.30	An end-for-end check on a surface: Level is accurate, surface is out	245	
7.31	An end-for-end check on a surface: Level is out, surface is level	245	
7.32	End-for-ending a spirit level to draw a long line	246	
7.33	Centring an arch—plumb bob	246	
7.34	Plumbing a wall—plumb bob in parallel	247	
7.35	Plumbing a wall—sighting a plumb bob	247	
7.36	Basic traversing	247	
7.37	Transferring heights with a spirit level and straight edge	248	
7.38	Laser line generator: Horizontal line at dado height for wall papering	248	
7.39	Laser line generator: Crosshair mode	249	
7.40	Plumbing a ceiling point to the floor	249	
7.41	Laser warning labelling	250	
7.42	Laser warning signs	250	
7.43	Types of benchmarks	252	
7.44	Taking a reading: Optical survey	252	
7.45	Line of collimation through an optical level	253	
7.46	Taking a reading: Laser receiver	253	
7.47	Line of collimation, projecting like a large, flat disc	253	
7.48	Finding the difference in height between two points using a laser level	254	
7.49	Maintaining or plotting a continuous height around a site	255	
7.50	Maintaining or plotting a height relative to a datum	256	
7.51	Ceiling being levelled in by laser	256	
7.52	Taking a reading on channelling	257	
7.53	Tribrach screws	257	
7.54	Blister level and mirror	258	
7.55	Imagining a 'T' over the screws	258	
7.56	Adjusting the first two screws for level	258	
7.57	Adjusting the last screw for level	259	
7.58	Lens and focusing array of an optical level	259	
7.59	Eyepiece focused in front of reticle	259	
7.60	Eyepiece focused behind reticle	260	
7.61	Survey staff increments	260	
7.62	Reading the staff	260	
7.63	Rise and fall booking sheet (column headings)	261	
7.64	Rise and fall	261	
7.65	Backsights, foresights and reduced levels	262	
7.66	Backsight, immediate sights and foresight	262	
7.67	Line of sight from Station 1	263	
7.68	The survey	263	
7.69	Station 1 backsight	264	
7.70	Booking the backsight	264	
7.71	Station 1 foresight	265	
7.72	Rise and fall booking sheet (column headings)	265	
7.73	Station 2 backsight—Peg NW	266	
7.74	Entering the backsight from Station 2	266	
7.75	Sighting the remaining pegs	267	
7.76	Entering the rise and fall developed from each reading	267	
7.77	Entering the rise and fall developed from each reading	268	
7.78	Entering the calculated reduced level for each peg	268	
7.79	Checking the booked results	269	
7.80	Stadia readings	269	
7.81	Setting up the instrument and locating pegs ready for testing	270	
7.82	Establishing the difference in height between the two pegs	270	
7.83	A correctly calibrated instrument establishing a level line height	270	
7.84	Level line height established with an upward-pointing instrument	271	
7.85	Level line height established with a downward-pointing instrument	271	
7.86	Locating Station 2	271	
7.87	Taking readings from Station 2	272	
7.88	The difference between Station 1 and Station 2 readings	272	
7.89	Level sighting up: Difference in heights is unequal	272	
7.90	Level sighting down: Difference in heights is unequal	272	

7.91	Correctly calibrated level: Differences in heights is equal	272
7.92	The four-peg test: Rotary laser levels	273
7.93	Booking the results	273
7.94	Sketch the procedure for finding the height difference	279
7.95	Find the height differences and RLs	283
7.96	Basic survey	284
7.97	Booking sheet	284
8.1	The open hand	290
8.2	Good stance	290
8.3	Good hand positions	291
8.4	Never hold tools like this!	291
8.5	Combination square	291
8.6	Steel square	291
8.7	(a) Compass, (b) trammel head, (c) trammel in use	292
8.8	Mallets: (a) Wooden, (b) plastic tipped, (c) mash	293
8.9	Claw hammer	293
8.10	Warrington or cross pein hammer	293
8.11	Sledge-hammer	293
8.12	Nail punch	293
8.13	Crowbar	294
8.14	Pinch or jemmy bar	294
8.15	Spanners	294
8.16	Sockets, extensions and ratchet	295
8.17	Handsaw	295
8.18	Hacksaw	295
8.19	Keyhole saw	296
8.20	Utility knife	296
8.21	Putty knife	296
8.22	Scrapers	296
8.23	Broad knives	297
8.24	Wood chisel	297
8.25	Bolster	297
8.26	Aviation snips	297
8.27	Bolt cutters	298
8.28	Fibro cutters	298
8.29	Spade, auger, masonry and jobber	299
8.30	Torque screwdriver, Allen key, Phillips (butt) screwdriver, flat (electrical) screwdriver, Phillips screwdriver, flat screwdriver	299
8.31	Multi-grips, side cutters and locking pliers	300
8.32	Sliding cramps in use	300
8.33	(a) Straw broom, (b) yard broom, (c) broad floor broom	301
8.34	Cleaning bin and brush	301
8.35	Wire brush	301
8.36	Shovels: (a) Square mouth, (b) long-handled square mouth, (c) round mouth, (d) long-handled round mouth	302
8.37	Hose and nozzle	302
8.38	Site skip	302
8.39	Rectangular trowel	303
8.40	Adjustable corner trowel	303
8.41	Hawk and knife	303
8.42	Taping knife	303
8.43	Adhesive knife	303
8.44	Small tool	304
8.45	Trowels and floats on display	304
8.46	Loading a tape box	304
8.47	Filling a jointing box	304
8.48	Using a jointing box	305
8.49	Sanding float	305
8.50	Plasterer's square	305
8.51	London-pattern trowel	306
8.52	American-pattern trowel	306
8.53	Gauge trowel	306
8.54	Brick hammer	307
8.55	Plugging chisel	307
8.56	Corner block and profile	307
8.57	Examples of line or mortar pins	307
8.58	Dutch pin	307
8.59	Round jointer	308
8.60	Raker	308
8.61	Profile	308
8.62	Profile clamp	308
8.63	Brick carrier	309
8.64	Just one rack of the many brushes available	309
8.65	Wall brush	309
8.66	Sash cutter	310
8.67	Rat-tailed brush	310
8.68	Oval wall brush	310
8.69	Fitches	310
8.70	Detail roller	311
8.71	Paint tray and roller	311
8.72	Extension handle	312
8.73	Multi-purpose tool in use as a paint scraper	312
8.74	Masking tool	313
8.75	Tile cutter	313
8.76	Tile cutter in use	313
8.77	Straight nipper	314
8.78	Parrot nippers	314
8.79	Applying adhesive with a notched trowel	314
8.80	Notched and gauge trowels	315
8.81	Grout float and scourer	315
8.82	Grout sponge	315
8.83	Suction grips	316
8.84	Caulking gun	316
8.85	Accessing domestic power boards	316
8.86	Typical builder's temporary power pole and board	317
8.87	Portable generator	318
8.88	RCD unit	318
8.89	Tested and tagged lead	319
8.90	Lead stand	319
8.91	Grinder (100 mm) on tile	320
8.92	Angle grinder—9 inch (230 mm)	320
8.93	Typical circular saw components	321
8.94	Bench-mounted wet saw	321
8.97	Cordless drill	322
8.96	Large drill used for mixing	322
8.97	Keyed chuck	322
8.98	Screw gun	323
8.99	Palm sander	323
8.100	Portable compressor	324
8.101	High-pressure low-volume air spray unit	325
8.102	Air spray gun with two feeds: Air and paint	325
8.103	Concrete mixer	326
8.104	Concrete mixer and wheelbarrow	326
8.105	Brick trolley	327
8.106	Brick saw	327
8.107	Industrial vacuum cleaner	328
8.108	Trestles, planks and trailer	329
8.109	Aluminium trestle	329
8.110	Typical extension ladder components	330
8.111	Ladder placement regulations	330
8.112	Stepladder	330
8.113	Scissor lift	331
8.114	The fourth tool is the one that matters	332

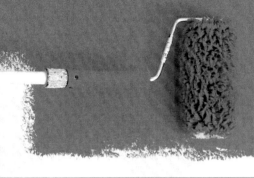

CHAPTER 1
OCCUPATIONAL HEALTH AND SAFETY

This chapter covers the requirements for safe work on a general construction site by outlining the requirements of the relevant state and territory OH&S Acts and Regulations, as they apply to all persons at work.

Areas addressed from the unit of competency include:
- following safe work practices;
- assessing risks; and
- following emergency procedures.

OH&S INTRODUCTION AND RESPONSIBILITIES

The term **occupational health and safety (OH&S) injuries**, making a healthier and safer working environment for all concerned. Building construction workers must be familiar with occupational health and safety requirements as these apply to the building and construction industry, and must understand the responsibilities of the employers and construction workers (employees).

Workplace statistics show that since 1997, approximately 300 Australians die every year from work-related **accidents**, with many more associated deaths from things such as silicosis, due to long-term exposure to dangerous substances such as silica in cement, tiles, stone and adhesives. In addition, almost 1000 persons a day suffer some type of work-related injury or illness. The importance of OH&S is evident in these statistics. Historically, the building industry is one of the most dangerous industries for workers.

The prevention of accidents in industry is the concern not only of experts: all workers must learn how to work without hurting themselves or endangering fellow workers. Every worker's own efforts in keeping the workplace safe, and reporting possible causes of injury, are most important.

Origin of modern occupational health and safety legislation

To understand fully any piece of **legislation** (law), it is important to appreciate the origin and reasons for the law first being introduced.

The industrial workplace in Australia is governed by either federal or state legislation, which varies from state to state and industry to industry. These laws and regulations provide a set of minimum standards of protection for the health and safety of workers.

On the federal level, the National Occupational Health and Safety Commission, known as Worksafe Australia, under the authority of the national *Occupational Health and Safety Act 1985*, assumes responsibility for developing national standards. Worksafe Australia seeks to build cooperation between the three groups involved—governments, business and unions—bringing them together to forge solutions and decide on policy. From this, the states develop their own legislation and policies.

South Australia was the first state to introduce legislation. In 1972 it introduced the *Industrial Safety and Welfare Act*. In 1986 it adopted the *Occupational Health, Safety and Welfare Act*, following the general form of Victorian state legislation.

In NSW the *Occupational Health and Safety Act* was proclaimed in 1983. It was enacted following the Williams Inquiry into health and safety practices in the workplace. The Inquiry was commissioned in 1979 as a result of pressure from trade unions and community groups over serious **hazards** faced by workers in NSW. The new **Act** amended and complements other legislation that previously covered workplace health and safety in NSW under seven separate Acts. In 1987 major changes were made to the Act.

Victoria introduced legislation in 1985 that became the model for the rest of the country. It had far-reaching social and industrial concepts incorporated into the legislation.

Other states and territories to introduce OH&S legislation are:

- Western Australia in 1984, amended in 2008;
- Tasmania in 1977, current Act is 1995;
- Queensland in 1989, current Act is 1995;
- Northern Territory in 1989, current Act is 2007;
- ACT in 1989, amended in 2009.

OH&S Acts

In Australia, states and territories have responsibility for making laws about occupational health and safety (OH&S) and for enforcing those laws. Each state and territory has a principal OH&S Act, setting out requirements for ensuring that workplaces are safe and healthy. These requirements spell out the duties of different groups of people who play a role in workplace health and safety (see Table 1.1).

OH&S Regulations and Codes of Practice

Some workplace hazards have the potential to cause so much injury or disease that specific Regulations or Codes of Practice are warranted. These Regulations and Codes, adopted under state and territory OH&S Acts, explain the duties of particular groups of people in controlling the risks associated with specific hazards. Note that:

- **Regulations** are legally enforceable.
- **Codes of Practice** provide advice on how to meet regulatory requirements. As such, Codes are not legally enforceable, but they can be used in courts as evidence that legal requirements have or have not been met.

Reasons for the introduction of occupational health and safety legislation

Cost of compensation

The total cost of work-related accidents and injuries on a national level was approximately $6 billion per year. Workers' compensation claims had increased by 10% since 1974.

Many workers not covered

Only about one-third of the workforce was covered by any occupational health and safety legislation.

Table 1.1 Relevant Australian state and territory OH&S Acts and Regulations, as of February 2009

State / Territory	Current OH&S Act	Current OH&S Regulation	OH&S Regulating Authority	Web link & contact numbers
WA	*Occupational Safety and Health Act 1984*	Occupational Safety and Health Regulations 1996	WorkSafe WA (a division of WA's Department of Commerce)	<www.slp.wa.gov.au/legislation/agency.nsf/docep_home.htmlx> Contact: 1300 307 877 or 08 9327 8777
Vic	*Occupational Health and Safety Act 2004*	Occupational Health and Safety Regulations 2007	WorkSafe Vic	<www.worksafe.vic.gov.au/wps/wcm/connect/WorkSafe/Home/Laws+and+Regulations> Contact: 1800 136 089 or 03 9641 1444
Qld	*Workplace Health and Safety Act 1995*	Workplace Health and Safety Regulation 2008	Department of Employment and Industrial Relations Qld	<www.deir.qld.gov.au/workplace> Contact: 1300 369 915 or 07 3225 2000
NSW	*Occupational Health and Safety Act 2000*	Occupational Health and Safety Regulation 2001	WorkCover NSW	<www.workcover.nsw.gov.au> Contact: 13 10 50 or 02 4321 5000
SA	*Occupational Health, Safety and Welfare (SafeWork SA) Amendment Act 2005*	Occupational Health, Safety and Welfare Regulations 1995	SafeWork SA	<www.safework.sa.gov.au> Contact: 1300 365 255 or 08 8303 0400
Tas	*Workplace Health and Safety Act 1995*	Workplace Health and Safety Regulations 1998	Workplace Standards Tas	<www.wst.tas.gov.au> Contact: 1300 366 322 or 03 6233 7657
ACT	*Occupational Health and Safety Act 1989*	Occupational Health and Safety (General) Regulation 2007	Department of Justice and Community Safety ACT	<www.ors.act.gov.au/workcover/WebPages/WorkSafe/ohs.htm> Contact: 02 6207 3000 or 02 6205 0200
NT	*Workplace Health and Safety Act 2007*	Workplace Health and Safety Regulations 2007	NT WorkSafe	<www.worksafe.nt.gov.au> Contact: 1800 019 115 or 08 8999 5010

National and overseas legal developments

On a national and international level there were efforts to update existing occupational health and safety legislation, to bring it into line with the 20th-century working environment. The states needed to keep abreast of these changes for social and economic reasons.

Too much occupational health and safety legislation

Although only one-third of the workforce was covered by any OH&S legislation, there were up to 26 different Acts in one state alone relating to occupational health and safety. Enforcement procedures for these Acts created a legal nightmare.

Self-regulation not working

Allowing organisations to regulate their own occupational health and safety programs was not working, at one stage, over 500 people were dying each year on a national basis due to work-related accidents, injuries and diseases.

Rights and responsibilities of employers and employees

In each state there are specific rights and responsibilities for employers and employees under the state's workplace OH&S legislation.

Occupational health and safety Acts aim to protect the health, safety and welfare of people at work. They lay down general requirements that must be met at places of work. All states have similar aims and regulations.

Requirements of state regulations

Each state has its own specific requirements, which may include any of the following.

The employers

Employers must provide for the health, safety and welfare of their employees at work. To do this, employers must:
- provide and maintain equipment and systems of work that are safe and without risks to health;
- make arrangements to ensure the safe use, handling, storage and transport of equipment and substances;
- provide the information, instruction, training and supervision necessary to ensure the health and safety of employees at work;
- maintain places of work under their control in a safe condition and provide and maintain safe entrances and exits;
- make available adequate information about research and relevant tests of substances used at the place of work.

Employers must not require employees to pay for anything done or provided to meet specific requirements made under the Acts or associated legislation. They must also ensure the health and safety of people visiting their places of work who are not employees.

The employees

Employees must take reasonable care of the health and safety of others. Employees must cooperate with employers in their efforts to comply with occupational health and safety requirements.

Employees must not:
- interfere with or misuse any item provided for the health, safety or welfare of persons at work;
- obstruct attempts to give aid or attempts to prevent a serious risk to the health and safety of a person at work;
- refuse a reasonable request to assist in giving aid or preventing a risk to health and safety.

Offences and penalties

Nationally, under the relevant OH&S Acts and Regulations pertaining to each state and territory, there are a number of offences, penalties and infringement systems in place. For example, certain states and territories adopted a system that applies a varying number of penalty units according to the severity of the offence, with each unit having a set monetary value. While the contravention value of offences varies, the underlying principles behind each state and territories infringement system are the same.

The harshness of fines issued is also influenced by whether an individual or a corporation is guilty of the offence and whether or not they are previous offenders. It should also be noted that, as well as the imposing of fines, courts may opt for a sentence of imprisonment; they also have the option to order offenders to undertake any or all of the following:
- to take steps to remedy or restore any matter caused by the offence;
- to pay WorkCover for the costs of the investigation;
- to publicise or notify other persons of the offence; or
- to carry out a project for the general improvement of health and safety.

It is recommended that readers become familiar with their relevant state or territory infringement systems by either contacting their respective OH&S Regulatory Authority, or accessing their website, as detailed in Table 1.1.

Site induction

Outlined within the relevant OH&S Acts and Regulations for Western Australia, Victoria, Queensland, New South Wales and South Australia are the duties and obligations of the principal contractor, employers and self-employed persons, in ensuring that all construction workers have undertaken mandatory **safety induction training**. Non-compliance could invoke a breach of the OH&S Act and subsequent fines may apply.

It is a requirement under the relevant OH&S Acts, enforced by each state's Regulating Authorities (i.e. WorkCover NSW, SafeWork SA, Department of Employment and Industrial Relations QLD, WorkSafe VIC and WorkSafe WA) that all workers carry out OH&S induction training to familiarise themselves with:
- the origins of the modern OH&S legislation;
- the rights and responsibilities of employers and employees in relation to the Act;
- identification of common workplace hazards;
- inspection of a workplace to assess risks;
- identification of quality control measures to control hazards;
- the purpose and use of work method statements;
- identification of essential **PPE (personal protective equipment)**;
- identification of barricades, hoardings and various signage to highlight site hazards and to protect workers.

Statement and proof of induction training

On successful completion and attendance of an OH&S induction training session a worker will be issued with a statement to outline the training and identify the training body, the training assessor and the date of the assessment.

In states where there is compulsory OH&S induction training, after the course each person is provided with a small plastic card.

Although there have been discussions for a national card, at the time of publication, each state of Australia differs. Examples of induction cards that you may come across are shown in Figures 1.1–1.4. They show the person's name, the date training was completed and the name of the group carrying out the training.

There has been a mutual agreement between the applicable state OH&S Regulating Authorities, to acknowledge acceptance for these OH&S Induction Cards from state to state. As long as the training meets existing standards for currency, as per the relevant state's requirements, and the construction worker can provide sufficient evidence that they hold either one of the general OH&S induction cards as shown in Figures 1.1–1.4, or similar, then they will be permitted to carry out work on a construction site without having to undertake the relevant general OH&S induction course pertinent to the state with which they are seeking employment.

The card should be carried on-site at all times and produced on demand for inspection.

Note: A statement of OH&S induction training may cease to be valid when a person has not carried out

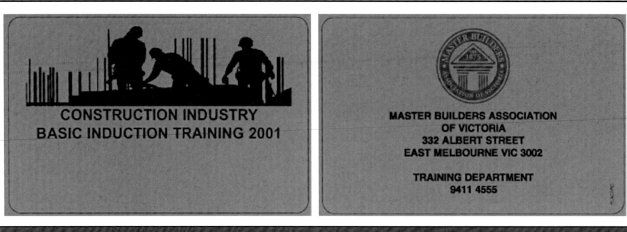

Figure 1.1 Red card sample as issued in Victoria (prior to 1 July 2008)

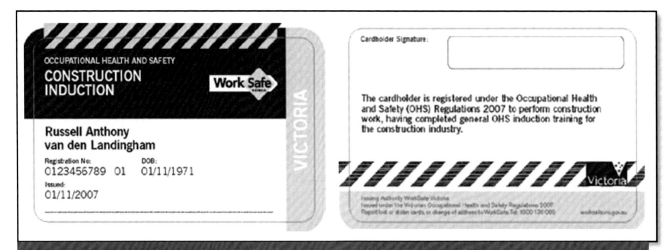

Figure 1.2 Construction induction card sample as issued in Victoria (post 1 July 2008)

Figure 1.3 White card sample as issued in Queensland

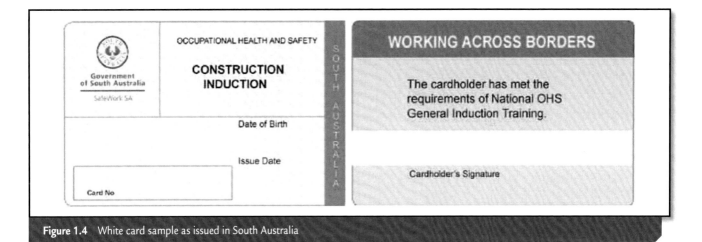

Figure 1.4 White card sample as issued in South Australia

construction work for a specific consecutive period of time. In some states this period may be as little as two years. This means that if you have not carried out construction work for the period stipulated by the applicable state's OH&S Regulating Authority, you must undergo safety induction training again. It is recommended that readers become familiar with their appropriate state's OH&S induction training programs, by either getting in contact with their respective OH&S Regulating Authority, or viewing their website, as detailed in Table 1.1.

What are Codes of Practice?

Codes of Practice are used in conjunction with the Act and Regulations, but they are not classified as law documents. However, while not mandatory, the Codes provide significant guidance and may be used as evidence in a court of law in cases relating to the *Occupational Health and Safety Act*.

The basic purpose of these Codes is to provide workers in the building industry with practical, commonsense, industry-acceptable ways of dealing with the OH&S Act and working safely.

They are put together and published by each state and territory OH&S Regulating Authority, and cover such areas as electrical safety, roof tiling, formworking, personal protective equipment (PPE), use of safety harnesses, construction and use of hoardings. A typical example is shown in Figure 1.5.

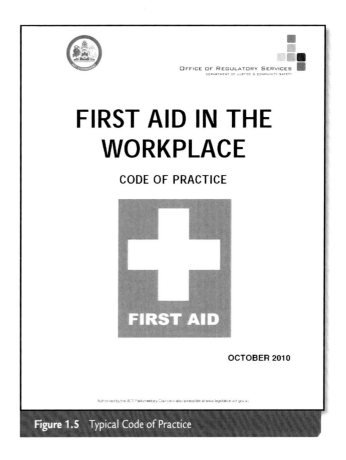

Figure 1.5 Typical Code of Practice

SAFE WORK PRACTICES

Manual handling

Manual handling is an activity requiring a person to use force to lift, lower, push, pull, carry, move or hold any type of object. As manual handling is the most common hazard in the building industry, it is important for all workers to understand, and to be fully trained in, correct manual handling techniques.

In 1990, Worksafe Australia provided standards for manual handling with the introduction of the National Standard for Manual Handling and the National Code of Practice for Manual Handling.

The National Standard aims to prevent injury and reduce serious injuries resulting from manual handling tasks at work. It requires employers, in consultation with their workers, to identify, investigate and control the risks coming from manual handling activities in the workplace.

States may adopt the National Standard and Code for manual handling in their OH&S Regulations. This means that both the Standard and Code will operate as law in that state.

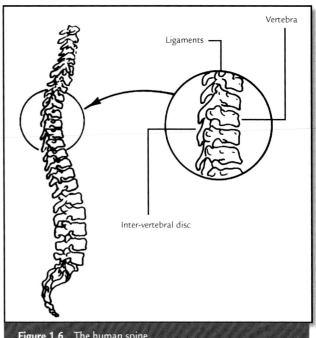

Figure 1.6 The human spine

Causes and effects of bodily injury

Back injuries

Most manual handling injuries are to a person's back. The spine consists of a series of vertebrae, separated by spongy discs or gristle. These discs are called inter-vertebral discs (Figure 1.6). They act as shock absorbers between the vertebrae. If the back is bent or twisted, the discs will be deformed by the vertebrae they support. Severe injuries occur when a load is so great that the disc ruptures (a slipped disc). However, painful injuries can occur without a rupture actually taking place.

Some experts believe that serious back injuries result from damage caused by years of bad practices, rather than from the single lift, twist or other movement that finally causes the injury to become apparent.

Fatigue

Fatigue caused by constant or heavy manual handling tasks can increase the chances of having an accident through loss of concentration.

Muscle injuries

Muscle injuries (musculoskeletal injuries) can be caused by strain to the legs, back, arms and tendons from overuse or by exceeding the capacity of the muscle to carry the load. These injuries may cause inflammation of the joints and surrounding nerves, spinal disc damage and hernias (rupturing of body tissue).

Heart and respiratory disease

Existing medical conditions can be aggravated by bad practices or excessive manual handling. The majority of these injuries are caused by:
- using incorrect techniques for lifting objects;
- being physically unfit;
- not using mechanical aids such as forklifts and conveyors to eliminate the need for manual handling;
- carrying out tasks without enough helpers;
- continuing on the same task for long periods of time;
- having work benches and tables at an unsuitable height for the particular work practice;
- prolonged exposer to dusts and noise.

To help prevent injuries resulting from the lifting and carrying of objects, one should:
- use suitable mechanical equipment whenever possible;
- redesign the task to minimise the risk of injury and eliminate the hazard;

- use the appropriate PPE;
- learn the correct methods of lifting and carrying.

Methods of manual handling

Lifting

Correct lifting methods require you to bend your knees, not your back. Never twist your body when lifting, carrying or moving a load. Protect your hands and feet with suitable PPE.

1. *Size up the load.* Consider the shape and size of the load, as well as the weight. If the load appears too heavy, get assistance.
2. *Position the feet.* Face the intended direction of travel. Place your feet comfortably apart, one foot forward of the other and as close as possible to the object to be lifted.
3. *Obtain a proper hold.* Get a safe, secure grip, diagonally opposite the object, with the whole length of the fingers and part of the palms of your hands.
4. *Maintain bent knees, straight back.* The knees should be bent before the hands are lowered to lift or set down a load. Keep the upper part of your body erect and as straight as possible.
5. *Keep the head erect, chin in.* Keep the head erect and chin in to help keep the back straight. Take a deep breath and begin to raise the load by straightening your legs. Complete the lift with your back held straight.
6. *Keep the arms in.* Keep your arms close to the body. Keep your elbows and knees slightly bent. Hold the load in close to your body. Maintain flexible control over the load with your arm and leg muscles.

Lowering

Setting down the load is the reverse of lifting. It is just as essential to keep the back straight and bend the knees while lowering the load as when lifting it.

Dual lifting

When more than one person is required to lift and carry a load, the correct lifting methods (as shown in Figure 1.7) must be practised, and coordinated team-lifting techniques should be applied.

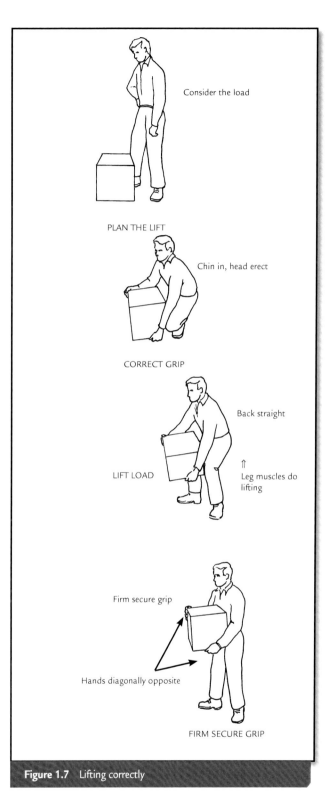

Figure 1.7 Lifting correctly

Coordinating team lifting

- One person only should give the orders and signals, and this person should be able to see what is happening.

- The movements of the team members should be performed simultaneously (all lift together).
- All persons involved in the lift should be able to see or hear the one giving the orders.
- To enable load sharing, lifting partners should be of similar height and build, or lifters should be graded by height along the load.
- Persons should be adequately trained in team lifting and preferably have been trained together.

Pushing and pulling

Tasks requiring the pushing or pulling of loads are more effectively carried out if the force is applied at or around waist level (Figure 1.8). When setting the load in motion, jerky actions should be avoided. Apply the force gradually to avoid overexertion and damage to the body.

Shovelling

The selection of the correct type of shovel for the job is important. In all cases it is essential that the length of the handle is suitable to reduce the strain and exertion on the body (Figure 1.9).

Long-handle shovelling
- Grasp the shovel with hands well apart.
- Place feet apart, one behind the other.

Figure 1.9 Correct posture when using a long-handled shovel

- Bend the forward knee.
- Use your body weight and pressure from the rear leg to drive the shovel forward and under the material.
- Lift the load by pressing down with the rear hand and straightening the front knee.
- Deliver the load by pivoting on the feet, using the front hand as a fulcrum.

Short-handle shovelling (see Figure 1.10)
- Grasp the shovel with hands well apart.
- Place the feet apart, one behind the other; bend both knees.
- Keep the back straight and inclined forward.
- Use your body weight in a forward and downward motion, with pressure from the rear leg; drive the shovel forward and under the material.
- Lift the load by straightening the front leg and back to a vertical position.

Mechanical aids

Mechanical aids reduce the amount of manual handling and the effort required for lifting and carrying tasks undertaken at the workplace. Some aids will give a mechanical advantage to the user to reduce the amount of effort required to carry out the task (e.g. the use of a manhole cover lifter, as in

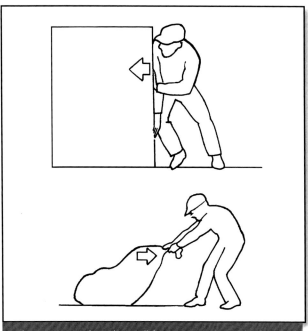

Figure 1.8 Apply push or pull force at around waist level

Figure 1.10 Correct posture when using a short-handled shovel

Figure 1.11 Manhole cover lifter

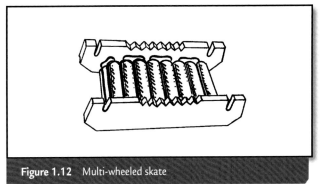

Figure 1.12 Multi-wheeled skate

- *Wheelbarrows*—the most common carrying aid used on building sites, used to cart concrete, bricks, tools etc. over all types of site conditions. They make it easy to negotiate tight situations due to the single wheel (Figure 1.13).

Figure 1.13 A typical wheelbarrow

- *Hand trucks, trolleys and wheelsets*—carrying aids that take most of the weight of the load (see Figures 1.14 and 1.15).
- *Cranes and hoists*—lift heavy loads without the use of manual handling. They may be hand- or power operated (Figure 1.16).
- *Jacks and lifting tackles*—lift heavy loads. Jacks may be hydraulic or mechanical. Lifting tackle may

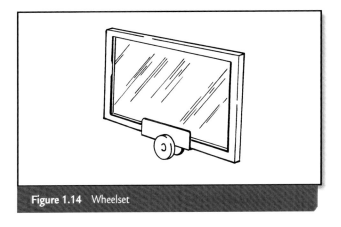

Figure 1.14 Wheelset

Figure 1.11). Some of the mechanical aids available are as follows:
- *Crowbars and levers*—give a mechanical advantage when lifting or moving an object, and can be used to set a heavy object in motion when using rollers or a crowbar to slide a heavy load forward.
- *Rollers*—placed under heavy loads to move them into position. The rollers may be simply pieces of water pipe or round rod, or may be more sophisticated, such as air bags, for use over rough terrain, or multi-wheeled skates to move very heavy loads (Figure 1.12). The larger the diameter of the roller, the easier the object is to move. The path of travel of the load must be cleared of all obstacles before commencing the move.

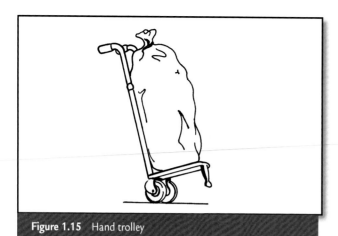

Figure 1.15 Hand trolley

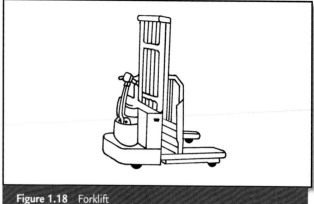

Figure 1.18 Forklift

Figure 1.16 Sheet lifter

be pulley blocks with rope tackle, chain blocks (Figure 1.17) or wire rope tackle.
- *Forklift trucks and pallet trucks*—hand- or power driven. These move large quantities of materials fast and safely. Materials are normally stacked on pallets for ease of handling (Figure 1.18).

- *Lifting grips*—used to allow safe lifting and carrying of awkward materials, e.g. suction grips for handling glass, carry grips for lifting and carrying sheet material (Figures 1.19 and 1.20).

Safe and responsible manual handling
- Compare the correct and incorrect ways of carrying a load in both arms. The worker must walk in an upright position and avoid bending the back either forward or backward.
- Do not carry a heavy load in one hand or under one arm, as this tends to bend the spine sideways. Distribute the load evenly so that the bone structure of the body can support the load without distortion. If this cannot be achieved on your own, get a helper or use carrying aids such as yokes or straps.
- Before attempting to move the load, check the route to be travelled. Make sure that there is

Figure 1.17 Chain block pulley system

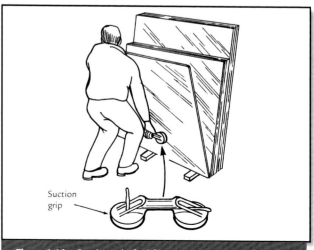

Figure 1.19 Suction grip for lifting

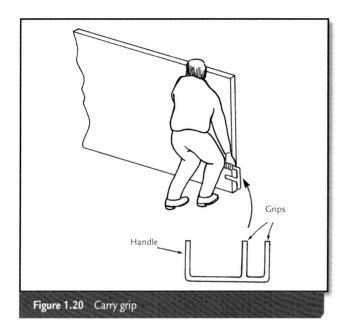

Figure 1.20 Carry grip

nothing in the way on which you could slip or trip; that it is clear of other obstructions; and that there is no overhead danger, or dangers from vehicular traffic.
- Check the area where the load is to be placed for space, and that packers (known as gluts or dunnages) are in place ready for stacking, before commencing to lift and carry the load.
- If supports are to be used to help carry the load, check that they are strong enough and are correctly placed to take the load.
- Dangerous substances or loads that are labelled corrosive or fragile should be handled with proper care.
- When it is necessary to change direction, move your feet and turn the whole body to avoid twisting your spine.
- Avoid manual handling in tight, constrictive positions.
- When carrying loads of separate units, divide the load evenly between both arms.

Clothing

Work clothes should be comfortable and allow freedom of movement. Clothing that is too tight will restrict movement and make safe manual handling more difficult.

Special note

Analysis of Industry Performance—Workplace Injuries may be accessed for workers' compensation statistics by contacting the respective OH&S Regulating Authority, or accessing its website, as detailed in Table 1.1 on p. 3.

HOUSEKEEPING AND WORKPLACE MAINTENANCE

Housekeeping of a building site involves the maintaining of the worksite in a *safe* and *clean* manner. This will improve:
- safety—by maintaining safety standards that will provide safe work areas;
- productivity—by allowing work to proceed faster, improving production times;
- access—by allowing safe access free of hazards to and from the work areas.

The tasks included under the term *housekeeping* involve:
- sorting and stacking of materials and equipment;
- removal of hazards;
- cleaning of work areas;
- disposal of rubbish (non-toxic waste);
- erecting/maintaining safety rails and barricades;
- maintaining safety equipment;
- removal of water hazards.

(**Non-toxic wastes** are all wastes created on a building site that do not produce either a toxic or poisonous health hazard or a toxic threat to the environment. They may, however, cause hazards to workers and the environment in other ways.)

Cleaning of work areas and removal of health hazards must be an ongoing operation on construction sites and other workplaces. It will help maintain a high standard of safety and a healthy environment in which to work. Before commencing the cleaning of a given work area, a planned approach must be formulated.

An inspection of the area should be carried out to determine the extent of the work.

The planned approach should consider:
- first, the removal of all hazards;
- the method of dust suppression to be used;
- designated material storage areas;
- cleaning and rubbish disposal methods to be used;
- use of drop sheets;
- safe access to and from the area;
- a systematic approach to the whole cleaning operation.

Housekeeping functions

- *Sort and stack* reusable and unused materials neatly and in a safe manner. A list should be placed on the stack giving any necessary details, such as the number of items and lengths of materials within the stack. This will avoid the unnecessary unpacking of the stacks by others to find required materials or items of equipment.
- *Remove hazards* that may cause people to trip, slip or be cut. Some of these hazards include broken tiles, bricks, paints, thinners, oil spills, water leakages, broken glass and sharp pieces of materials.
- *Transfer waste materials* to designated waste bins or rubbish stockpile areas. Special areas should be provided for hazardous materials found on the building site, e.g. asbestos, flammable liquids, oxy-acetylene bottles, cleaning materials and solvents. Hazardous materials must be removed or isolated to avoid dangers to workers and delays to work schedules. If left lying about the site they may cause deterioration to material finishes and plant and equipment by staining and corrosion.
- *Safety rails and barricades* should be erected around the edges of floor areas, openings in floors, stairways and trenches. Place and fix safety covers over holes where people could trip or fall.
- *Maintain safety equipment* in good condition so that it is ready to use. This should include cleaning and stocking first aid kits, making sure fire extinguishers are in place and charged, and seeing that safety signs are in place.
- *Electrical leads* should be kept clear of work and access areas by the provision of stands or hooks to keep them above the ground or floor.
- *Water hazards* should be drained, or barricaded off, to eliminate slippery conditions caused by spreading mud over walk areas.
- *PPE* should be worn at all times when carrying out housekeeping functions.
- *Correct lifting techniques* must be used.
- *Dangerous situations* can occur from lack of good housekeeping if:
  - combustible materials are left in areas where welding and grinding are being carried out;
  - spilt liquids are left on walk areas causing slippery conditions;
  - materials are stacked in an unsafe manner;
  - timbers and **formwork** materials are not de-nailed and stacked as they are dismantled;
  - unused materials are not stacked in a safe manner.

Hazard rectification

When inspecting the site prior to commencing cleaning operations, a hazardous or dangerous situation may be found. If the problem cannot be rectified immediately, the following steps should be taken:

1 Barricade off the area, warn others in the area of the danger and erect hazard warning signs.
2 Report the situation to your immediate supervisor or site safety officer.

Tools and equipment

Tools and equipment will be selected to suit the cleanup job to be carried out. In most cases the items required include the following:
- wheelbarrows—to provide a safe method of moving materials and rubbish;
- shovels and brooms—to sweep up and transfer rubbish to containers;
- cleaning equipment—to remove spills and stains;

- rubbish bins—for storage of rubbish until it can be removed from the site;
- rubbish chutes—to allow for a safe and easy method of transferring rubbish to ground level;
- vacuum cleaners—for the safe collection of hazardous dusts;
- pallets and pallet trucks—for the stacking of reusable and unused materials so that they can be moved with safety to the required areas;
- gluts—packers used to keep materials off the ground, as spacers between materials, or to allow space for forklift tines to get underneath. Also known as dunnage;
- ropes—to lift or lower materials or equipment from one level to another and for stabilising and tying loads;
- personal protective equipment—safety equipment for cleaning operations, which may include:
  - safety boots or footwear
  - hard hats
  - safety goggles
  - ear muffs/plugs
  - protective gloves
  - protective clothing
  - respirators
  - dust protective masks.

  Wearing these items of PPE equipment will minimise hazards to health and safety.

At the completion of the cleanup, all tools and equipment should be cleaned and returned to their correct storage places.

Dust suppression

Dusts in the workplace can cause:
- chemical hazards
- respiratory problems
- explosive hazards.

Therefore it is important to reduce to a minimum the amount of dust in the air. This is most important when cleanup operations are taking place, as large volumes of dust can be generated if care is not taken. Dust masks and eye protection may not keep all dust from entering the body tissue. Silica, asbestos, synthetic mineral fibre, cement and wood dusts are of particular concern on building sites.

The three most common methods of dust suppression are:
- *Wetting down.* This is a form of wet sweeping. The area to be swept is sprayed with a fine mist of water to dampen the dust particles before sweeping commences. This dampening of the dust stops it from floating in the air when disturbed. Care must be taken not to cause a hazardous area by the application of excessive amounts of water.
- *Damp sawdust.* This is another form of wet sweeping. Dampened sawdust is spread over the area and, when swept up, the fine dust particles cling to the sawdust, preventing them from floating in the air.
- *Vacuum cleaners.* Vacuuming is a very effective way of collecting hazardous dusts. It is particularly useful for those places that are difficult to reach with a broom. Ordinary household vacuum cleaners will not effectively trap the very fine dust particles, and are prone to clogging after a short time. You may need an industrial-quality vacuum cleaner with a HEPA (high effective particle air) filter. Where there is an explosion hazard, flame-proofed vacuum cleaners must be used. Some industrial-quality vacuum cleaners are of a 'wet and dry' type: these will pick up water, allowing you to wet down the area before commencing vacuuming.

Personal cleaning procedures

Cleaning operations bring workers into contact with many harmful substances and microorganisms that are harmful to a person's health. It is important to maintain personal hygiene at work.

The following minimum standards should be followed by all workers at all times:
- Wash hands and other exposed parts of the body before handling food or drink, before smoking and at the completion of the day's work.

- Wear proper clothing and footwear, which can be removed before leaving the work area, so that hazardous materials will not be spread away from the site. Clothing should be cleaned regularly.
- When working with hazardous dusts it is important to shower before leaving the site.
- Wash hands before leaving the toilet block. Use the soap and towels provided.
- Use rubbish bins provided for the disposal of food scraps.
- Don't spit.
- Apply a barrier cream to exposed areas of skin before handling harmful substances. This will prevent the absorption of the material into the skin and make it easier for you to wash it off.

Safety precautions

When working on-site, think *safety first* at all times:
- Always look up and check what is overhead.
- Do not infringe any safety rules or regulations.
- Take the safest, most direct route from one place to another.
- Keep access routes clear of obstructions.

PERSONAL PROTECTIVE EQUIPMENT (PPE) AND CLOTHING

PPE is the last line of defence to protect your health and safety from workplace hazards. It is the employer's responsibility to provide the PPE, clothing and training to protect the worker. It is the worker's responsibility to wear and look after the equipment provided.

PPE must be appropriate to each particular hazard. It must fit properly, and must be properly cleaned and maintained. It is designed and manufactured to provide protection from a specific hazard to a particular part of the body. No single design can be expected to provide protection from all types of hazards in the workplace.

PPE can be grouped according to the part of the body it will protect:
- *head*—safety helmets, sun hats;
- *eyes/face*—safety spectacles, goggles, face shields;
- *hearing*—ear muffs, ear plugs;
- *airways/lungs*—dust masks, respirators;
- *hands*—gloves, barrier creams;
- *feet*—safety boots and shoes, rubber boots;
- *body*—clothing to protect from sun, cuts, abrasions and burns; high visibility safety garments.

Identify hazards

To decide what PPE and clothing is required, you must first be able to identify the hazards involved. Types of hazards commonly identified where PPE and clothing are a suitable means of protection are:
- **physical hazards**—noise, thermal, vibration, RSI, manual and radiation hazards;
- **chemical hazards**—dusts, fumes, solids, liquids, mists, gases, and vapours.

Once the hazards have been identified, suitable equipment and clothing must be selected to give the maximum protection.

Head protection

Safety helmets

Wearing safety helmets on construction sites may prevent or lessen a head injury from falling or swinging objects, or through striking a stationary object.

Safety helmets must be worn on construction sites when:
- it is possible that a person may be struck on the head by a falling object;
- a person may strike his/her head against a fixed or protruding object;
- accidental head contact may be made with electrical hazards;
- carrying out demolition work;
- instructed by the person in control of the workplace.

Safety helmets must comply with AS/NZS 1801:1997, 'Occupational protective helmets', must carry the AS or AS/NZS label, and must be used in accordance with AS/NZS 1800:1998, 'Occupational protective helmets—Selection, care and use' (Figures 1.21 and 1.22).

When wearing a helmet the harness should be adjusted to allow for stretch on impact. No contact should be made between the skull and the shell of the helmet when subjected to impact.

Sun shades

The risk of skin cancer for building workers is increasing. The neck, ears and face are particularly exposed. Workers should wear sun protection at all times when working outdoors (including winter time).

Sun shades include wide-brimmed hats and foreign legion-style sun shields fixed to the inner liner of safety helmets, or safety helmet 'foreign legion sun brims' (Figure 1.23).

Figure 1.23 Fabric sun brim accessory for a safety cap

Eyes/face protection

The design of eye and face protection is specific to the application. It must conform to AS/NZS 1337:1992, 'Eye protectors for industrial applications'. The hazards to the eyes are of three categories:

- *physical*—dust, flying particles or objects, molten metals;
- *chemicals*—liquid splashes, gases and vapours, dusts;
- *radiation*—sun, laser, welding flash.

The selection of the correct eye protection to protect against multiple hazards on the job is important. Most eyewear is available with a tint for protection against the sun's UV rays, or may have radiation protection included (see Figures 1.24 and 1.25).

Face shields

Face shields give full face protection, as well as eye protection. They are usually worn when carrying out grinding and chipping operations. The shield may come complete with head harness or be supplied for fitting to a safety helmet (Figure 1.26).

Figure 1.21 Safety cap

Figure 1.22 Full welding helmet

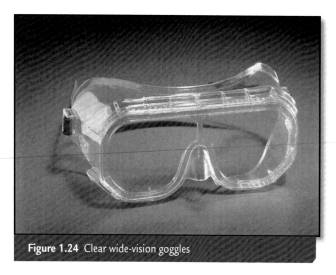
Figure 1.24 Clear wide-vision goggles

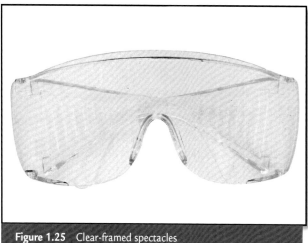
Figure 1.25 Clear-framed spectacles

Goggles are important for use with spray painting and overhead paintrollers. They are also used when power mixing paints and adhesives (Figure 1.24).

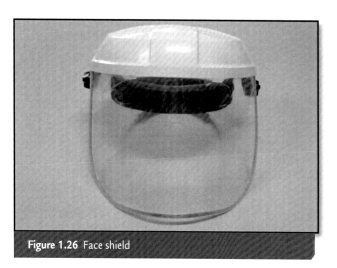
Figure 1.26 Face shield

Hearing protection

You should always wear ear protection in areas where loud or high-frequency noise operations are being carried out, or where there is continuous noise. Always wear protection where you see a 'Hearing protection must be worn' sign, and when you are using or are near noisy power tools.

The two main types of protection available for ears (Figure 1.27) are:
- *ear plugs*—semi- and fully disposable;
- *ear muffs*—available to fit on hard hats where required.

Choose the one that best suits you and conforms to AS/NZS 1270:2002, 'Acoustic—Hearing protectors'.

Airways/lungs

The greater use of mechanical equipment and the use of chemicals for building construction work has increased the need for personal respiratory protection. Breathing contaminated or oxygen-deficient air creates a health hazard that can range from mild discomfort to chronic or acute poisoning, or even death.

The type of cartridge in a filter will define the type of protection it will give. Cartridge types can be identified by the classification ratings from AS/NZS 1716:2003, 'Respiratory protective devices',

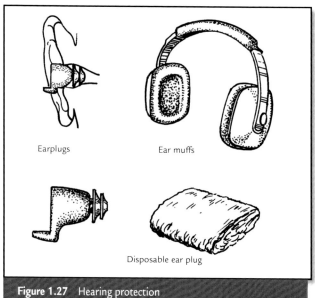
Figure 1.27 Hearing protection

e.g. a P2 class particle dust filter. It is important to learn about the wide range of face masks and respirators for various conditions.

Dust masks

Only high quality dust masks labelled P1 or P2 should be used in the construction industry (Figure 1.28).

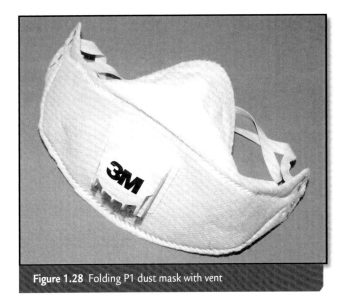

Figure 1.28 Folding P1 dust mask with vent

Respirators

Half-face and full-face respirators can have filters designed to keep out dusts, smoke, metal fumes, mists, fogs, organic vapours, solvent vapours, gases, acids etc. These respirators may contain a combination of dust and gas filters to give full protection.

Respirators fitted with P2 class dust filters (formerly class M; Figure 1.29) are suitable for use with the general low-toxic dusts and welding fumes that are commonly found on construction sites.

Further information on respirators and dust masks should be obtained from the suppliers. It is very important to be trained in the correct methods of selecting, fitting, wearing and cleaning of the equipment in accordance with AS/NZS 1715:1994, 'Selection, use and maintenance of respiratory protective devices'. Respirators and masks must be close-fitting to ensure that all air entering your respiratory passages has been fully filtered.

Hand protection

Hands require protection from both physical and chemical hazards.

Gloves

Gloves can be used to give protection from both physical and chemical hazards. Stout gloves are required when handling sharp or hot materials. Chemical-resistant gloves are used when handling hazardous chemical substances. Gloves should conform to AS/NZS 2161.1:2000, 'Occupational protective gloves—Selection, use and maintenance' (see Figure 1.30).

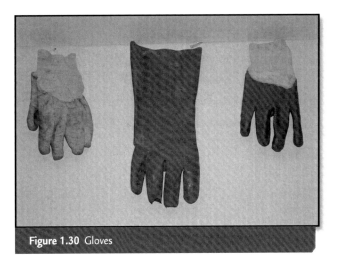

Figure 1.30 Gloves

Creams

Barrier creams may be used when gloves are too restrictive, to protect the hands from the effects of cement and similar hazards (see Figure 1.31).

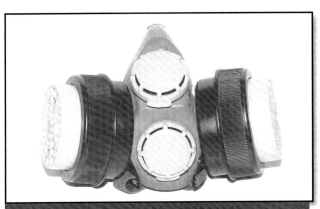

Figure 1.29 Half-face respirator with P2 class dust filters fitted

Figure 1.31 Barrier cream

Figure 1.32 Foot protection

Foot protection

It is mandatory to wear protective footwear at the workplace at all times. *Thongs are not permitted at any time*. Footwear should conform to AS/NZS 2210.1:1994, 'Occupational protective footwear—Guide to selection, care and use'.

All safety footwear must have:
- stout soles or steel midsoles to protect against sharp objects and protruding nails;
- good uppers to protect against sharp tools and materials;
- reinforced toecaps to protect against heavy falling objects.

Safety boots should be worn in preference to safety shoes on construction sites to give ankle support over the rough terrain.

Safety joggers may be required when carrying out roof work or scaffold work; they must have reinforced toecaps.

Rubber boots should be worn when working in wet conditions, in wet concrete, or when working with corrosive chemicals. They must have reinforced toecaps.

See Figure 1.32 for examples of foot protection.

Body protection

Clothing

Good-quality, tough work clothing is appropriate for construction work. It should be kept in good repair and cleaned regularly. If the clothing has been worn when working with hazardous substances it should not be taken home to launder but sent to a commercial cleaning company; this will prevent the hazards from contaminating the home and the environment.

A good fit is important, as loose-fitting clothing is easily caught in machine parts or on protruding objects. Work pants should not have cuffs or patch pockets, as hot materials can lodge in these when worn near welding or cutting operations.

Rings, bracelets and neck chains should not be worn at work.

Clothing should give protection from the sun's UV rays, cuts, abrasions and burns.

Industrial clothing for use in hazardous situations should conform to AS/NZS 4501.2:2006, 'Occupational protective clothing—General requirements'.

Sun protection

Sun protection, in the form of a 15+ minimum and preferably a 30+ sunscreen, should be provided for workers on construction sites. It should be applied on a regular basis to all areas of the body exposed to the sun.

PPE equipment

Points to consider when assessing workplace PPE requirements are:

- *The workplace.* Could it be made safer so that you don't need to use PPE?
- *PPE selection.* Is the PPE designed to provide adequate protection against the hazards at your workplace?
- *PPE comfort and fit.* Is the PPE provided comfortable to wear? Even if the equipment theoretically gives protection, it won't do the job if it doesn't fit properly or is not worn because it is too uncomfortable. For example, close-fitting respirators give protection only if the person is clean-shaven. Those with a beard or 'a few days' growth' will need to use a hood, helmet or visor-type respirator.

Cleaning and maintenance

It is crucial that cleaning and maintenance be carried out on all PPE on a regular basis. This must be done by someone who has been trained in inspecting and maintaining the equipment.

GUARDS FOR TOOLS AND EQUIPMENT

A guard on a power tool, static machine or any equipment having a moving blade is another form of protective equipment. Its purpose is to prevent material and/or material waste from being projected towards the operator, as well as preventing the fingers or hands from being drawn into moving parts or blades. Guards are also used to prevent pieces of shattered blade or abrasive disc from striking the operator when they are faulty or disintegrate due to being jammed.

A guard is fitted as the last line of protection for an operator, and therefore should **never** be removed or tied back while the tool or machine is in use. In fact, the only time the guard should be allowed to move from its safety position is when the tool is in use and it retracts as it is fed into the material or when the power has been disconnected from the tool to allow the blade or disc to be removed.

Tools and equipment with guards

There are many portable hand-held power tools and static machines that have guards fitted; for example:
- circular power saw (Figure 1.33);
- drop or drop/slide saw;
- tile saw;
- angle grinder;
- brick saw;
- mixing attachment for a drill.

Figure 1.33 Circular saw fitted with a guard

CHEMICAL HAZARDS IN THE CONSTRUCTION INDUSTRY

A recent survey of the construction industry in NSW revealed that very large numbers of chemicals are being used on construction sites, and are causing major hazards for the workforce. The possible effects associated with these chemicals are a major threat to construction workers' health and safety. This situation would be similar in all states.

Some of the effects of exposure to chemicals found on these sites are well known, such as dermatitis from cement and epoxy resins. The hazards of other materials are less well known, and the attitude in the industry is that most materials in use, such as adhesives, grouts, mastics and powders, 'are not really chemicals'. This is wrong; the majority of these products are chemicals. This incorrect assumption is causing most workers to use no protection, or very little protection, when using hazardous materials.

The chemical hazards identified include:
- *compressed gases*—e.g. nitrogen;
- *flammable gases*—e.g. acetylene;
- *oxidising gases*—e.g. oxygen;
- *flammable liquids*—e.g. petrol, solvents, thinners and paints;
- *poisonous substances*—e.g. two-pack products and isocyanate-containing paints;
- *harmful substances*—e.g. amine adduct adhesives;
- *irritant substances*—e.g. peroxide catalysts;
- *corrosive substances*—e.g. hydrochloric acid;
- *many other hazards*—e.g. aerosols.

The number of hazardous products that workers are coming into contact with, which were identified in the survey, varies from trade to trade (e.g. builders 64, painters 22, plumbers 12, and tile layers 10). It was found that all construction workers are coming into contact with at least two hazardous products on a regular basis.

Material safety data sheets (MSDS)

Different chemicals can cause different health problems. They also have different safe use requirements. It is important to know about the chemical products used at your workplace.

Material safety data sheets (MSDS) are prepared by the manufacturer of the products used at your workplace (Figure 1.34). They are available from the manufacturer or supplier.

Any hazardous material being delivered to a construction site should have an MSDS for the product provided at the site before it is delivered. These sheets should be kept on-site and should not be more than three years old. If you have not read the MSDS for a substance that you are going to use or will be exposed to when someone else uses it, make sure you obtain a copy and read it first. If any special training is required you should complete the training before the material is handled.

Manufacturers and suppliers may be required by law to provide information 'about any conditions necessary to ensure that the substances will be safe and without risk to health when properly used'; this information is included on the MSDS for that product.

Employers also may be required by law to provide such information, instruction, training and supervision as is necessary to ensure the health and safety of employees.

MSDS give advice on:
- the ingredients of a product;
- the health effects of a product and first aid instructions;
- precautions for use;
- safe handling and storage information.

MSDS will also help you to:
- be aware of any health hazards of a product;
- check that the site emergency equipment and procedures are adequate;
- store the chemicals properly;
- check that a chemical is being used in the right way for the right job;
- decide whether any improvements or changes should be made to machinery or work practices;
- decide whether any environmental monitoring should be done.

When you first view an MSDS, ensure that the following headings (at least) are complete and you understand the information stated:
1 Identification
2 Health hazards
3 Precautions for use
4 Safe handling.

Health hazards should include at least two subheadings:
1 Acute health effects;
2 Chronic health effects.

Acute health effects relate to short-term exposure to the chemical. For example, if you swallow a poison, you

Product Name   **GENERAL PURPOSE CEMENT**

1. IDENTIFICATION OF THE MATERIAL AND SUPPLIER

Supplier Name	BLUE CIRCLE SOUTHERN CEMENT LIMITED
Address	Clunies Ross Street, Prospect, NSW, AUSTRALIA, 2148
Telephone	(02) 9033 4000
Fax	(02) 9033 4055
Emergency	1800 033 111
Website	www.bluecirclesoutherncement.com.au
Synonym(s)	BERRIMA SL • GENERAL PURPOSE CEMENT • GP CEMENT • HE CEMENT • KOORAGANG GP • LOW HEAT PORTLAND CEMENT • MALDON GP • OFF WHITE CEMENT • PORTLAND CEMENT • SHRINKAGE LIMITED CEMENT • SL CEMENT • SOUTHERN WHITE CEMENT • TYPE GP • TYPE HE • TYPE SL • WINTERMENT
Use(s)	BINDING AGENT • CEMENT
MSDS Date	01 May 2006

2. HAZARDS IDENTIFICATION

CLASSIFIED AS HAZARDOUS ACCORDING TO ASCC CRITERIA

RISK PHRASES

R36/37/38	Irritating to eyes, respiratory system and skin.
R40	Limited evidence of a carcinogenic effect.
R43	May cause sensitisation by skin contact.

SAFETY PHRASES

S22	Do not breathe dust.
S24/25	Avoid contact with skin and eyes.
S36/37	Wear suitable protective clothing and gloves.

NOT CLASSIFIED AS A DANGEROUS GOOD BY THE CRITERIA OF THE ADG CODE

UN No.	None Allocated	DG Class	None Allocated	Subsidiary Risk(s)	None Allocated
Packing Group	None Allocated	Hazchem Code	None Allocated	EPG	None Allocated

3. COMPOSITION/ INFORMATION ON INGREDIENTS

Ingredient	Formula	CAS No.	Content
SILICA, CRYSTALLINE - QUARTZ	$Si-O_2$	14808-60-7	<1%
HEXAVALENT CHROMIUM	Not Available	Not Available	<0.1%
PORTLAND CEMENT	Not available	65997-15-1	>60%
FLY ASH	Not Available	68131-74-8	<5%
GROUND BLAST FURNACE SLAG	Not Available	65997-69-2	<5%
GYPSUM	$Ca-S-O_4-2(H_2-O)$	13397-24-5	<5%
LIMESTONE	$Ca-CO_3$	1317-65-3	<5%

CHEM ALERT

Figure 1.34 Page 1 of a sample MSDS
Note: The form consists of five pages; please visit <www.chemalert.com> to see other pages.

will be either dead or very ill within 48 hours. Similarly, if you have acid splashed onto your skin, you might suffer burns immediately or within the next 48 hours.

Chronic health effects relate to the long-term effects of exposure to a chemical. In general use, these effects may take years to become apparent. The chronic health effects may be just as serious as the acute health effects in the long term (e.g. exposure to cancer-causing agents may take 20 years to become apparent, but may still end up killing you).

Disposal of chemicals

It is most important to dispose of chemicals safely, as prescribed in the MSDS for the product. Always wash carefully with soap after handling any chemical.

If you spill any chemical on your clothing, remove the clothing and wash the body part affected. If you experience skin problems or difficulty in breathing, seek medical advice immediately. In addition, the environment must be considered when disposing of materials. Septic systems likewise do not respond well to chemicals.

Dangerous goods

Many of the chemicals used on building sites are classified as **dangerous goods**. Australia has adopted a system of classification and labelling for dangerous goods based on the United Nations' system used in other countries. The system helps people recognise dangerous goods, their properties and dangers quickly. The system for classification is detailed in the Australian Dangerous Goods Code.

Dangerous goods can be identified by a diamond sign or label (Figure 1.35).

The nine classes of dangerous goods under this system are:
- explosives;
- gases;
- flammable liquids;
- flammable solids;
- oxidising substances;
- poisonous and infectious substances;
- radioactive substances;
- corrosives;
- miscellaneous dangerous substances.

The diamond-shaped sign or label shows which of the nine classes the dangerous substance belongs to. These signs have distinctive symbols and colouring. Not all hazardous substances have dangerous goods labels because the dangerous goods diamond indicates only an immediate hazard, not necessarily a hazard that has only long-term health risks.

Figure 1.35 Examples of dangerous goods labels
Note: Refer also to the coloured Appendix at the end of this book.

Not all dangerous goods have safe handling and storage instructions printed on them; they may have only warning diamonds. The safe handling and storage instructions can be obtained from the MSDS for the product, or in the relevant Australian Standard for the substance.

Beware: If you find a product or substance with a 'diamond' on the container, obtain and read the safety instructions for the material before storing, opening or using it.

Signs and labelling

The details of the design and selection of signs and labels for the nine classes of dangerous goods are contained in the Australian Standard AS 1216-2006, 'Class labels for dangerous goods'.

Correct labelling means dangerous goods don't have to be any more dangerous than they already are. You should not only learn to recognise the various symbols but also learn about the actual properties of the substances you may be exposed to.

It is important to know which goods can produce toxic gases, which are highly flammable, which are dangerous when wet, which are dangerous on contact with air, and which are harmful when they come into contact with your skin. What you don't know could hurt you.

Each label has a distinctive colour, has a symbol to make it easily recognisable, states the hazard in words and gives the classification number.

Storage

With the withdrawal of AS 2508, there is currently no Australian Standard for the safe handling and storage of hazardous material in the building industry. There are, however, Codes of Practice (available from your relevant state or territory Regulating Authority), and these will reference you back to the material safety data sheets (MSDS) for each substance. It is critical that you adhere to the advice contained with an MSDS.

In all cases, storage must be in properly constructed containers or storage areas. Correct signposting of the area should be carried out to warn of the hazards of the chemicals stored.

Stored chemicals can be dangerous to outsiders, such as rescue workers and firefighters. In incidents where chemicals are spilt or involved in fires, toxic fumes and/or gases could be emitted. These situations can present a hazard to members of the public, as well as to the emergency personnel in attendance. Storage of specified quantities of substances that are classed as dangerous goods must be licensed in some states.

The purpose of licensing stored dangerous goods is to provide greater protection to people handling them, as well as to the public. The information obtained from the licensing is placed in a stored chemicals information database. This database can be accessed by emergency workers and government authorities in the case of an emergency, to find out what is being stored on premises or transported in a vehicle.

Licensing of the storage of hazardous chemicals on building sites has not worked well due to the short nature of the storage and the mobility of the industry. New regulations are being prepared to cover building sites.

Good work practices

In hazardous chemical areas, you should:
- change clothes daily;
- shower after work;
- isolate or enclose dust-producing machines;
- check the workplace to ensure that:
  - adequate personal protective equipment (PPE) is being used;
  - there is medical monitoring of persons exposed for damaging effects;
  - ventilation and other safety and health systems are effective.

SAFETY SIGNS AND TAGS

Note: Refer also to the colour Appendix at the end of this book.

Safety signs are placed in the workplace to warn of hazards or risks that may be present. They can also give information on how to avoid that hazard or risk, or how to avoid its effects. If a safety sign is required to be placed on a piece of equipment it may be in the form of a *tag*.

Safety signs:
- prevent accidents;
- signal health hazards;
- indicate the location of safety and fire protection equipment;
- give guidance and instruction in emergency procedures.

Standards Australia has three Standards covering the use of safety signs in industry:
- AS 1216-2006, 'Class labels for dangerous goods';
- AS 1318-1985, 'Use of colour for the marking of physical hazards and the indentification of certain equipment in industry (known as the SAA Industrial Safety Colour Code) (incorporating Amdt 1)';
- AS 1319-1994, 'Safety signs for the occupational environment'.

The three main types of safety signs covered in AS 1319 are as follows.

Picture signs use symbols (pictures) of the hazard, equipment, or the work process being identified, as well as standard colours and shapes, to give a message to people (see Figure 1.36).

Signs with word-only messages are made more recognisable by the use of standard colours and shapes (see Figure 1.37).

Figure 1.36 Picture signs—smoking prohibited

Figure 1.37 Word-only messages

Picture signs can be made clearer with a short written message (see Figure 1.38).

Figure 1.38 Combined picture and word signs

To make sure the message reaches everyone at the workplace, including workers from non-English-speaking backgrounds and workers with low reading skills, picture signs should be used wherever possible. When this is not possible it may be necessary to repeat the message in other languages.

Colour and shape

There are seven categories of safety signs identified by colour and shape:
- prohibition (don't do) signs;
- mandatory (must do) signs;
- restriction signs;
- hazard warning signs;

- danger hazard signs;
- emergency information signs;
- fire signs.

Prohibition (don't do) signs
These signs indicate that this is something you must not do: *red circle border* with *line* through it, *white* background and *black* symbol (see Figures 1.39, 1.40, 1.41 and 1.42).

Figure 1.39 Digging prohibited

Figure 1.40 No pedestrian access

Figure 1.41 Water not suitable for drinking

Figure 1.42 Fire, naked flame and smoking prohibited

Figure 1.43 Eye protection must be worn

Figure 1.44 Head protection must be worn

Figure 1.45 Hearing protection must be worn

Figure 1.46 Face protection must be worn

Figure 1.47 Restriction sign

Mandatory (must do) signs
These signs tell you that you must wear some special safety equipment: *blue solid circle*, *white* symbol, no border required (see Figures 1.43, 1.44, 1.45 and 1.46).

Restriction signs
These signs tell of the limitations placed on an activity or use of a facility: *red circular border*, no cross bar, *white* background.

Limitation or restriction signs normally have a number placed in them to indicate a limit of some type (e.g. a speed limit or weight limit) (Figure 1.47).

Hazard warning signs
These warn you of a danger or risk to your health: *yellow triangle* with *black* border, *black* symbol (Figures 1.48, 1.49, 1.50 and 1.51).

(Danger) hazard signs
These signs warn of a particular hazard or hazardous condition that is likely to be life-threatening: *white rectangular* background, *white/red* DANGER, *black* border and wording (Figure 1.52).

Figure 1.48 Fire risk

Figure 1.49 Toxic hazard

Figure 1.53 First aid

Figure 1.54 Emergency (safety) eye wash

Figure 1.50 Electric shock risk

Figure 1.51 Forklift hazard

Fire signs

These tell you the location of fire alarms and firefighting facilities: *red solid square*, *white* symbol (Figures 1.55 and 1.56).

Figure 1.55 Fire alarm

Figure 1.56 Fire hose reel station

Safety signs and tags for electrical equipment

Electrical wires and equipment that are being worked on or are out of service, or are live or may become live, must have 'Warning' or 'DANGER' safety tags fixed to them to help prevent accidents. Once the hazard has been removed, only the person who put the tag in place should remove it or authorise its removal.

Any standard safety sign may be made smaller and used as an accident protection tag (Figure 1.57). If words are to be used, this will generally be in the form of a danger sign. A tag should be at least 80 mm × 50 mm, plus any area required for tying or fixing the tag in place.

The background colour of the tag should be *yellow* for *warning* signs and *white* for *danger* signs.

Figure 1.52 Danger signs

Emergency information signs

These show you where emergency safety equipment is kept: *green solid square*, *white* symbol (Figures 1.53 and 1.54).

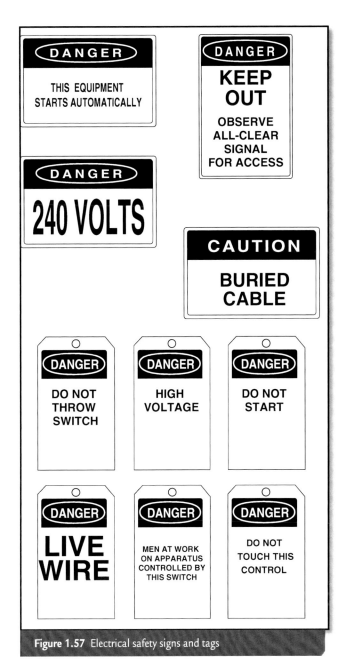

Figure 1.57 Electrical safety signs and tags

Placement of safety signs

Signs should be located where they are clearly visible to all concerned, where they can easily be read, and so that they will attract attention. If lighting is not adequate, illuminated signs can be used.

Signs should *not* be located where materials and equipment are likely to be stacked in front of them, or where other obstructions could cover them (e.g. doors opening over them). They should *not* be placed *on* movable objects such as doors, windows or racks so that when the object is moved they are out of sight or the intention of the sign is changed.

The best height for signs is approximately 1500 mm above floor level. This is at the normal line of sight for a standing adult. The positioning of the sign should not cause the sign itself to become a hazard to pedestrians or machine operators.

Regulation and hazard-type signs should be positioned in relation to the hazard to allow a person plenty of time to view the sign and take notice of the warning. This distance will vary: for example, signs warning against touching electrical equipment should be placed close to the equipment, whereas signs on construction work may need to be placed far enough away to permit the warning to be understood before the hazard is reached.

Care should be taken where several signs are intended to be displayed close together. The result could be that so much information is given in one place that little or no notice is taken of it, or that it creates confusion.

For more picture safety signs and their meanings, see:
- WorkCover NSW booklet: *Picture safety signs for the workplace*;
- Standards Australia: AS 1319-1994, 'Safety signs for the occupational environment';
- Safety sign manufacturers' catalogues: see *Yellow Pages*—'Signs–Safety &/or Traffic'.

RISK ASSESSMENT

In the building industry, electricity, falls, collapsing trenches and melanoma often kill. Chemicals, corrosives, noise and dust inhalation can result in blindness, deafness, burns and injuries to lungs. Back problems or other serious strains or sprains can slow workers down and put them out of action for weeks or even permanently.

Accidents

An accident is when something happens unexpectedly, without design, or by chance. An accident may cause a person pain and injury.

Hazards

A hazard is a potential source of harm where a person is exposed to a dangerous or harmful situation, physical or otherwise, which may affect the health and safety of that person.

Acute hazards

An acute hazard is one where short-term exposure to the hazard will cause an injury or sickness (e.g. being burnt in an explosive fire).

Chronic hazards

A chronic hazard is one where long-term exposure to the hazard will cause an injury or sickness (e.g. melanoma from extended exposure to the sun, or slow poisoning from chemicals building up in the body's system over a long period of time).

If every person is aware of the hazards to health and safety they may be exposed to, and follows the commonsense safety rules, both employers and employees will benefit. Now is the time to make the building industry safer for everyone.

Workplace hazards

To be aware of potential dangers at work we must be able to identify workplace hazards. We come into contact with these hazards every day. Some of the more common hazards that influence health and safety in the workplace are:
- lifting and handling materials;
- falls—of objects and people;
- machinery—power and handtools;
- chemicals and airborne dust;
- noise;
- vibration;
- thermal discomfort;
- illumination (visibility);
- potential fire hazards.

Hazard groups

Hazards can be described as being:
- safety hazards;
- health hazards.

Health hazards can be further subdivided into:
- physical;
- chemical;
- biological;
- stress.

See Figure 1.58 for the major groupings under which to define the various *types* of hazards found in the workplace.

Common types of workplace hazards

The following is a list of the more common hazards found on building and construction sites.

Safety hazards

- *Poor housekeeping*—untidy sites, lack of guardrails, inadequate walkways and rubbish left in work areas.
- *Electricity*—underground and overhead supply cables, site wiring and power leads.
- *Water*—presence of water in a work area can cause slips and falls, electrocution, drowning and caving in of excavation works.
- *Ladders and scaffolding*—slips and falls from ladders and scaffolding.
- *Hand tools*—can cause serious injuries if not used and maintained correctly.
- *Electrical tools*—can cause serious cuts through contact with unguarded moving parts, electrocution, or falls and burns from electrical shock.
- *Heavy equipment*—heavy motorised equipment moving around construction sites is a major hazard.
- *Sharp materials*—materials with sharp edges can cause cuts and lacerations.
- *Airborne materials (projectiles)*—one of the most common causes of injury in the industry is falling or flying materials.
- *Airborne dusts and mists*—from spray paints, mixing adhesives, cements, grouts, etc.

Physical hazards

- *Noise*—exposure to excessive noise can cause temporary hearing loss, stress and annoyance.
- *Heat and cold*—cause reduction in concentration and heat-related medical conditions.

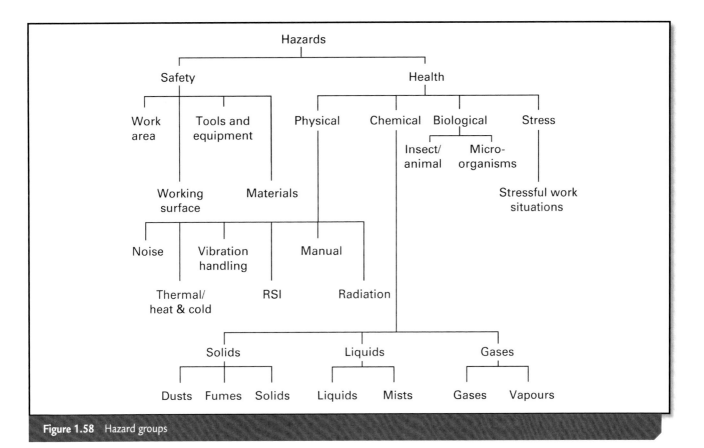

Figure 1.58 Hazard groups

- *Vibration*—whole-body vibration injuries from working with heavy equipment; and hand and arm vibration injuries caused by the use of vibrating tools.
- *Manual handling*—these injuries (lifting, carrying, pushing, pulling) are probably the most common injuries in the building industry.

Chemical hazards

Many different chemicals that may be harmful to health are used in the construction industry in the forms of:
- *solids*—dusts, fumes and solid materials;
- *liquids*—liquid materials and mists;
- *gases*—gases and vapours.

Some of these chemicals can cause acute or chronic injuries and severe medical conditions.

Toxic chemicals

A large number of chemicals are toxic or poisonous. The effects of chemicals on your health are difficult to identify at the source of exposure. Often the symptoms appear in parts of the body far removed from the point of absorption or perhaps do not appear until some years later.

Biological hazards

- *Insect/animal*—poor sanitation in dining areas or toilet areas and poor practices of rubbish disposal increase the chances of the spread of disease.
- *Micro-organisms*—when health and welfare facilities are not cleaned often, bacteria can grow and spread, causing disease. Some work areas may also harbour dangerous bacteria.

Stress hazards

Stress is normally experienced as fatigue, anxiety and depression. Factors that may lead to stress disorders are:
- workload;
- work pressure—excessive pace;
- excessive overtime;
- monotonous work;
- uncertainty of responsibilities;
- poor physical environment.

Hazard control procedures

In many cases a number of control methods must be used to control a hazard. Various methods should be considered. The order of preference for controlling hazards is known as the 'Hierarchy of Hazard Controls'.

The emphasis in the list is on controlling the hazard at the source. This is done by giving preference to the first five controls, which are engineering controls.

Engineering controls

- *Design*—ensuring that hazards are 'designed out' when new materials, equipment and work systems are being planned for the workplace.
- *Eliminate* the hazard or *substitute* less hazardous materials, equipment or substances.
- *Change the process*—alterations to tools, equipment or work systems can often make them much safer.
- *Enclose or isolate the process*—e.g. through the use of guards and the use of drop sheets to cover floors and openings.
- *Provide effective ventilation*—through local or general exhaust ventilation systems.

Other controls

- *Establish suitable administrative procedures* such as job rotation to reduce exposure, or by timing the job so that fewer workers are exposed.
- *Provide training* on hazards and safe working procedures.
- *Establish routine housekeeping and maintenance procedures*.
- *Provide* suitable and properly maintained *personal protective equipment*, provide training in its use, and ensure the equipment is used.

Selection of controls

The methods used should be those that are less likely to fail due to human failure. Personal protective equipment should not be uncomfortable to the wearer or disrupt the activities being performed by that person.

The capacity of the control used to provide safety to the persons affected should be checked regularly.

Eliminate the hazard

The posting of safety signs, like the issue of personal protective equipment, should be seen as a last resort, where it is not possible to eliminate the hazard at its source. These signs and personal protective equipment are not an alternative to the elimination of the hazard or risk.

Effects on the human body

Injuries to the human body, and diseases caused by accidents or exposure to an unhealthy environment through work-related hazards, will vary from hazard to hazard. The most common effects on the body are:

- broken limbs, cuts, lacerations, crushing and bruising—caused by slips, trips and falls, misuse of tools, contact with sharp objects; being hit by falling or flying materials, hand tools or equipment; and accidents through poor concentration;
- electrocution and electrical burns—from coming into contact with live electrical equipment;
- burns—from hot objects, flames or contact with corrosive chemicals;
- crushing and suffocation—through excavation cave-ins, being crushed by materials or equipment, lack of oxygen;
- hearing loss (temporary or permanent)—through exposure to excessive noise;
- work-related stress—from working under stressful work conditions;
- heat stroke, cramps, heat exhaustion and rashes—from exposure to excessive heat;
- heart conditions and high blood pressure—from exposure to harmful chemicals and body stress and strain;
- bone damage, slipped disks, hernias—from body stress and strain;
- stomach and digestive problems—from exposure to harmful chemicals and body stress and strain;

- nervous system problems—from exposure to harmful chemicals and body vibration;
- white finger—from excessive hand vibration;
- poisoning (acute and chronic), allergies, irritations—from exposure to harmful chemicals;
- cancers, dermatitis and respiratory irritations—from exposure to harmful chemicals and radiation;
- diseases—from exposure to harmful chemicals and micro-organisms;
- death—in the worst cases, all workplace hazards could cause death.

Hazard report forms

Controlling the risks arising from hazards offers the greatest area of opportunity for reducing injury and illness in the workplace. All workplaces should have a hazard reporting system in place for workers to report potential hazards. This will bring problem areas to the attention of management as soon as the hazard has been identified.

A means of providing this system is to have standardised **hazard report forms** readily available to the workforce (Figure 1.59). Workers should complete a form and give it to their immediate supervisor as soon as a potential hazard is identified. This will allow control measures to be put in place to remove the hazard at the earliest possible time.

Workplace inspections

These are regular inspections of the workplace to determine what hazards exist by observation. They are made by management and representatives of the workforce. Reports and recommendations from these inspections will allow hazard control measures to be undertaken.

At workplaces with more than 20 employees, permanent safety **committees** are normally established, with representatives from management and the workforce. These safety committees are then responsible for carrying out the inspections.

MANAGEMENT OF SITE HAZARDS

Responsibility and duty of care

Employers have a 'duty of care' in relation to the health, safety and welfare of their employees. An employer must identify any foreseeable hazard that may arise from the work carried out by the company or partnerships.

To identify these risks and state how the employer will safeguard employees and all other persons at work, the employer is required to prepare an OH&S site management plan, in accordance with the relevant state or territory OH&S Regulation (refer to Table 1.1 on p. 3).

This management plan is site-specific: that is, it relates to individual sites being worked on at any one location only, and must be maintained and kept up to date during the course of the construction work. Non-compliance could invoke breach of the OH&S Act and subsequent fines may apply.

Safe work method statements (SWMS)

Like employers, employees have a duty of care to ensure that they work in a safe manner, obey site rules and do not interfere with any safety item or system. Therefore, it is a requirement under the relevant state or territory OH&S Act that all persons on a construction site adhere to safety requirements and take part in the formation and application of safety plans, which include safe work method statements for risky work.

Definition (extract from OH&S Regulation 2001):

> **Safe work method statement** *means a statement that:*
> *describes how work is to be carried out, and*
> *identifies the work activities assessed as having safety risks, and*
> *describes the control measures that will be applied to those work activities, and*

HAZARD/INCIDENT REPORT FORM

To be completed by supervisor

1. **Who reported the hazard/incident?** _____

   Time: _____   Date: _____

2. **What is the hazard/incident?** _____

   _____

   _____

   _____

3. **What has been done to rectify the hazard/incident?** _____

   _____

   _____

   _____

   Time: _____   Date: _____

4. **What further action needs to be taken?** _____

   _____

   _____

   _____

5. **Referred to manager/owner for information or action (date)** _____

   Signed: _____   Date: _____

To be completed by manager/owner

6. **Action taken by manager/owner** _____

   _____

   _____

Figure 1.59 Sample hazard report form

includes a description of the equipment used in the work, the standards or codes to be complied with and the qualifications of the personnel doing the work.

Subcontractors' responsibilities

The principal contractor must ensure that each subcontractor provides a written safe work method statement for the work to be carried out by the subcontractor, which should include an **assessment of the risks** associated with that work.

If the subcontractor does not comply with what has been outlined in the safe work method statement, the principal contractor has the right and responsibility to direct the subcontractor to stop work and not resume work until the safe work method statement requirements are met. This could result in the subcontractor losing money due to the implementation of safety requirements and/or equipment, or receiving a fine under the OH&S Act.

If the subcontractor neglects to maintain or modify the work method statement or fails to inform the principal contractor of the changes, then the subcontractor may receive a fine for breaching the OH&S Act.

If the principal contractor neglects to enforce the requirements of the Act on the subcontractor, then the principal contractor may receive a fine for breaching the OH&S Act.

It is therefore essential that all parties on the site comply with the requirements of their relevant state or territory OH&S Act to create, implement and monitor safe work method statements to reduce risk and possible loss of income.

Format of safe work method statements

These statements should be kept as simple as possible and show the greatest amount of information (Figure 1.60). Contractors don't want to spend an excessive amount of time on the preparation and maintenance of these statements, so a standard format should be developed and used by individuals to suit the type of work being carried out.

Subcontractors may reuse the same statement with modifications to suit new sites and/or conditions, as they tend to specialise in one particular area and carry out the same type of work on every job.

ACCIDENT REPORTING

All workers at a place of work have a responsibility to report any illness, accident or near miss they are involved in or see. They should report these **incidents** to their immediate supervisor, site supervisor, member of the workplace OH&S committee, a union delegate or first aid officer. If there is any **plant** or equipment involved they must make the operator aware of the problem immediately.

Once the supervisor becomes aware of an illness, accident or near miss, he or she must determine whether it is a reportable accident (see below). If so, the necessary forms must be completed and forwarded to the appropriate authorities (see Figure 1.61). This reporting procedure allows for steps to be taken on a state and national level to help reduce workplace accidents and illnesses in the future. It also allows inspections of the workplace at which the accident or illness occurred to be made by the state's OH&S authority inspectors, to see that steps are taken to prevent the problem from happening again.

Reportable accidents

Standards Australia's AS 1885.1-1990, 'Measurement of occupational health and safety performance—Describing and reporting occupational injuries and disease (known as the National Standard for workplace injury and disease recording)', provides a national Standard for describing and reporting occupational injuries and disease. From this national Standard each state develops an accident reporting register and reporting procedures to suit its requirements under its own OH&S legislation.

Work Method Statement (Side 1)		(Principal Contractor/Site Supervisor)
Contractor/Sub-contractor: _____		Signed off: _____
Project/Site Address: _____		Date: _____
Trade/Work to be carried out: _____		Accepted: YES / NO

TASK/PROCEDURE	POTENTIAL HAZARDS	SAFETY CONTROLS
Describe the individual tasks or jobs step by step and include any details where an action is required, e.g.: Cut timber to length with power saw, Lay ant capping, Drill 10 mm diameter holes in concrete, Lift steel beams into place, etc.	Identify all potential hazards related to the task or sections of the task description where the greatest risk is anticipated, e.g.: Dust, noise and injury caused by saw, Cuts from sharp edges, Noise and vibration from drill, Manual handling, back injuries from lifting, etc.	List all safety controls to avoid injury and/or lower risk of accidents or injury caused by the action or equipment being used, e.g.: Wear PPE such as respirator, ear muffs, use both hands on the saw, Wear long pants and use gloves, Wear PPE such as ear muffs and gloves when using the drill, Lift using legs not back, dual lift heavy objects, etc.

(Side 2)

PERSONAL QUALIFICATIONS/EXPERIENCE	PERSONNEL DUTIES/RESPONSIBILITIES	SPECIAL TRAINING REQUIRED
List all details of sub-contractors and employees, qualifications and experience required for the task or project. List any certificates, licences or previous training.	List the duties and responsibilities of all workers, e.g. who will be in charge of first aid, who will carry out regular safety inspections, who is responsible for training, who is licensed to carry out specific work, etc.	List any training or special instruction required to carry out specific tasks or projects, e.g. wearing safety harnesses for roof work, asbestos removal, scaffolding above 4 m high, etc.

ENGINEERING DETAILS/CERTIFICATES/APPROVALS	CODES OF PRACTICE/LEGISLATION
Provide any special requirements for demolition, precast member erection, formwork erection and/or dismantling, working times in busy commercial settings, etc.	List the Codes of Practice being used for particular work areas, e.g. Safe work on roofs - Part 2 - Residential buildings, overhead protective structures, safety line systems, etc.

PLANT/EQUIPMENT	MAINTENANCE CHECKS
List the type of plant and equipment being used for the task of project, e.g. portable power tools, elevated work platforms, scaffolding, ladders, hammer drills, etc.	List the requirements for equipment maintenance or checking including regular tagging of electrical equipment and tools, including 3-monthly tagging of tools for residential sites and monthly tagging for commercial/multistorey sites.

Read and signed by: Principal Contractor _____

Sub-contractor _____

Sub-contractor Employees _____

Figure 1.60 Typical SWMS format

HAZARD/INCIDENT/ACCIDENT REPORT FORM

Who uses this form?
Two people – the worker and his or her supervisor (from the host employer).
Purpose?
When a hazard, incident or accident occurs, record what happened, what investigations occurred, and what was done to prevent future injury or illness in relation to this incident or accident.
What should happen?
The host employer keeps the original and a copy is to be given to the labour-hire agency, to be kept in a file with the host employer's name on it.

PART A – To be completed by employee

Name of employee:		Date:	01/01/05
Time of incident / accident:			
Supervisor:		Work Area:	

1. Describe the hazard / detail what happened – include area and task, equipment, tools and people involved.

2. Possible solutions / how to prevent recurrence – Do you have any suggestions for fixing the problem or preventing a repeat

PART B – To be completed by supervisor

3. Results of investigation – Determine whether the hazard is likely to cause an injury and explain what factors caused the event.

PART C – To be completed by supervisor

4. Action taken – Supervisor to identify actions to prevent injury or illness.

	ACTION	RESPONSIBILITY	COMPLETION DATE
4.1			
4.2			
4.3			
4.4			
4.5			

Feedback has been provided to person who reported the hazard / incident / accident.

Employee representative (health and safety representative)	Name Surname	Date:	01/01/05
Business Manager	Name Surname	Date:	01/01/05

WORKSAFE VICTORIA / SAFETY MANAGEMENT SYSTEMS GUIDE FOR LABOUR HIRE AGENCIES (2nd Edition, October 2005).

Figure 1.61 Sample accident report form

The Regulating Authority may require serious work-related illnesses, injuries or dangerous occurrences to be reported on an **accident report form** in Victoria. These accidents became known as **reportable incidents**, and are reportable under that state's OHS Act of 2004.

Accident

This is any event that causes human injury or property damage. These events may occur as a result of unsafe acts, which may include the following:
- practical jokes;
- using tools and/or equipment in a manner for which they were not designed;
- rushing and taking short cuts;
- not using PPE or other safety devices;
- throwing materials/rubbish from roofs or upper floors.

Other causes

These include unsafe conditions, such as:
- little or no training in safety and the proper use of tools and equipment;
- poor housekeeping;
- poor management of site safety issues;
- poorly maintained tools and/or equipment;
- damaged tools and/or equipment;
- inadequate PPE, or PPE not supplied or not used;
- poor site conditions and congestion due to lack of preparation.

A reportable incident that takes the form of an illness is defined as involving:
- a person requiring medical treatment within 48 hours of exposure to a substance; or,
- a person requiring immediate treatment as an in-patient in a hospital.

An injury that becomes a reportable incident is defined as involving a person requiring immediate medical treatment for any of the following:
- the amputation of any part of their body;
- a serious head injury;
- a serious eye injury;
- the separation of their skin from an underlying tissue;
- electric shock;
- a spinal injury;
- the loss of a bodily function;
- serious lacerations.

Other incidents that must be reported are those that, while not actually causing an injury, could have done so. These are sometimes referred to as 'near misses'. Examples include (but are not limited to):
- where there is damage to any boiler, pressure vessel, plant, equipment or other thing that endangers or is likely to endanger the health and safety of any person at a workplace;
- where damage to any load-bearing member or control device of a crane, hoist, conveyor, lift, escalator, moving walkway, plant, scaffolding, gear, amusement device or public stand occurs;
- where any uncontrolled explosion, fire or escape of gas, dangerous goods or steam occurs, or where there is a risk that any of these events is likely to occur at any moment;
- where a hazard exists that is likely to cause an accident at any moment and may cause death or serious injury to a person (e.g. an electric shock) or substantial damage to property.

Employers or persons in control of a workplace are normally required to send an accident report form to the state authority even if the person injured or killed is not one of their employees (e.g. is a subcontractor or visitor to the site).

Near misses

These are basically accidents that didn't quite happen. They occur when the conditions are right for an accident but people don't get hurt and equipment is not damaged.

Near misses usually indicate that a procedure or practice is not being carried out correctly or site conditions are unsafe. By reporting these near misses the problems may be looked at and rectified before someone is hurt or killed or equipment is seriously damaged.

Accident report forms

The accident report normally requires the following information to be given, where appropriate.

Information about the employer or workplace:
- name of company;
- office address;
- address of site where accident happened;
- main type of activity carried out at the workplace (e.g. building construction);
- major trades, services or products associated with this activity;
- number of people employed at the workplace and whether there is an OH&S committee at the workplace.

Information about the injured or ill person:
- name and home address;
- date and country of birth;
- whether the person is an employee of the company;
- job title and main duties of the injured person.

Information about the injury/illness:
- date of medical certificate;
- type of illness as shown on the medical certificate;
- whether the injury resulted in death;
- particulars of any chemicals, products, processes or equipment involved in the accident.

Information about the injury or dangerous occurrence:
- time and date that it happened;
- exact location of the event;
- details of the injury;
- type of hazard involved;
- exactly how the injury or dangerous occurrence was caused;
- how the injury affected the person's work duties;
- details of any witnesses;
- details of the action taken to prevent the accident from happening again.

Details of the person signing and the date of signing the accident report are also required. When the report is completed, copies are sent to the state authority and a copy is kept by the employer.

Injury management

The loss or disruption a company can experience as a result of a hazardous incident may be multiplied tenfold when that incident leads to a worker being injured.

A comprehensive risk management system should include a well-thought-out plan to maximise the opportunity for injured workers to remain at work. This allows the worker to be productive in some capacity and assists with the recovery and rehabilitation process.

Therefore, the risk management system should cover the following points:
- early notification of the injury;
- early contact with the worker, their doctor and the builder's insurance company;
- provision of suitable light duties as soon as possible to assist with an early return to work; and
- a written plan to upgrade these duties in line with medical advice.

Payment of compensation

Workers' compensation

Workers' compensation insurance is a system that provides payment benefits and other assistance for workers injured through work-related accidents or illnesses. It may also provide their families with benefits where the injury is very serious or the worker dies.

The current workers' compensation system in NSW is known as WorkCover and was established under the *Workers' Compensation Act 1987* and the *Workplace Injury Management and Workers' Compensation Act 1998*. Employers must take out workers' compensation insurance to cover all workers considered by law to be their employees. Each state has its own system, which should be investigated by the building student.

Eligibility for compensation

To be able to claim workers' compensation a worker must have suffered an 'injury' or disease. The injury or disease must be work-related: that is, it must have happened while working, during an allowed meal break or on a work-related journey.

The injury or disease must result in at least one of the following:
- the death of the worker;
- the worker being totally or partially unable to perform work;

- the need for medical, hospital or rehabilitation treatment;
- the worker permanently losing the use of some part of the body.

A claim for compensation is made by:
- informing the employer and lodging a claim as soon as possible;
- seeing a doctor and obtaining a medical certificate.

If any problems arise with the compensation claim the worker should contact the state compensation board or authority or the worker's own union.

A record of *all* injuries that occur at a workplace must be entered in an **injuries register book** kept at the site.

EMERGENCY PROCEDURES

An emergency may develop due to a number of reasons, such as a fire, gas or toxic fumes leak, improper use of flammable materials, partial collapse of a building (Figure 1.62), a bomb threat, a crane overturning, unstable ground, materials improperly stored, or a trench collapse.

Therefore, every organisation is required to have an emergency procedure in place and personnel appointed to control the safe exit of persons at the workplace.

Responsible personnel

On a large building site the responsible personnel may include the head contractor, safety officer, head foreperson or site supervisor.

Figure 1.62 Emergency caused by a partial building collapse

Small-building-site personnel may include the builder, foreperson, leading hand or a nominated tradesperson. Whichever the case, these persons are responsible for following set procedures to get all other persons on-site out of the danger area to a predetermined collection point so that in the event of an incident all persons may easily be accounted for. The nominated responsible personnel are indemnified against liability resulting from practice or emergency evacuations from a building, where the persons act in good faith and in the course of their duties.

On large sites, these persons may be identified by a coloured helmet they would wear, which would be determined as part of the organisation's emergency plan.

Roles
Emergency coordinator
- Determine the nature of the emergency and the course of action to be followed.
- Set off any alarm or siren to warn persons of an emergency.
- Contact appropriate services such as police, fire or ambulance.
- Initiate the emergency procedure and brief the emergency services when they arrive.

Warden or controller
- Assume control of the occupants/workers until the emergency services arrive.
- Notify all persons regarding the nature of the emergency.
- Give clear instructions and make a record of what was carried out.
- Report all details to the coordinator as soon as possible.

Casualty control
- Attend to casualties and coordinate first aid.
- Coordinate the casualty services when they arrive.
- Arrange for further medical or hospital treatment.

First aid

An employer must ensure that employees have access to first aid facilities that are adequate for the immediate treatment of common medical emergencies; if more than 25 persons are employed at the workplace there should be trained first aid personnel. Trained first aid personnel may include a person with a current approved first aid certificate, a registered nurse or a medical practitioner.

An employer must also ensure that there are first aid facilities, a suitable first aid kit and sufficiently trained first aid personnel at their place of work.

Note: Although the following first aid kit guidelines are specific to the requirements as stipulated under the New South Wales OH&S Regulation 2001, on a national level, all other state and territory obligations are very similar. It is therefore highly recommended that readers refer to their relevant OH&S Regulations in order to get an exact understanding of the legislation specific to their location. It is additionally recommended that readers access the St John Ambulance Australia website <www.stjohn.org.au>, to research what first aid kits are available to accommodate their specific state or territory needs.

There are three main types of kit, which are typically used on different types of construction sites as described below:

First Aid Kit A—Construction sites at which 25 or more persons work or other places of work at which 100 or more persons work;
First Aid Kit B—Construction sites at which fewer than 25 persons work or other places of work at which fewer than 100 and more than 10 persons work;
First Aid Kit C—Places of work (*other than construction sites*) at which 10 or fewer persons work.

Most residential building sites would fall into the 'Kit B' category, as it would be rare to have 25 or more persons on the site at any one time.

Note: 'Kit C' is *not* suitable for construction sites of any kind, as it lacks many of the items required for first aid treatment of common building site injuries.

Contents of a first aid kit type B

The contents of a type B kit are listed in Table 1.2 and the first aid kit and stores are shown in Figures 1.63 and 1.64.

Table 1.2 Contents of a type B first aid kit

CODE: FKN405
WORKPLACE RESPONSE KIT 4 (NSW)
METAL WALL CABINET
CONTENTS LIST

CODE	DESCRIPTION	QTY
FRD005	ADHESIVE PLASTIC STRIPS (50) 72 X 19MM	50
FRS022	ANTISEPTIC CREAM 25G TUBE	1
FRS023	ANTISEPTIC LIQUID 50ML SPRAY RAPAID	1
FRC450	ANTISEPTIC WIPE STERILE NON-STING	10
FLB001	BOOKLET FIRST AID GUIDE FASTAID	1
FRH003	BURNAID GEL 3.5G SACHET	8
FRH500	COLD PACK INSTANT MEDIUM	1
FRD237	COMBINE DRESSING PAD 10 X 20CM STERILE	1
FRB210	CONFORMING BANDAGE 10CM WHITE	1
FRB205	CONFORMING BANDAGE 5CM WHITE	1
FLC999	CONTENTS LIST	1
FRC260	COTTON BALLS - LARGE (60)	60
FRC100	COTTON TIPS 7.5CM DOUBLE ENDED PLASTIC (100)	100
FRB405	CREPE BANDAGE HEAVY 5CM BROWN	1
FRA202	CUPS PLASTIC 200ML	3
FRD026	DRESSING STRIP FABRIC ADHESIVE 75MM X 1M	1
FRA111	EMERGENCY SHOCK/RESCUE BLANKET	1
FRD107	EYE PADS STERILE PK1	2
FRS100	EYEWASH 15ML SODIUM CHLORIDE	6
FRG500/L	FINGER COTS LATEX POWDER FREE LARGE	1
FRG500/M	FINGER COTS LATEX POWDER FREE MEDIUM	1
FRG500/S	FINGER COTS LATEX POWDER FREE SMALL	1
FRI140	FORCEP SPLINTER 12.5CM STAINLESS STEEL	1
FRI142	FORCEPS PLASTIC DRESSING	1
FRC400	GAUZE SWAB 7.5 X 7.5CM PK 3 STERILE	2
FRG004	GLOVES LARGE (4)	4
FLI501	LEAFLET "REGISTER OF INJURIES" SAMPLE	1
FRD306	NON-ADHERENT DRESSING 10 X 7.5CM STERILE	2
FRA005	PLASTIC BAG LARGE RESEALABLE	1
FRA004	PLASTIC BAG MEDIUM RESEALABLE	1
FRA003	PLASTIC BAG SMALL RESEALABLE	1
RCF035	RESUSCITATION FACE SHIELD DISPOSABLE	1
FRI012	SAFETY PINS (12)	12
FRI100	SCISSORS STAINLESS STEEL 12.5CM SHARP/BLUNT	1
FRI036	SPLINTER PROBE 4CM DISPOSABLE	5
FRT002	TAPE PAPER 2.5CM X 9M HYPO-ALLERGENIC WHITE	1
FRB620	TRIANGULAR BANDAGE N/W DISP. 110CM LARGE	4
FRD515	WOUND DRESSING #15 LARGE STERILE	3

Figure 1.63 Type B first aid kit

Figure 1.64 Stored contents of the kit

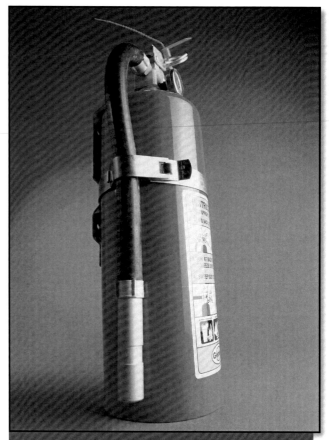

Figure 1.65 Use a fire extinguisher to control small outbreaks

Fire orders

The initial action may be undertaken by a single person or with the assistance of others available at the time; however, any person discovering a fire should:
- rescue any person in immediate danger, if it is safe to do so;
- alert other people in the immediate area;
- take action to extinguish the fire if equipment is available before it takes hold (Figure 1.65);
- confine the fire by closing any doors;
- dial 000 for the fire brigade to attend;
- contact the emergency coordinator, or warden, as soon as possible.

Procedures for emergencies

An employer has the responsibility to ensure that in the event of an emergency arrangements have been made for the following:
- the safe and rapid evacuation of persons from the place of work or to a designated collection area on the site;
- emergency communications, such as a landline phone or mobile with emergency phone numbers clearly visible and accessible; and
- provision of appropriate medical treatment of injured persons by ambulance, medical officer or access to a suitable first aid kit.

Note: If a person is seriously injured the relevant state or territory Regulating Authority must be notified. If a person is killed, the relevant state or territory Regulating Authority and the police must be notified.

Firefighting equipment

Learning how to prevent and fight fires is part of every worker's responsibility. It is important for the safety of every worker on a job to understand the procedures to follow in the event of a fire.

Large construction sites and buildings should have firefighting teams responsible for each floor or the whole building. The firefighting team must be specially trained staff members who can direct the evacuation and firefighting operations until the fire brigade arrives.

Fire hazards

The elements necessary before there can be a fire are shown in Figure 1.66:

FUEL + HEAT + OXYGEN

Figure 1.66 The elements necessary for a fire

Fuel can be any combustible material; that is any solid, liquid or gas that can burn. **Flammable materials** are any substances that can be easily ignited and will burn rapidly.

Heat that may start a fire can come from many sources, such as flames, welding operations, grinding sparks, heat causing friction, electrical equipment, hot exhausts.

Oxygen comes mainly from the air. It may also be generated by chemical reactions.

If any one of the three elements is taken away, the fire will be extinguished.

Common causes of fires on-site

- Burning off rubbish—site rubbish should be cleared away regularly and no burning off should take place, in accordance with EPA requirements.
- Electrical fires—due to overloading equipment, faulty equipment, faulty wiring etc. All equipment should be carefully checked, maintained and used correctly.
- Contractors using naked flames—such as plumbers, structural steel workers etc. These contractors must ensure that they do not carry out naked flame operations within the vicinity of stored rubbish, paints, sawdust or any other highly flammable material.
- Smokers—carelessly disposing of cigarettes, matches etc. Butane lighters may also be a source of ignition and should not be exposed to naked flames or other situations where ignition could occur.
- Spark generation—use of angle grinders and the like in extreme weather conditions or in highly flammable environments.

How to prevent fires

Don't give them the chance to start.

- Remove unwanted fuel from the workplace, e.g. rubbish and waste materials.
- Store fuels and combustible materials carefully; use safety carrying and pouring cans.
- Use only approved electrical fittings, keeping them in good order.
- Don't overload electrical circuits.
- Don't smoke at the workplace.
- Take special care if working with flammable liquids or gases.
- Be careful of oily rags, which can ignite from spontaneous combustion, e.g. turps- or linseed oil-soaked rags.
- Avoid dust hazards. Many types of dust are so highly flammable that they can explode when mixed with air or when they are exposed to flame or sparks.

In the event of a fire

- Don't panic—keep calm and think.
- Warn other people in the building.
- Those not needed should leave the building at once and assemble at the designated fire-assembly area.
- Arrange for someone to phone the fire brigade.
- Have the power and gas supplies turned off if it is appropriate (some lighting may still be required). Close doors where possible to contain the fire.
- Stay between a doorway and the fire.
- Be aware of containers of explosive or flammable substances. Remove them from the area only if it is safe to do so.
- If it is safe to fight the fire, select the correct type of extinguisher, having others back you up with additional equipment.
- Know how to use the extinguisher; be confident and attack the fire energetically.

- If the fire is too large for you to extinguish, get out of the building and close all doors. Assemble at the designated area.

Classes of fires and extinguishers

Extinguishers have been grouped according to the class of fire on which they should be used. The class is determined by the type of material or equipment involved in the fire. The four main classes of fires are as follows.

Class A fires

Class A—ordinary combustible materials, e.g. wood, paper, plastics, clothing and packing materials (Figure 1.67).

The correct extinguishers to use are:
- water type—most suitable;
- any other type (except: B(E) dry chemical powder).

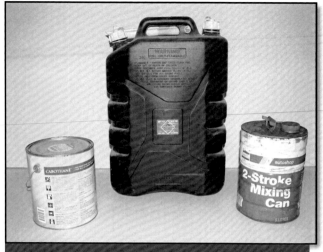

Figure 1.68 Class B—flammable and combustible materials: paint, oil and petrol

Class C fires

Class C involve flammable gases, such as LPG, acetylene, gas-powered forklifts (Figure 1.69).

The correct extinguisher to use is:
- dry chemical powder.

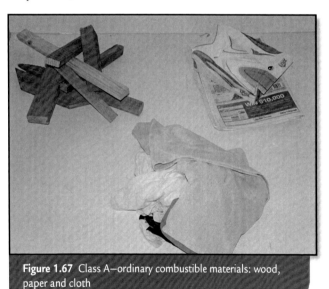

Figure 1.67 Class A—ordinary combustible materials: wood, paper and cloth

Figure 1.69 Class C—flammable gases: LPG

Class B fires

Class B fires involve flammable and combustible liquids, e.g. petrol, spirits, paints, lacquers, thinners, varnishes, waxes, oils, greases, petrol or diesel-driven motor vehicles, and many other chemicals in liquid form (Figure 1.68).

The correct extinguishers to use are:
- foam type;
- carbon dioxide (CO_2);
- dry chemical.

Class E fires

Class E fires involve 'live' electrical equipment, e.g. electric motors, power switchboards, computer equipment (Figure 1.70).

The correct extinguishers to use are:
- only extinguishers displaying an (E); these will not conduct electricity; **never** use water or foam extinguishers;
- dry chemical powder;
- carbon dioxide.

CHAPTER 1 OCCUPATIONAL HEALTH AND SAFETY 45

Figure 1.70 Class E—electrical switchboards

Identification and operation of extinguishers

Note: Refer also to the colour Appendix at the end of this book.

Water extinguisher (Figure 1.71)

Operation:

- Three types of water extinguisher available are: gas pressure, stored pressure, and soda–acid (being phased out).
- Range of operation: up to 10 metres.
- Methods of activating the extinguishers are different for each type; it is therefore important to read the instructions on the container before attempting to use it.
- Once activated, the jet of water should be directed at the seat of the fire.

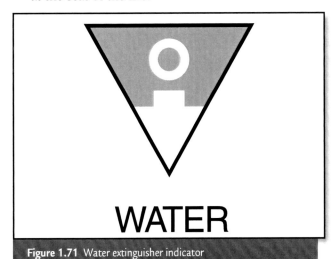

Figure 1.71 Water extinguisher indicator
Identifying colour: All RED

Foam extinguisher (Figure 1.72)

Operation:

- Three types of foam extinguisher available are: gas pressure, stored pressure, and chemical (being phased out).
- Range of operation: up to 4 metres.
- Methods of activating the extinguishers are different for each type; it is therefore important to read the instructions on the container before attempting to use it.
- Once activated, the jet of foam is directed to form a blanket of foam over the fire. This stops oxygen from getting to the fire long enough to allow the flammable substance to cool below its reignition point.

Figure 1.72 Foam extinguisher indicator
Identifying colour: Pre 1999—All BLUE; Post 1999—RED container with BLUE band

Dry chemical powder—AB(E) (Figure 1.73)

Operation:

- Small sizes have a range of 3 metres, with larger types to 6 metres.
- This is most effective for the extinguishing of large areas of burning liquid or free-flowing spills.
- Powder is discharged through a fan-shaped nozzle. It should be directed at the base of the fire, which should be covered with a side-to-side sweeping action.

Figure 1.73 Dry chemical powder indicator
Identifying colour: RED container with WHITE band

Carbon dioxide (CO_2) (Figure 1.74)

Operation:
- Small sizes have a range of 1 metre, with larger types to 2.5 metres.
- These are useful for penetrating fires that are difficult to access.
- Use as close to the fire as possible. Aim the discharge first to the rear edge of the fire, moving the discharge horn from side to side, progressing forward until flames are out.

Warning: Exposure to carbon dioxide in a confined area could cause suffocation. Move clear of the area immediately after use and ventilate the area once the fire is out.

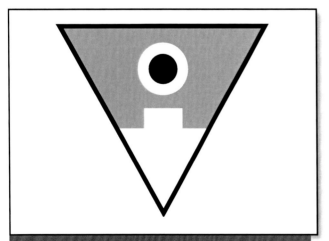

Figure 1.74 Carbon dioxide indicator
Identifying colour: RED container with BLACK band

Note: Halon-type extinguishers (all yellow) have been found to have ozone-depleting potential, and their use is being phased out by new government regulations.

Operating instructions for any extinguisher can be found on the label fixed to that extinguisher.

Other firefighting equipment

Fire blankets

Fire blankets can be used on most types of fires (Figures 1.75 and 1.76). The action of the blankets is to smother the fire. These are very useful on burning oils and electrical fires. Keep the blanket in place until the fire is out and long enough to allow the flammable substance to cool below its reignition point.

Figure 1.75 A typical fire blanket packet

Figure 1.76 A fire blanket in use

Hose reels

Where hose reels are provided they are a very effective means of fighting Class A fires (Figure 1.77). Extreme care is required to ensure that no live electrical equipment is in the area. They should not be used on liquid fires, as water may spread the burning liquid.

Figure 1.77 Hose reel

Worksheet 1

Student name: _____

Enrolment year: _____

Class code: _____

Competency name/Number: _____

To be completed by teachers:	
Student competent	☐
Student not yet competent	☐

Task: Read through the sections beginning at *OH&S introduction and responsibilities* and up to and including *What are Codes of Practice?*, then complete the following.

Q. 1 OH&S self-regulation wasn't working. Approximately how many people were dying per year throughout Australia before the Act came into being?

_____

Q. 2 Employers must provide for the health, safety and welfare of their employees at work. Briefly describe three areas that employers must address:

1. _____
_____
_____

2. _____
_____
_____

3. _____
_____
_____

Q. 3 Complete the following statement: 'Employers must not require employees to _____ for anything done or provided to meet specific requirements made under the Acts or associated legislation. They must also ensure the _____ and _____ of people _____ their places of work who are not _____.'

Q. 4 Employees must take reasonable care of the health and safety of others. Employees must cooperate with employers in their efforts to comply with OH&S requirements. Briefly describe two areas that employees must address:

1. _____
_____
_____

continued ▶

2. _____
_____
_____

Q. 5 What may be incurred if a construction worker is working on a construction site, without first undertaking the mandatory OH&S induction training?

_____
_____
_____

Q. 6 List three areas that should be addressed as part of a site induction training session:

1. _____
2. _____
3. _____

Q. 7 What are Codes of Practice?:

_____
_____
_____

Worksheet 2

Student name: _____

Enrolment year: _____

Class code: _____

Competency name/Number: _____

To be completed by teachers:	
Student competent	☐
Student not yet competent	☐

Task: Read through the sections beginning at *Safe work practices* and up to and including *Personal cleaning procedures*, then complete the following.

Q. 1 Briefly describe the aims of the National Standard for Manual Handling and the Code of Practice for Manual Handling:

_____
_____

Q. 2 State the four main bodily injuries that occur due to poor or incorrect manual handling techniques:

1. _____
2. _____
3. _____
4. _____

Q. 3 List the six main steps to follow for correct lifting:

1. _____
2. _____
3. _____
4. _____
5. _____
6. _____

Q. 4 State the importance when shovelling materials of the length of the handle of the shovel being used:

_____
_____

continued ➤

Q. 5 State a suitable mechanical aid to use for the following situations:

1. To carry concrete, bricks, tools, rubbish etc. around a building site.
   _____

2. To lift loads too heavy for manual lifting techniques.
   _____

3. To lift awkward or heavy sheet materials.
   _____

Q. 6 Maintaining a worksite in a safe and clean manner will improve what three main areas?

1. _____
2. _____
3. _____

Q. 7 What are three areas of concern dusts in the workplace may cause?

1. _____
2. _____
3. _____

Q. 8 State one method of suppressing dust on-site:
_____

CHAPTER 1 OCCUPATIONAL HEALTH AND SAFETY — 53

Worksheet 3

Student name: _____

Enrolment year: _____

Class code: _____

Competency name/Number: _____

To be completed by teachers:
Student competent ☐
Student not yet competent ☐

Task: Read through the sections beginning at *Personal protective equipment (PPE) and clothing* and up to and including *Signs and labelling*, then complete the following.

Q. 1 Briefly describe the main purpose and function of PPE:

_____

_____

_____

Q. 2 List two items of PPE suitable to protect the following body areas:

Body area		
Head	_____	_____
Eyes/face	_____	_____
Hearing	_____	_____
Airways/lungs	_____	_____
Hands	_____	_____
Feet	_____	_____
Body	_____	_____

Q. 3 PPE is designed to protect against specific hazards. Briefly describe each of the following hazard groups:

1. Physical hazards _____

_____

2. Chemical hazards _____

_____

Q. 4 What is the name given to the stretching impact barrier placed inside a safety helmet between the skull and shell of the helmet?

_____

Q. 5 How is the back of the neck protected from sunburn when wearing a safety helmet?

_____

_____

continued ➤

Q. 6 List the three hazard categories that eye protection is designed for:

1. _____

2. _____

3. _____

Q. 7 State the two main PPE methods used to protect hearing:

1. _____

2. _____

Q. 8 State a suitable use for P2 (class M) class filters:

_____

_____

Q. 9 State the main requirements of all safety footwear to provide maximum protection:

_____

_____

_____

Q. 10 What is the minimum protection rating of sunscreen for use by construction workers?

_____

CHAPTER 1 OCCUPATIONAL HEALTH AND SAFFTY   55

Worksheet 4

Student name: _____

Enrolment year: _____

Class code: _____

Competency name/Number: _____

To be completed by teachers:	
Student competent	☐
Student not yet competent	☐

Task: Read through the sections beginning at *Safety signs and tags* and up to and including *Placement of safety signs*, then complete the following.
(Also see the coloured Appendix at the end of the textbook.)

Q. 1 Briefly describe the three main categories of common safety signage used in the building industry:

1. Picture signs _____
_____
_____

2. Word-only messages _____
_____
_____

3. Combined picture and word signs _____
_____
_____

Q. 2 Briefly describe the shape, colours and detail/symbol found on a sign used to indicate a toxic hazard:

1. Shape _____
2. Colours _____
3. Detail/symbol _____

Q. 3 State the colours used for a sign that indicates a danger or risk to your health:

1. Background _____
2. Symbol _____
3. Border _____

Q. 4 Signs should be placed in a position that allows them to be clearly seen. State the preferred position for safety signs:
_____
_____

continued ▶

Q. 5 Identify the following signs, stating what they represent:

Figure 1.78 Various signs

SUGGESTED ACTIVITY **one**

Video: View the video produced by TAFE and C&T ESD related to signs and tags.
Title: *Safety Signs and Tags for the Occupational Environment*
Running time: 10 minutes
Available from: Resource Distribution Unit, 3/61–71 Rookwood Road, Yagoona, NSW 2199
Ph: 9793 3347 Fax: 9793 3242

Task: Engage in a class discussion, directed by your teacher/instructor, to identify important issues.

CHAPTER 1 OCCUPATIONAL HEALTH AND SAFETY 57

Worksheet 5

Student name: _____

Enrolment year: _____

Class code: _____

Competency name/Number: _____

To be completed by teachers:	
Student competent	☐
Student not yet competent	☐

Task: Read through the sections beginning at *Risk assessment* and up to and including *Workplace inspections*, then complete the following.

Q. 1 Briefly describe the difference between an 'acute hazard' and a 'chronic hazard':

_____

_____

_____

_____

_____

Q. 2 List the five main common workplace hazard areas:

1. _____

2. _____

3. _____

4. _____

5. _____

Q. 3 In relation to hazard control procedures, list two methods of providing engineering controls to reduce the risk of the hazard:

1. _____

2. _____

Q. 4 Injuries to the human body and diseases caused by accidents or exposure to an unhealthy environment are wide and varied. List and describe any three effects or injuries to the body caused by work-related hazards:

1. _____

_____

_____

continued ▶

2. _____

3. _____

Q. 5 Who would normally carry out workplace inspections at a workplace with more than 20 employees?

_____

Q. 6 List four areas that safety signs in the workplace set out to address or make workers aware of:

1. _____
2. _____
3. _____
4. _____

Worksheet 6

Student name: _____

Enrolment year: _____

Class code: _____

Competency name/Number: _____

To be completed by teachers:	
Student competent	☐
Student not yet competent	☐

Task: Read through the sections beginning at *Management of site hazards* and up to and including *Format of safe work method statements*, then complete the following.

Q. 1 Name the two documents that govern health and safety in the workplace of your relevant state or territory:

1. _____

2. _____

Q. 2 What are the three main tasks required to be completed on the front page of a safe work method statement?

1. _____

2. _____

3. _____

Q. 3 Safe work method statements are a compulsory part of any site safety management plan and should form part of the planning for any risky site task. These statements should also form part of student practical task preparation for you to safely plan and prepare for practical jobs in the workshop or within the college grounds.

Using the blank proforma overleaf, complete the three columns to identify the tasks, risks/hazards and safety controls required to carry out the construction of a *Timber Framing Joints* project.

continued ➤

	Safe Work Method Statement	Signed off: _____ (Teacher)

Student name: _____

Workshop/college grounds: _____

Practical project: _____

Signed off: _____ (Teacher)
Date: _____
Accepted: YES / NO

TASK/PROCEDURE	POTENTIAL HAZARDS	SAFETY CONTROLS

SUGGESTED ACTIVITY two

Video: View the video produced by Safetycare related to accidents and reporting.
Title: *Accident Investigation*
Running time: approx. 10 minutes
Available from: Safetycare Australia Pty Ltd, Ph: (03) 9569 5599

Task: After discussing the section on *Accident reporting*, view the video and complete the following questions.

Q. 1 What is the main aim of accident investigation?

Answer:

_____

Q. 2 What is an accident?

Answer:

_____

Q. 3 State the three main types of accidents.

Answer:

1. _____

2. _____

3. _____

Q. 4 Which is the main group that results in the accident being investigated?

Answer:

_____

Q. 5 What are the two things that all accidents have in common?

Answer:

1. _____

2. _____

Q. 6 What are the four main contributing factors to accidents?

Answer:

1. _____

2. _____

3. _____

4. _____

continued ➤

PAINTING AND DECORATING, AND MORTAR TRADES

Q. 7 Who are the five personnel who should be involved in an accident investigation?

Answer:

1. _____
2. _____
3. _____
4. _____
5. _____

Q. 8 List the four main points that form a strategy for the investigation procedure.

Answer:

1. _____
2. _____
3. _____
4. _____

Q. 9 What is the most important factor in relation to accurate recording and recollection of the events of the accident?

Answer:
_____

Q. 10 What are the five main benefits of an accident investigation?

Answer:

1. _____
2. _____
3. _____
4. _____
5. _____

Worksheet 7

Student name: _____

Enrolment year: _____

Class code: _____

Competency name/Number: _____

To be completed by teachers:	
Student competent	☐
Student not yet competent	☐

Task: Read through the section beginning at *Accident reporting* up to and including *Payment of compensation*, then complete the following.

Q. 1 What is the responsibility of every worker at a place of work in relation to illness, accident or near miss?
_____

Q. 2 What action should a supervisor take when made aware of illness, accident or near miss?
_____

Q. 3 Define a reportable incident.
_____

Q. 4 List 5 unsafe acts which can lead to accidents.

1. _____   2. _____
3. _____   4. _____
5. _____

Q. 5 List 7 posssible types of unsafe conditions.

1. _____   2. _____
3. _____   4. _____
5. _____   6. _____
7. _____

Q. 6 In what circumstances can illnesses be reportable incidents?
_____

Q. 7 Define a near miss.
_____

Q. 8 Give examples of a near miss.

1. _____   2. _____
3. _____   4. _____

Q. 9 How can a claim for compensation be made?
_____

Worksheet 8

Student name: _____

Enrolment year: _____

Class code: _____

Competency name/Number: _____

To be completed by teachers:	
Student competent	☐
Student not yet competent	☐

Task: Read through the sections beginning at *Emergency procedures* and up to and including *Other firefighting equipment*, then complete the following.

Q. 1 State why it is necessary for an organisation to have an emergency procedure in place:

_____

_____

Q. 2 Apart from giving clear instructions, recording what was carried out and reporting details to the emergency coordinator, state two other responsibilities of a warden or controller:

1. _____

_____

2. _____

_____

Q. 3 With regards to first aid, what are three areas of concern that an employer must ensure are sufficiently covered at their place of work?

1. _____

2. _____

3. _____

Q. 4 In an emergency situation where a person is killed, who must be contacted?

_____

Q. 5 List three common causes of fires on-site:

1. _____

2. _____

3. _____

Q. 6 State the three elements required to start and sustain a fire:

1. _____

2. _____

3. _____

CHAPTER 1 OCCUPATIONAL HEALTH AND SAFETY   **65**

Q. 7 List three ordinary combustible materials that may be used as fuel for a Class A fire:

1. _____

2. _____

3. _____

Q. 8 State the source or fuel and most suitable types of fire extinguishers for use on the following classes of fires:

Class A

Source/Fuel _____

Suitable extinguishers _____

_____

Class B

Source/Fuel _____

Suitable extinguishers _____

_____

Class E

Source/Fuel _____

Suitable extinguishers _____

_____

Q. 9 What type of extinguishers should never be used for Class E fires?

_____

Q. 10 Fire extinguishers come in a variety of easily identifiable colours. State the common colour used to identify a foam-type extinguisher:

_____

SUGGESTED ACTIVITY **three**

Video: View the video produced by Safetycare related to workplace safety committees.
Title: *You Can't Do It Alone—Making Safety Committees Work*
Running time: Part 1, approx. 12 minutes
Available from: CFMEU and Master Builders Association

Task: Review and discuss as a whole-class group.

SUGGESTED ACTIVITY four

Video: View the video produced by Safetycare related to fires and their containment.
Title: *Fire Awareness*
Running time: approx. 16 minutes
Available from: Safetycare Australia Pty Ltd, Ph: (03) 9569 5599

Task: After viewing the video, complete the following questions.

Q. 1 State the three main fire safety management elements.
Answer:

1. _____
2. _____
3. _____

Q. 2 What is the name given to the chemical reaction that results in fire?
Answer:

_____

Q. 3 What are the two forms of energy created by fire?
Answer:

1. _____
2. _____

Q. 4 What are the main elements of fire?
Answer:

1. _____
2. _____
3. _____

Q. 5 State four sources of heat for fires:
Answer:

1. _____
2. _____
3. _____
4. _____

Q. 6 List the five main principles of fire:
Answer:

1. _____
2. _____
3. _____
4. _____
5. _____

Q. 7 Why should you close windows and doors behind a fire?
Answer:

_____

Q. 8 State the three essential fire management elements.
Answer:

1. _____
2. _____
3. _____

Q. 9 What is the most common method of reducing the heat of a fire?
Answer:

_____

Q. 10 What are the three main causes of death from fire?
Answer:

1. _____
2. _____
3. _____

REFERENCES AND FURTHER READING

Acknowledgment
Reproduction of the following *Resource List* references from *DET, TAFE NSW C&T Division (Karl Dunkel, Program Manager, Housing and Furniture) and the Product Advisory Committee* is acknowledged and appreciated.

Texts

Graff, D.M. & Molloy, C.J.S. (1986), *Tapping group power: A practical guide to working with groups in commerce and industry*, Synergy Systems, Dromana, Victoria

National Centre for Vocational Education Research (2001), *Skill trends in the building and construction trades*, National Centre for Vocational Education Research, Melbourne

NSW Department of Education and Training (1999), *Construction industry: Induction & training: Workplace trainers' resources for work activity & site OH&S induction and training*, NSW Department of Education and Training, Sydney

NSW Department of Industrial Relations (1998), *Building and construction industry handbook*, NSW Department of Industrial Relations, Sydney

TAFE Commission/DET (1999/2000), 'Certificate 3 in General Construction (Carpentry) Housing', course notes (CARP series)

WorkCover Authority of NSW (1996), *Building industry guide: WorkCover NSW*, Sydney

Web-based resources

Regulations/Codes/Laws

<www.austlii.edu.au/databases.html> Laws database

<www.workcover.nsw.gov.au> Codes of Practice etc.

<www.workcover.vic.gov.au> Victorian WorkCover

Resource tools and VET links

<www.resourcegenerator.gov.au> ANTA resource generator

<www.ntis.gov.au> National Training Information Service

<http://hsc.csu.edu.au/construction> NSW HSC Online—Construction

Industry organisations' sites

<www.citb.org.au> SA Construction Industry Training Board

<www.btgda.org.au> Building Trades Group Drug & Alcohol

<www.citab.com.au> NSW CITAB home

<www.W.A.gov.au> manual handling website WA Government OH&S site—manual handling

<www.workcover.nsw.gov.au/Publications/Industry/Construction/demolition.htm> WorkCover NSW (1999) Hazard Profile for Demolition

Audiovisual resources

Title: *Safety Signs and Tags for the Occupational Environment*

Running time: 10 minutes

Author: TAFE Construction and Transport Division

Available from: Resource Distribution Unit 3/61–71 Rookwood Road, Yagoona, NSW 2199 Ph: 9793 3347 Fax: 9793 3242

Title: *Accident Investigation*

Running time: approx. 10 minutes

Author: Safetycare Australia Pty Ltd

Available from: Safetycare Australia Pty Ltd Ph: (03) 9569 5599

- (c1990), *Easy Guide to Worksite Safety*, Pt 2, Workplace Video Productions

- TAFE NSW (2000), *Safety Signs and Tags*, C&T Division, Sydney
- TAFE NSW (2000), *Temporary Power to a Job Site*, C&T Division, Sydney

Australian Standards

AS/NZS 1269.0:2005, 'Occupational noise management—Overview and general requirements'

AS 1470-1986, 'Health and safety at work—Principles and practices'

AS/NZS 1800:1998, 'Occupational protective helmets—Selection, care and use'

AS/NZS 2210.1:2010, 'Safety, protective and occupational footwear—Guide to selection, care and use'

AS 2436-2010, 'Guide to noise and vibration control on construction, demolition and maintenance sites'

State Occupational Health and Safety Acts

Occupational Health and Safety Act 2000 (NSW)

Workplace Health and Safety Act 1995 (Qld)

Occupational Safety and Health Act 1984 (WA)

Occupational Health, Safety and Welfare Act 1986 (SA)

Workplace Health and Safety Act 1995 (Tas)

Occupational Health and Safety Act 2004 (Vic)

WorkCover NSW Publications

'Applying the new safety regulations', Cat. no. 229 (also 100, 110, 1008, 2001)

'Back watch industry profile—Construction trades', Cat. no. 718

'Manual handling', Cat. no. 9020

'Occupational protective gloves', Cat. no. 3017

'Personal protective equipment', Cat. nos 032, 208, 310, 3003, 3010, 3012, 3017, 3019, 3029, 4005, 4007, 4500

'Protection from UV radiation in sunlight', Cat. no. 9017

'Protective helmets standard', Cat. no. 3012

'Reading labels on material safety data sheets', Cat. no. 400

'Safety helmets', Cat. no. 4500

'Skin cancer and outdoor workers', Cat. no. 116, 117

'Work method statements', Cat. no. 231

CHAPTER 2

WORKING EFFECTIVELY IN THE GENERAL CONSTRUCTION INDUSTRY

This chapter looks at the historical background and underpinning knowledge required for preparation to work effectively within the general construction industry. It covers the identification and scope of the general construction industry work context, employment conditions, responsibility of individuals, working effectively within a team, career path options and how to participate in site meetings.

Areas addressed from the unit of competency include:
- the industry work context and setting;
- organising and accepting responsibility for own workload;
- working in teams;
- participating in identifying and pursuing own development needs; and
- participating in site meetings.

HISTORICAL INDUSTRY BACKGROUND

Human beings have not always built. In ancient times they used caves and makeshift shelters while moving from place to place in search of food. It was not until they were able to grow their own food that they were able to settle in one place.

Human needs over time have not changed, as satisfying the physiological and safety requirements is still the first priority for all civilisations, according to Maslow, a renowned humanist psychologist (Figure 2.1). As shelter was one of the most basic needs, humans developed the principles of construction, and invented tools to build using the materials available.

Early buildings were not built to last, but the establishment of more permanent settlements allowed a change in building materials to take place from mud and straw to bricks, stone, timber, concrete and steel.

The principles of construction have developed from the simple post and lintel—seen at Stonehenge, England; arches—seen in the Roman aqueducts; domes—seen in the Roman Pantheon and Byzantine domes on a square-based pendentive; barrel vaults—seen in Roman, Romanesque and Renaissance structures; buttresses—seen on English and French cathedrals (e.g. Notre Dame Cathedral); the reinforced concrete slab technique—seen in the architect Le Corbusier's reinforced concrete house; steel-framed construction—seen in the Eiffel Tower, Paris; to the reinforced and prestressed concrete constructions of today—as seen in the Sydney Opera House and the Sydney Centrepoint Tower (Figures 2.2 to 2.11).

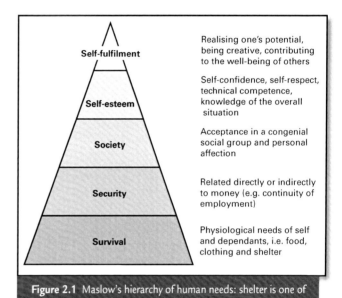

Figure 2.1 Maslow's hierarchy of human needs: shelter is one of our most basic needs

Figure 2.2 Primitive post and lintel construction as seen at Stonehenge, Salisbury, England, c. 1800-1400 BC

Figure 2.3 Roman aqueduct lined with lead and supported by stone arches, built in Segovia, Spain, nearly 2000 years ago

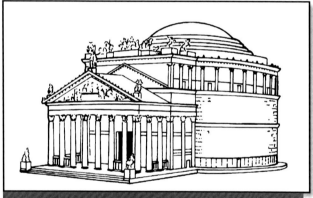

Figure 2.4 The Pantheon, Rome, 120–124, showing dome roof

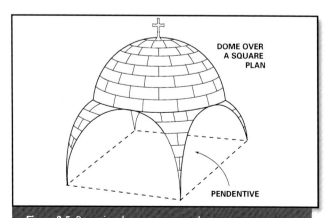

Figure 2.5 Byzantine dome on a square base

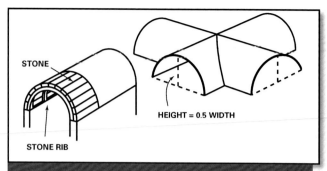

Figure 2.6 Barrel or tunnel vault and barrel groin vault. Vaults were used in Roman, Romanesque and Renaissance structures from 500 BC to c. 1500 AD

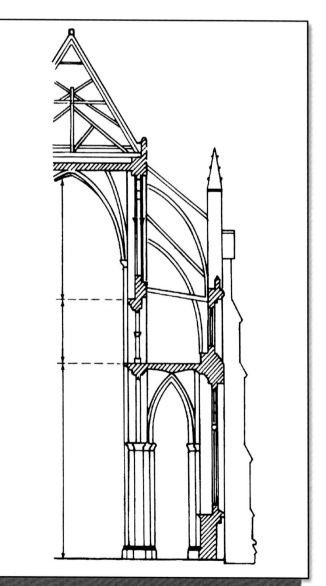

Figure 2.7 Gothic buttresses similar to those found on Notre Dame, 1163–c. 1250

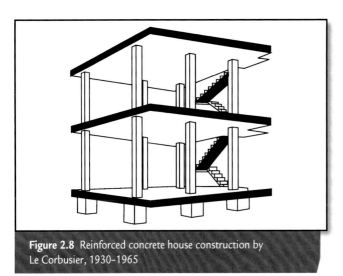

Figure 2.8 Reinforced concrete house construction by Le Corbusier, 1930–1965

Figure 2.10 Sydney Opera House, designed by Danish architect Jørn Utzon and built 1959–1973

Figure 2.9 Steel-framed Eiffel Tower, stands 300 m tall. Designed and built by Gustave Eiffel in 1889

Figure 2.11 Sydney Tower, completed in 1981, stands 304.8 m above street level, including a 30 m spire

Function

Buildings were used not only for shelter but also as tombs for burials, monuments, palaces from which to govern, shrines where gods could be worshipped, and for the storage of materials. The relationship between the materials used and the function of the building was a close one, especially if the building was to last for centuries.

Only since the Industrial Revolution during the second half of the 18th century has technology begun to catch up with building, which now appears to have been left far behind. In the past 200 years building operations have become largely mechanised through the use of machines such as excavators, concrete mixers and power tools.

Our creative thinking about building is still rooted in the 18th century—our walls, floors and roofs are not very different from those of 200 years ago. The virtual explosion of scientific development during the past 40 years or so has left many aspects of building largely untouched. For example, the automobile has been around for over 100 years but we no longer use cars of that age, whereas we still live in and renovate houses that are over 100 years old and these are still considered to be comfortable and functional.

The Australian building industry

Early forms of construction in Australia commenced shortly after Captain Arthur Phillip landed in Sydney Cove in 1788 with his charge of convicts and soldiers. Phillip foresaw the need for housing. Tents were used, but as these were designed for a more temperate climate they were quickly discarded in favour of more permanent structures, in keeping with the climate of eastern Australia.

So commenced the Australian style of construction, which enhanced and further improved more traditional methods brought to the colony from England. The early form of construction was mainly timber frame with earthen floors. Wall and roof construction consisted of small-diameter trees cut and shaped for the various components.

External wall claddings consisted of a range of materials, such as vertically fixed split timber, which were brought to a much more acceptable form of finish by thatching branches of the plentiful wattle tree, weaving them onto wall framing and covering them with several coats of rendered mud (Figure 2.12). Roof coverings for these early buildings consisted of bark stripped from large-diameter mature trees, or reeds woven into a matrix, commonly known as thatch.

Figure 2.12 A wattle-and-daub hut typical of those constructed by early settlers

Early colonial building

With the advent of locally manufactured clay bricks in the early colonies, the style of constructed buildings resembled the more traditional forms existing in the homeland of the first European settlers. Walls were made of either thick, solid brick or locally quarried sandstone, as found at Pyrmont and Woolloomooloo circa 1800. Roofs were covered with either split timber shingles or, as in later developments, a locally pressed and fired clay roofing tile (Figure 2.13).

As the various colonies prospered, so did the various forms of construction. Features such as the damp-proof course in wall construction, properly constructed chimneys and fireplaces, and more elaborate joinery inclusions enhanced the habitable qualities of the early homes, most of which were built originally for merchants, land owners and senior officers of the military.

Figure 2.13 Elizabeth Farm, Parramatta, Australia's oldest surviving building, c. 1794

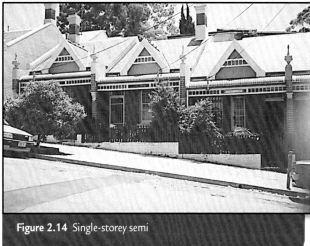

Figure 2.14 Single-storey semi

With the requirement to procure further penal accommodation and relieve the overcrowding in Sydney Cove, other states such as Tasmania (then known as Van Diemen's Land) and Queensland were established as outlying annexes to accommodate further shipments of prisoners from England. By then the Australian climate had started to influence methods of construction and styles of **architecture**.

Progress in building styles

Queensland pioneered the development of the all-timber form of construction, as the local climate conditions could not justify heavy masonry construction. Early and successful examples of these timber buildings were often designed and built solely by landowners, and were outside the influence of architects and traditional builders. This gave birth to the traditional homestead featuring wide verandas. The overall profile of the structure was kept low to the ground to reduce the ingress of the morning and afternoon sun.

Construction processes then further improved, with Sydney and Melbourne adopting the terrace form of construction. These houses were English in style and utilised maximum coverage of available land.

A further development from the terrace was the semi (Figure 2.14), a pair of adjoining buildings separated by a common dividing wall of masonry construction. These styles are at present being revived to make more efficient use of the available land in Australian capital cities.

As housing styles changed, so did people's expectations. This resulted in the creation of the 'Australian dream', a quarter-acre (1000 m^2) block with a two- or three-bedroom cottage.

The typical cottage built in any one of Australia's states or territories would rank among the best in the world when affordability and quality were considered. The purchase of a home is the single largest investment most people make in a lifetime; therefore, price and quality are important.

The Federation period

Federation architecture appeared around 1900 with the introduction of a style known as Edwardian. The design consisted of a red-brick house with a steeply pitched roof covered with orange terracotta roof tiles, and decorative finials and saddles on the ridges. Most of the interior fixings and fitments echoed the **Art Nouveau** style, which was based on organic or plant-like forms, giving the walls, ceilings, painted and papered surfaces, furniture, carpets and so on, a very decorative finish.

During the 1920s a new style of art and architecture arrived in Australia, known as the **Art Deco** style. This was based on symmetrical and, to a large extent, geometrical shapes. The style was strongly influenced by Egyptian papyrus motifs, as it was at this time that the mummified body of the boy king Tutankhamen was found by an archaeologist in an Egyptian tomb. A good example of this architecture may be found on the exterior and interior of the Chrysler Building in New York.

During this time the styles began to change from being predominantly European to very American, as typified by the bungalow style of cottage. The bungalow style took off in Australia around 1914, and continued to be popular through to the 1940s. At first the style consisted of a narrow, partly enclosed front veranda and steeply pitched roof line, and was known as the Federation bungalow. Gradually it changed to resemble the wide-fronted, partly enclosed front veranda and lower-pitched roof line of the American bungalow. Other typical aspects of this style include a roughcast skirt around the base of the cottage and the veranda piers, while the gable ends would have exposed ends and raked eaves.

Another significant style of the period between 1930 and 1940 was the Spanish mission style, which had similar lines to the bungalow but with a more Mexican pueblo look, made obvious by the inclusion of semicircular arched windows, twisted supporting columns at the front, and a light-coloured render or white-paint wall finish.

Construction developments—cavity walls

Another important change in construction technique came with the introduction of cavity walling. The first recorded types of cavity wall were known as *proto-cavity* and found in Victoria in the 1850s. The true cavity wall was first noted in 1868 in Bendigo and Stawell, Victoria. These walls were of a masonry construction, with both internal and external skins being either brick or brick and stone, and with a cavity of approximately 50 mm. The skins were linked by metal ties such as hoop iron straps, which were tarred and sanded to prevent rusting. Later, wrought and cast iron ties were introduced from overseas, and the galvanised wire or Morse tie was invented and introduced from America around the 1890s. It is still used today.

Brick veneer construction is virtually unknown overseas—and even in some areas of Australia—but it has been adapted to suit our conditions. It consists of a timber frame which supports the load and forms the real structure, surrounded by a leaf of brickwork to give the appearance of a solid brick structure. Experimental versions date back to 1903 and can be found in country areas, while houses made partly of brick were built in Melbourne from about the time of the First World War. It was only around 1928 that brick veneer became widely accepted by municipal councils and lending institutions.

Our buildings have further improved due to the availability of a variety of materials, technological advances, refined techniques and the working climate. Changes in the approach to work and improvements in the industry have largely occurred due to the building workers' trade union movement. Trade unions took a leading role in the battle for the 'eight-hour day', which occurred between 1855 and 1858, and the building industry was the first in the world to have a 48-hour working week, which has since been modified to a 38-hour week. The Australian Builders Labourers Federation was established in 1910 and has campaigned tirelessly for improved wages and conditions for its workers. It even fought a successful campaign to retain green spaces in Sydney's inner-city area during the late 1960s and early 1970s. The union was deregistered during the 1980s after misconduct was proved against its leading officials. The present building workers' union is known as the CFMEU (Construction, Forestry, Mining and Energy Union).

Apart from unions, employers may belong to associations that may act on their behalf in disputes, and which are also responsible for training and the promotion of new technologies to a large number of building workers. The two main associations are the HIA (Housing Industry Association) and the MBA (Master Builders Association).

Industry structure

The building industry is basically divided into two groups: the commercial or large business sector, and the residential or smaller domestic sector.

Commercial buildings can be classified as:
- shopping centres;
- factories;
- warehouses;
- multistorey buildings;
- carparks;
- hospitals;
- schools;
- hotels;
- offices.

Residential buildings can be classified as:
- single dwellings;
- boarding houses;
- flats;
- villas;
- townhouses;
- guest houses;
- hostels.

Traditional trade areas

Building workers on either commercial or domestic sites usually fit into one of the following trade categories:
- carpenter and joiner;
- bricklayer;
- stonemason;
- metalworker;
- plumber;
- drainer;
- plasterer (render and plasterboard);
- wall and floor tiler;
- roof tiler;
- painter and decorator;
- glazier;
- concreter;
- electrician;
- rigger;
- scaffolder;
- labourer;
- trades assistant;
- machine operator.

Current trends and technological development

Due mainly to the loss of skilled people and the shortage of materials after the end of the Second World War in 1945, the design of cottages was simplified, and more timber-framed structures clad with asbestos cement sheets—the wonder product of the day—were built, a trend that continued right up into the 1960s. During the 1960s roofs began to be covered with concrete roof tiles and walls were lined with plasterboard instead of fibrous plaster. Towards the end of the 1960s, brick veneer construction came back into vogue and is still the most commonly used method of construction throughout major centres of Australia.

Technological improvement has allowed a wider range of materials and fabrics to be used in the construction of residential cottages, including a variety of patent-type floor systems, such as suspended steel framing and reinforced concrete slab-on-ground, precast tilt slab walling and aerated block walling, lightweight timber trusses and a variety of improved roof coverings such as concrete tiles, terracotta tiles and Colorbond® metal sheeting.

Twentieth-century advances in technology have been notable: for example, multistorey skyscrapers could not have been built without the invention of the elevator early in the last century, or the use of steel reinforced concrete. More recent advances include the use of computers to design, calculate and solve engineering problems, such as those encountered on Jørn Utzon's Sydney Opera House, the Sydney and Melbourne football stadiums and the new Parliament House in Canberra. Computer-controlled or programmed machines make light work of tedious jobs such as pre-cutting timber frames and trusses, polishing or grinding precast concrete facings to buildings, and programming for laser cutting equipment. CAD, or computer-aided design, has allowed the designer the luxury of seeing the building or components of the building in three-dimensional form before the designs are finalised. Other advances include prefabrication of many elements, the use of new versatile materials such as the autoclaved concrete Hebel blocks, improved adhesives and premixed plastering compounds.

Apart from new materials and technology there has been a change in building trade culture, which traditionally has been a male-dominated area. In recent years, female building workers have begun to take their place alongside their male counterparts and now make up 8%–10% of the trade areas. Commonwealth incentives are increasing this combined intake, which should make the Australian building industry a versatile and progressive area for the future.

THE ROLE OF EMPLOYERS, EMPLOYEES AND WORKPLACE COMMITTEES

Building and construction sites are complicated workplaces involving a wide range of issues, including people working close together and fitting into a hierarchical system where line management is used to control the day-to-day running of the site. Each person has an important role to play within this hierarchy, and if the site is to run smoothly there needs to be good communication, civility and respect between workers and management.

The types of issue encountered on any building site, whether on a large, multistorey construction or a single residential dwelling, to some degree will include the following:
- industrial relations—people working and interacting together in a workplace;
- awards—covering pay and conditions for employees;
- resolution of disputes—how disputes occur and how they are resolved;
- grievance procedures—the methods used to settle disputes;
- trade unions—representatives of employees, e.g. CFMEU;
- employer associations—representatives of employers, e.g. HIA, MBA;
- changes affecting the building industry—award restructuring, enterprise bargaining, technological improvements etc.;
- training—introduction of national, competency-based training packages;
- workplace committees—employee negotiators;
- site safety induction—mandatory training for all new recruits to the industry;
- roles of management—position and responsibility within the site hierarchy;
- career paths—choosing a specific job pathway in the building industry;
- building sectors—the main differences and skills between these sectors.

All these issues affect every individual either directly or indirectly while working in the industry as part of management, a supervisor, an employee or a subcontractor. Therefore, it is essential that everybody has a basic understanding of these issues to allow for workplace harmony and individual improvement.

Industrial relations

Industrial relations is about people and organisations working together within the social and political systems of our society. Employment makes up a large part of our lives and determines our living standards, while the industrial relations process determines the employment conditions of the **environment** in which the employee works.

Industrial relations issues include occupational health, safety and welfare, childcare, new technology, social welfare, unemployment, illness, redundancy, wages, award restructuring, multi-skilling, career paths and early retirement.

There are two industrial relations systems in Australia: the federal system and the state system. The federal system is independent of all the state systems and is intended to cover industrial problems that are larger and more far-reaching than those in any state. The system is regulated by the *Conciliation and Arbitration Act 1904*.

Each state has its own industrial relations system and there are no constitutional limitations. As a result, state systems have a broader scope of operation than the federal system.

In NSW, the *Industrial Arbitration Act 1904* established the Industrial Commission and conciliation committees. These bodies are responsible for making variations to state awards. Matters of importance and public interest are referred from the committees to the Industrial Commission.

Awards

An award is the law that establishes the wages and conditions of employment in defined industries or occupations. An award is made after a dispute is registered by a decision of, for example, the Industrial Commission of NSW for state awards in NSW, and the Conciliation and Arbitration Commission for federal awards. An award provides for minimum wages and conditions (e.g. overtime, sick leave, annual leave loading and occupational health and safety).

Resolution of disputes

A dispute exists when conflict arises out of a disagreement over the rights and interests of two parties. A dispute can occur when a job done by one member of a union should in fact belong to a member of another union. This is commonly referred to as **demarcation**.

Industrial action taken by employees includes:
- overtime bans;
- work-to-rule campaigns;
- go-slows;
- picketing;
- strikes.

Industrial action taken by employers includes:
- blacklisting;
- lockouts.

Employees are paid for striking only when this is directed by their union and if there are sufficient funds. Office workers cannot afford to stay away from work, because strikes always result in loss of wages.

Grievance procedure

It is usually in the interests of the worker, the employer and the government to settle disputes as quickly as possible. If unions and employers cannot settle their differences by discussion among themselves, they can make use of conciliation and arbitration procedures.

One of the principal ways of resolving disputes or disagreements in the workplace is through grievance or dispute-settling procedures. These include:
- negotiation;
- collective bargaining;
- enterprise bargaining;
- mediation;
- conciliation;
- arbitration;
- other tribunals.

The resolution of industrial disputes involves changes to the wage and/or non-wage aspects of the employment relationship.

Trade unions

A trade union is an association formed by employees to act for them inside and outside the workplace. Historically, trade unions arose in Australia as organisations for the defence and improvement of the conditions of various sections of the workforce. Unions also exert influence in environmental issues and in the pressure for public facilities. Unionism is a useful tool in the area of negotiations between worker and employer.

Unions usually employ organisers whose job covers promoting membership, contact with shop stewards, assisting shop stewards in difficult negotiations at the workplace, representing workers in negotiations, court appearances on award matters, and liaising with employer associations, other unions, the state labour councils, the ACTU, the media and politicians.

Employer associations

Employer associations are organisations formed by management to act for them outside the workplace and to provide information and advice. The Australian Chamber of Commerce and Industry (ACCI) is the largest single organisation representing industry and commerce in Australia.

Rights and obligations of trade union members

Union members are obliged to:
- pay regular dues;
- abide by union rules;
- pay levies for specific purposes;
- encourage other employees to join the union;
- support fellow members discriminated against or victimised by employers;
- press for improved working conditions where necessary;
- ensure that wages are paid;
- support any stop-work meetings or strikes that may be called in members' interests;
- encourage fellow unionists to attend union meetings off the job.

Union members have the following rights:
- access to all services provided by the union;
- legal aid and funeral benefits;
- protection from unfair dismissal;
- long-service leave;
- improved amenities at the workplace;
- improved safety measures at work;
- improved job security;
- higher wages.

In general, members should take an interest in issues not directly connected with the workplace, such as conservation, politics, and the spread and development of technology.

Changes in the building industry

The National Training Reform Agenda was introduced by the federal government, in response to the Carmichael Report of 1989, to increase productivity and competitiveness. It is a collection of government policies aimed at improving vocational education and training from an educational perspective, and at making it more responsive to the needs of industry.

These improvements include:
- training and skills development at all levels of the workforce;
- a diverse and efficient training market;
- an emphasis on competence rather than time served;
- more flexible approaches to training;
- nationally consistent arrangements for standards and qualifications;
- improved access for target groups;
- improved articulation arrangements within and across sectors.

Components of the National Training Reform Agenda are:
- competency-based training;
- competency standards;
- recognition of training;
- curriculum, delivery and assessment;
- entry-level training;
- training market;
- funding training;
- access and equity.

Entry-level training offers a number of training schemes in building and construction, with some new school-based apprenticeships being made available. The Australian New Apprenticeships System offers apprenticeships in over 500 occupational areas, including bricklaying, carpentry, joinery and construction carpentry, through to stonemasonry, wall and floor tiling within the building industry sector. All are aligned to a current award or industrial agreement. NSW has retained the existing system of trades and callings under the *Commonwealth Industrial Training Act 1989*.

Further details may be obtained from a New Apprenticeships Centre (NAC) (at <www.newapprenticeships.gov.au>).

Award restructuring

To improve Australia's economic performance, we need to rely less heavily on primary products and to provide high-quality goods and services that can compete successfully in the global marketplace. Unions, employers and governments are finding common ground and working together to improve industry and workplace efficiency.

Measures that are taking place to do this include:
- linking wages to skills;
- broadening the range of skills of individual workers;
- developing career paths for workers—even those previously classed as unskilled;
- restructuring awards.

Enterprise bargaining

In its broadest sense, **enterprise bargaining** involves an employer negotiating directly with its employees with regard to wages, conditions and work practices for that particular workplace. The end result of enterprise bargaining is an enterprise agreement.

Enterprise agreement

An **enterprise agreement** is a contract between an employer and employees on wages and conditions of work in the employer's business. Enterprise agreements, or workplace agreements, fall into one of the following categories:
- a collective agreement involving a group of employees and an employer;
- an individual agreement or contract between an employer and an employee covering employment matters on an individual basis;
- independent contractor agreements or contracts, which are a special form of individual contract covering independent contractors who are sometimes engaged by a business to do work that otherwise would be done by employees.

Mechanism for obtaining agreements

If negotiations result in agreement on the content of an enterprise agreement, the following steps must then occur:
1. If there is a negotiator or negotiating team representing employees, agreement by the employees is required.
2. An agreement containing the agreed conditions should be drawn up on the appropriate form.
3. The agreement should be made official by lodging it for registration with the appropriate organisation.

Relevant industrial awards

Industrial awards will generally continue to have relevance, as enterprise bargaining will operate in tandem with the award system for the foreseeable future. A registered enterprise agreement forms part of a common law contract of employment for employees who are bound by it.

The biggest change that has occurred recently in the building construction industry is the National Training Reform Agenda. This recognises the centrality of work to people's lives and the needs of industry to develop adaptable workforces with a wide and diverse range of skills.

The reforms include:
- the introduction of a competency-based training system for vocational education and training (VET) that is focused on outcomes;
- a national framework for the recognition of training and qualifications;
- national consistency between training systems, allowing greater transferability of skills and qualifications based on National Industry Competency Standards;
- increased industry participation in VET;
- wages relative to acquisition of a demonstrated skill or competence.

The National Training Reform Agenda proposed wage structuring system, shown in Table 2.1, is one which the construction industry has looked at adopting.

Table 2.1 Wages relative to skills level, with proposed percentage

Construction worker level		Wage relativity
CW1	(a) On entry	85%
	(b) After three months	88%
	(c) After 12 months	90%
	(d) After skill assessment	92.4%
CW2		96%
CW3		100%
CW4		105%
CW5		110%
CW6		115%
CW7		120%
CW8		125%

To implement these reforms, the following pathways for education and training are being offered. Apprenticeships and traineeships are a good way to expand a business, increase its skills base and keep up with technological changes so that existing customers' needs are met and the demands of new and emerging markets are addressed. Apprenticeships and traineeships are jobs that combine work and structured training. Although they vary from one industry to another, all apprenticeships and traineeships include the following:
- paid employment under an appropriate industrial arrangement (e.g. an award or enterprise agreement);
- a training agreement/training plan or registration that is signed by both the employer and apprentice, or trainee, in conjunction with an RTO (registered training organisation), and then registered with the relevant state authority, e.g. a NAC (New Apprenticeship Centre);
- a training program, delivered by a registered training organisation, that meets the requirements of a declared apprenticeship or traineeship and leads to a nationally recognised qualification.

The duration of an apprenticeship is normally four years and is available for young people, as well as mature-age and special target groups such as women, or workers with disabilities. Subsidy levels are available for employers.

Training packages

Nationally endorsed training packages, developed by national industry training advisory bodies (ITABs), are being introduced into the building and construction sector. The training package forms the foundation of vocational education and training in the industry, and provides a range of flexible training options that can be used by employers and registered training organisations to train apprentices and trainees.

Training packages consist of three components: the qualification, the endorsed components, and the non-endorsed components.

The qualification includes: the final nationally recognised title and certificate of diploma which the training package sets out to achieve; for example, the final qualification for carpentry would be a Certificate III in General Construction—Carpentry.

The endorsed component includes:
- the relevant competency standards for the industry;
- the qualifications that can be achieved;
- the way in which competency standards are assessed.

The non-endorsed component includes:
- a range of resources to support learning and assessment;
- the professional development of teachers and trainers.

Note: These training packages and national competency standards are updated regularly and should be checked for currency through the National Training Information Service (NTIS) website at <www.ntis.gov.au>.

Workplace committees

Consultative committees

These committees consist of employees and middle management, and consider problems and make suggestions for policies in areas such as safety, health, social activities and amenities.

Works committees (employee negotiation teams)

It may be appropriate to have a team of employee negotiators representing different areas of the workplace. The team might consist of a full-time union official as team leader, supported by employees from different work areas. If a union is not involved, the team leader and other team members should be elected from among the employees to be represented by the team.

Occupational health and safety (OH&S) committees

All Australian states and territories now legislate to provide for workplace consultation via health and safety representatives, or health and safety committees. The legislation provides a systematic approach by which management and employee representatives can regularly discuss occupational health and safety issues, with the objective of preventing and resolving occupational health and safety problems in a self-regulatory manner.

A general rule expressed within the relevant Acts and Regulations for a majority of the states and territories across the nation, is that an OH&S committee should be established at a place of work if there are 20 or more persons employed.
Note: It is recommended that readers become familiar with their relevant state or territory guidelines for the establishment of OH&S committees, by either getting in contact with or visiting their respective OH&S Regulating Authority's website, as detailed in Chapter 1 (see Table 1.1 p. 3: Relevant Australian state and territory OH&S Acts and Regulations).

Safety induction training

Employers and principal contractors must ensure that persons carrying out the work have relevant training, including occupational health and safety training. This includes identifying the occupational health and safety training requirements of those who are carrying out any construction work activities. This is an existing obligation outlined within the relevant OH&S Acts and Regulations for Western Australia, Victoria, Queensland, New South Wales and South Australia. It applies to all persons carrying out work in the residential, commercial and high-rise sectors. It is also a requirement under the relevant Acts and Regulations that principal contractors and employers must not direct or allow a person to carry out construction work unless that person has completed occupational health and safety (OH&S) induction training as follows:

- general OH&S induction (meeting the criteria for the card system relevant to your state), consisting of a broad range of safety awareness instruction;
- work activity OH&S induction;
- site-specific OH&S induction.

The Acts and Regulations also requires that self-employed individuals must not carry out construction work until they have personally completed OH&S induction training. This induction training is to be provided for all workers at no charge.

The person responsible for the induction training must also provide a written statement for each person being inducted, stating that the person concerned has satisfactorily completed the training; listing the activities covered in the training; specifying the dates on which the training occurred; specifying the name and qualifications of the person who conducted the training; and signed by the person who conducted the training. Each person who has successfully completed the induction should keep the written statement until such time as they receive their plastic induction card (see Figures 1.1–1.5 in Chapter 1), which is to be produced if a safety officer asks for it.

Site induction must be provided for every construction site, as there may be site-specific safety issues or hazards that need to be identified. This means that each worker must undertake site safety induction prior to entering that site. (For further details and information, it is recommended that readers either contact their respective OH&S

Regulating Authority, or access their website, as detailed in Table 1.1.

Career paths for building personnel

Rather than having specific job titles such as carpenter, plumber or tiler, since the beginning of 1996 tradespeople have been uniformly referred to as construction workers. This generic title refers to all those entering a particular skill stream. There are five skill streams that make up the building and construction industry training framework. These are:
- civil operations;
- general construction (made up of the previously separate groups known as structures, and fit-out and finish);
- heavy engineering;
- services;
- off-site.

Civil operations

Civil operations include work using plant, bulldozers, excavators, graders, scrapers, front-end loaders, backhoes, skidsteer loaders and non-plant. This area involves the alteration and development of the physical environment. The work is generally carried out prior to building construction or formation of structures.

General construction

General construction includes demolition work, site preparation, foundation work, materials handling, structural corework, structural steelwork, structural cladding, structural finishing and structural framework—the work that needs to be done to ensure that the building is structurally sound.

It also includes dry skills for housing and commercial/industrial, wet skills and final finishes. These are the final trades for the internal finishing of buildings.

Heavy engineering

Heavy engineering includes:
- power stations, oil refineries, terminals and depots;
- chemical, petrochemical and hydrocarbon plants; and associated plant, plant facilities and equipment;
- major industrial and commercial undertakings and associated plant and plant facilities;
- plant and facilities in connection with extraction, refining and/or treatment of minerals, chemicals etc.;
- transmission and similar towers;
- transmission lines;
- associated plant, plant facilities and equipment.

The construction activities covered by the heavy engineering stream overlap with metal trades.

Services

Services include plumbing, mechanical services such as air-conditioning, fire protection, electrical/electronics, lifts and escalators. These services are needed for the building to function.

Off-site

Off-site includes joinery, shopfitting, maintenance, pre-cast concrete manufacture, prefabricated buildings and stonemasonry. This work does not need to be done on-site.

Levels of training

The levels of training may vary depending on the skill stream being undertaken within the qualifications framework applicable to that stream. For example, the General Construction Stream consists of a number of training package qualifications, which include:
- Certificate I in General Construction;
- Certificate II in General Construction;
- Certificate III in General Construction (Bricklaying/Blocklaying);
- Certificate III in General Construction (Carpentry);
- Certificate III in General Construction (Carpentry—Formwork/Falsework);
- Certificate III in General Construction (Concreting);
- Certificate III in General Construction (Demolition);
- Certificate III in General Construction (Dogging);
- Certificate III in General Construction (Painting and Decorating);

- Certificate III in General Construction (Rigging);
- Certificate III in General Construction (Roof Tiling);
- Certificate III in General Construction (Scaffolding);
- Certificate III in General Construction (Steel Fixing);
- Certificate III in General Construction (Wall and Ceiling Lining);
- Certificate III in General Construction (Wall and Floor Tiling);
- Certificate III in General Construction (Waterproofing).

Note: There are no Certificate I or II qualifications embedded within any of these qualifications. Other stream training package qualifications should be checked out by visiting either:
- ITAB: Industry Training Advisory Bodies; or
- DEST: Department of Education, Science and Training.

Competency-based training

Skills are very important in the construction industry: there are skills you already have and skills you need to acquire, and these can be assessed on-site. Competency-based training refers to the concept of making sure people have the right skills to do a particular job competently. It does not matter if they spend six months or three years in training, as long as they are competent at the job. This is where recognition of prior learning (RPL) becomes important. Skills assessment is central to this concept. Skills can be assessed on-site—that is, actually on the job by an accredited workplace assessor. This is occurring on many sites. Skills can also be assessed off-site. There are a number of centres, or RTOs, throughout the individual states that offer recognition of current skills through an 'assessment pathway only'. For example, in NSW, numerous institutes offer a Skill Assessment Recognition Program, commonly referred to as 'Skills Express', which allows adult learners with no formal training to obtain a recognised qualification in a number of trade areas. Other similar organisations are in place to perform skills assessment for licensing purposes, but there is no formal qualification attached. See Figure 2.15 for career path options.

Building and construction industry sectors

There are three main sections of industry, which incorporate a wide range of skills carried out in an equally wide range of environments. The main sectors are:
- domestic construction;
- commercial construction;
- industrial construction.

Domestic construction

This sector is involved with the construction of residential buildings. These may include dwellings of one to three storeys, duplex dwellings, villas, townhouses and home units. Generally, work involves the use of concrete, lightweight framing in timber and steel, cladding in brick, timber or composite sheeting, and fitting out with linings, cupboards, ceramic tiles and basic furnishings. Jobs may involve specialisation, subcontracting or multi-skilling.

Commercial construction

This sector is involved with construction of shopping centres, schools, office blocks, hospitals and multistorey construction. Work may involve reinforced concrete-framed construction, precast concrete panels, pre- and post-tensioned concrete, steel-framed construction etc. Construction involves mainly large buildings and structures that involve a wide variety of skills areas such as special hydraulics, lifts, telecommunications, electronics and a vast variety of designs. Jobs range from simple and repetitive to difficult and complex.

Industrial construction

This sector is mainly involved with construction of factories, storage areas and industrial complexes. Work may involve precast tilt-up concrete panels, portal frame steel construction and combinations of

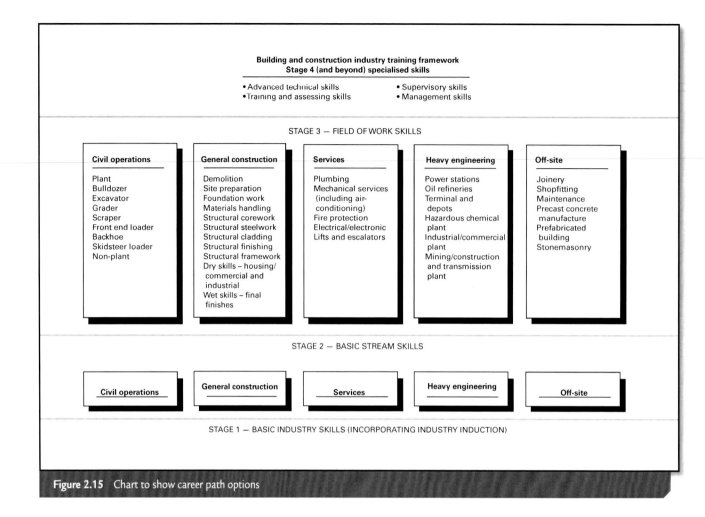

Figure 2.15 Chart to show career path options

these. This also requires many skills, which may range from labour-based tasks to highly mechanised tasks. Formwork and steel-tying skills are highly regarded in both industrial and commercial works. Another area of construction is the civil sector, which is involved with large projects such as bridge building, road building, airport construction and dam building. Again, a variety of jobs and skills are required, from driving or operating heavy earth-moving equipment to detailed construction of formwork.

A final career path choice may be made using the job descriptions and field or work areas previously outlined. From the chart in Figure 2.16, a career path may be mapped according to the learner's preferences.

WORKPLACE STRUCTURE

When many people are together in an organisation, there is a need to divide the work among them and be sure that they understand what they have to do and to whom they are responsible. This requires planning and controlling to ensure harmony and efficiency in the organisational structure.

An organisation chart is drawn up to give a pictorial representation of the business structure. Senior positions are shown at the top, descending to the lower positions; they can extend horizontally with multiple positions of equal status (Figure 2.17).

There are four main levels of authority, with each person being responsible to the person or persons above them on the next level.

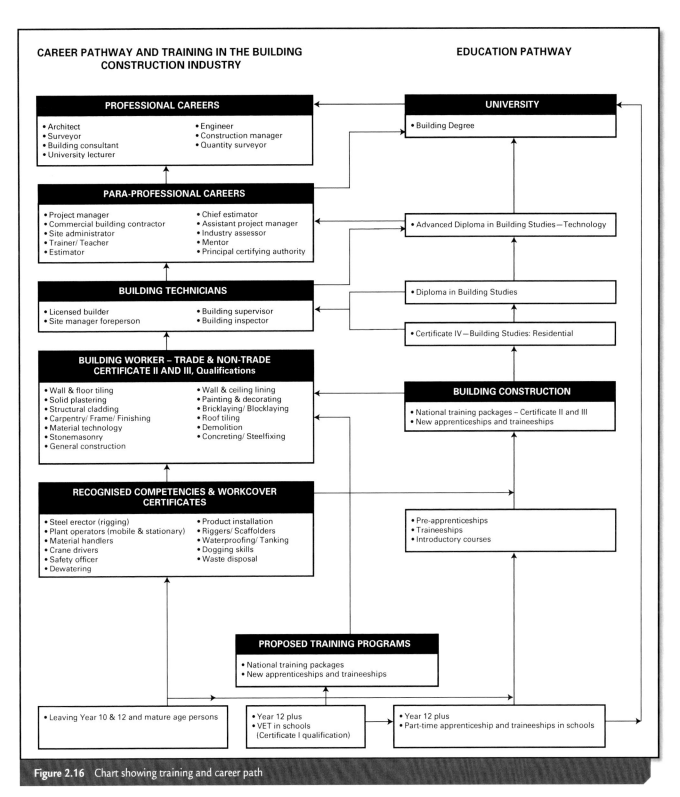

Figure 2.16 Chart showing training and career path

This arrangement of site personnel can be referred to by a number of descriptions, such as:
- site hierarchy;
- site lines of authority;
- site organisation diagram.

A large building firm, such as would be involved with multistorey construction, civil construction and so on, is structured in the same way as the individual sites: that is, with the person having the most authority at the top of the firm and the people with

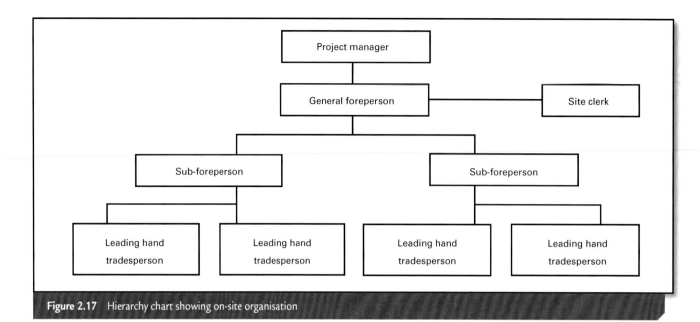

Figure 2.17 Hierarchy chart showing on-site organisation

the least authority at the bottom. It is also possible to have several people with the same authority or responsibility at the one level.

Roles

Client—finances the project. This may be the developer, or persons involved in a joint venture who require a building for a specific purpose.

Architect—engaged by the client to design and control the building on the client's behalf. Other responsibilities include lodging the development application with designs or models and supervising the drafting of plans and the specification, preparing tender documents to engage a builder, engaging consultants for the structural, mechanical and other specialist areas.

Quantity surveyor—responsible for the preparation of a bill of quantities for the architect, to enable the building to be priced by an estimator. Also required to prepare quantities for variations to the contract and assessing progress claims.

Land surveyor—responsible for site setting out and control of the vertical alignment of a building.

Draftsperson—engaged by the architect to prepare plans and details before and during construction.

Builder—the main contractor selected by a tender process to carry out construction as per the plans and specifications. The builder will also engage subcontract labour to deal with specific parts of the construction.

Construction manager—responsible for all building contracts the company has won or is tendering for. Usually working from the main office, the construction manager forms tender documents and controls the building activities for the company.

Project manager—usually employed on very large sites to control the running of the site and to make sure the project or projects run to schedule and budget.

Site manager—runs the day-to-day activities and liaises directly with the foreperson and the project manager.

General foreperson—responsible for the delegation of duties to workers on-site and also responsible for site safety. Other responsibilities are the coordination of subcontractors and the progress of work.

Quality assurance officer—engaged to control the quality of construction on-site and ensure that the work meets the standards laid down in the specification and tender documents.

Building inspector—for domestic construction, the inspector visits the site on behalf of the council to check structural work and pass the completed stages of a job.

Principal certifying authority (PCA)—for domestic construction the PCA carries out similar duties to the local government building inspector, but is engaged directly by the client to conduct the inspections.

TIME MANAGEMENT

Planning methods

Time is very important in building and construction projects, simply because it relates directly to money. It is therefore the building manager's responsibility to manage this very valuable resource to the best of his or her ability. The planning needed to manage this resource is of the utmost importance, and should be carried out by experienced managers using sound management and planning practices.

Planning suggests an orderly and natural action of doing things, together with putting methods and systems in place. Construction planning applied to these principles would involve many sound decisions made by the building manager, from the decision to tender, to the final completion of a project. Adequate time for planning is extremely important and must begin before site operations commence so that the proper methods and equipment are decided on and the correct materials are ordered. Planning has a direct effect on the whole job.

The output of subcontractors and tradespeople will determine the programming of a job. Stages of construction and sequencing of trades should commence as soon as possible, without necessarily waiting for the completion of the preceding work, and should continue without interruption. Sufficient labour and time should be proportional to the extent of work being performed. Brickwork is a labour-intensive trade, as is the erection of formwork for reinforced concrete. Buildings that have these trades will require detailed planning for the job to be done.

Time charts

Planning will involve charting programs, to look at when subcontractors will be needed, when orders are to be placed, and when supplies are to be delivered; and to show the sequence of operations and provide a guide for progress and costing. These charts can be used to show time, progress and financial position. They can be weekly or monthly charts, or for the duration of the total contract. The charts should be adjusted to take into account any change in the contract, or any other disruption to the on-site work. They are used to compare the planning prior to the commencement of the contract with current work, and also for the delivery of materials and commencement of different stages of the job.

Table 2.2 shows a typical sequence of events, with the time given in days. It also shows the relationship of activities to each other. Figure 2.18 on p. 92 shows the same information but graphically highlights the overlaps in activities.

Forecasting

To be able to stay in business requires long-term planning: you can't rely on the future to look after itself. Planning for the future, as well as for the ongoing commitments of a building company, is the manager's responsibility. When deciding on the type of contracts that will be undertaken it is necessary to consider the financial outlay and the risks involved for financial gain. The success of a competitive company is dependent on its personnel and its equipment. This must be kept in mind during the planning process. Although it is difficult to predict financial and government policies, these also have to be considered when planning a long way ahead.

Anticipation of delays

With the planning process it is not possible to predict delays that may happen on a job, but it is common for a delay to be caused by an industrial or safety dispute, or because of a change to the design. With some contracts a delay will allow the contractor to have an extension of time, while with others it may not. The construction plan should be adjusted when these delays are experienced.

Any delay should be documented, as all delays are expensive: the normal output and progress

Table 2.2 Schedule for a brick veneer cottage on a concrete slab

CONSTRUCTION OF BRICK VENEER COTTAGE ON A CONCRETE SLAB

Ref	Activity	Time (days)	Preceding activity
1	Site establishment	2	. .
2	Setting out of house	1	1
3	Excavate for slab	1	2
4	Concrete slab	3	3
5	Drainer sewer/storm water	2	4
6	Carpenter frame/roof	2	5
7	Metal worker fascia/gutter	1	6
8	Roofer	2	6,7
9	Carpenter windows	1	8
10	Bricklayer perimeter course	1	8
11	Termite protection	1	10
12	Bricklayer main	4	9,11
13	Electrician rough-in	3	6
14	Plumber rough-in	3	6
15	Carpenter eaves	2	12
16	Insulation	1	8,10,13,14
17	Plasterer	3	16
18	Waterproofing	2	17
19	Carpenter internal joinery	5	17
20	Tiler wall and floor	4	18
21	Painter	4	17,19,20
22	Concrete paths and driveway	3	5
23	Landscaping/fencing	5	22
24	Electrical final fit	2	21
25	Plumbing final fit	2	21
26	Floor coverings	1	24,25
27	Window dressings	1	26
28	Internal clean	1	26,27
29	External clean	1	23
30	Hand over	1	28,29

of the job is affected and any lost time is difficult to catch up.

Costing

As time is money, the construction plan takes into account the duration of the job, and the costing is linked to the bill of quantities, the working drawings and the contract documents. Once the contract is signed, the project is managed financially and a close watch is kept on the budget for the job. Progress is matched to budget and costing on a monthly basis to keep track of the overall contract.

Causes of delays

Delays are expensive on building sites, and this adds to the problem of planning. The manager must overcome the problem, whether it be the weather, industrial disputes, material shortage, or government and council requirements.

Hot weather can be a problem when placing concrete, as it dries out the concrete too quickly. Wet weather can cause water stain, the swelling of timber and the flooding of excavations. All of this requires extra work that has not been allowed for in planning, for replacement or rectification.

Industrial disputes can cause a site to close down, resulting in no production on-site at all, and there may be a financial outlay for labour. This can have a disastrous effect if the increase in labour has not been allowed for in the contract.

The supply of material can be interrupted at any time for various reasons, and unless other supply lines can be obtained the output of labour at different stages of the job is seriously affected. The supply of materials could be held up on the wharves in containers, due to a breakdown in the manufacturing stage or because of a transport strike. Any such cause is difficult to foresee. Planning could therefore involve stockpiling materials or obtaining alternative sources of supply.

Government and council requirements can cause delays at the beginning of a job, for example by holding up the approval to commence work. Complying with demands for occupational health and safety from the state WorkCover authority, for example, pollution control from the Environmental Protection Authority, and traffic control from the Roads and Traffic Authority, can all take time and can cause delays when waiting for approvals or when complaints arise.

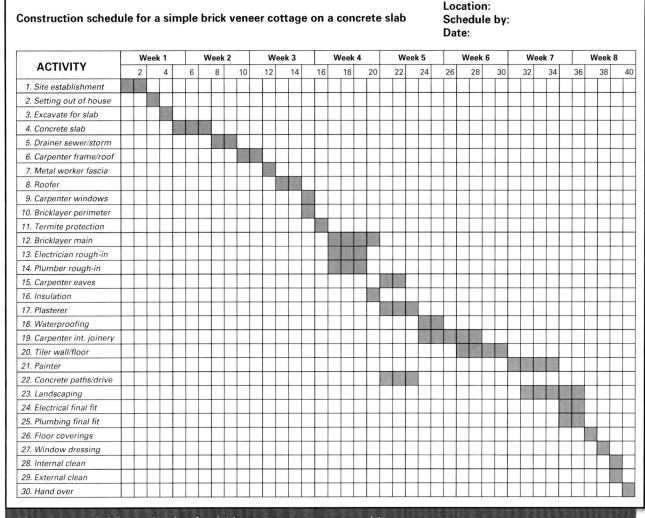

Figure 2.18 Typical Gantt or bar chart for a brick veneer cottage on a concrete slab

Cost factors

One of the manager's greatest responsibilities is to plan for a building to be constructed under budget. The financial control of a contract is most important, and all materials, labour and time are related to money. A manager with financial expertise will mean financial gain for the company.

Labour is the major cost for a building, and intensive planning will be required to cover this. Statutory and award costs can change over the life of a contract. It will depend on the type of contract whether it will be a lump sum and whether an allowance has been included for the rise and fall of material and labour costs. The costing of the building during its construction will also keep a check on the financial state of the company and must be done on a monthly basis.

The choice of plant and equipment and its operators is also a great concern for management. Breakdowns and the wrong choice of equipment can result in delays and lower than expected outputs. Considerable planning is needed for this, as these choices are linked to the costing of the job, and if mistakes are made at this stage the company will have difficulty with its budget.

During the progress of the job it is necessary to secure the materials stored on-site. This is even more important when the job is nearing completion. If valuable materials or equipment go missing, it disrupts the progress of the job and adds to costs.

Therefore, some sort of security has to be considered, even if it is only for the night and weekend hours. Security fencing, lighting or a surveillance system are some options.

Insurance can be taken out to provide compensation for unexpected happenings, such as fire, burglary or accidents. This is short-term, and premiums may vary from year to year. Building companies usually deal with insurance brokers to get the best deal and coverage. During the planning stage the insurance cover is another budgetary consideration.

Use of bonuses and incentives

The greatest asset of an organisation is its labour resource. The expected output from labour can be the basis for costing a job, but this can change due to many factors. Labour can be utilised to perform to its best advantage when a methods study is employed. The best way to do a job is to maximise output and job satisfaction. To guarantee this, some sort of incentive scheme needs to be put in place.

Some factors that contribute to job satisfaction or motivation for work are: achievement, recognition, the work itself, responsibility, advancement and money. This can be achieved by giving bonuses, profit-sharing schemes, payment in kind, and non-financial incentives.

SEQUENCING OF MAJOR BUILDING ACTIVITIES

The sequencing of major building activities will depend on the type of building to be constructed. It might be an upper-storey conversion, an addition to an existing building, or it may involve a new building on a vacant site. With multistorey buildings it might begin with hoardings or excavations for basements, or the removal or importing of foundation material. On other sites it might be the stockpiling of materials or the erection of compounds for security. Services may have to be diverted, extended or capped off before work commences on the actual building. Therefore, the site and the type of building will present their own sequence of building activity.

The sequence of events will be started by the client, who will provide the finance (it could be the financier who provides the funds for the owner/builder).

Method of sequencing

There are several possible combinations of sequence relating to the need and reason for the building to be constructed. The development application and building application have to be approved by local council before the work commences on-site.

The architect, together with the draftsperson and consultant, prepares the drawings and specification for the building to present to the council for a building approval. The quantity surveyor prepares a bill of quantities so that the building can be costed. The land surveyor sets up levels, reference points and set-out points to enable the builder to construct the building, and assists in checking the vertical alignment for tall buildings.

The amount of preplanning a builder does before commencing on-site will depend on the type of building, as mentioned above. Once a contract is won it can be carried out. The builder will then sequence the order of trades, tradespersons, plant and equipment for the construction of the building. Depending on the size of the job, the builder will employ either a project manager or a site manager to look after the on-site construction management.

Staff at head office will include a construction manager, who controls the entire building operations for the company. A chief estimator assists the construction manager, and is responsible for costing the building works.

Smaller construction sites and residential construction may not need the same force of professionals to run the job. It may simply be the builder who controls the site and coordinates the activities.

A typical **sequence** of events in construction for a simple brick veneer cottage on a concrete slab may be as follows:

1. Site establishment
   a. Arrange temporary electricity supply.
   b. Arrange temporary water supply.
   c. Provide site amenities.
   d. Erect site security, fences, signs.
   e. Erect necessary sedimentation barriers.
2. Contact subcontractors
   a. Excavator.
   b. Plumber and drainer.
   c. Electrician.
   d. Concreter.
   e. Carpenter/joiner.
   f. Bricklayer/blocklayer.
   g. Roofer.
   h. Plasterer.
   i. Tiler (wall and floor).
   j. Painter.
3. Organise materials
   a. Steel (reinforcement/structural).
   b. Formwork.
   c. Concrete.
   d. Windows.
   e. Timber/steel frames and roof.
   f. Bricks, sand and cement.
   g. Roofing.
   h. Internal lining.
   i. Joinery/hardware.
4. Site setting out
   a. Locate and inspect building site (check survey pegs).
   b. Clear and level site.
   c. Set out for building alignments.
5. Excavation
   a. Excavate/clear site.
   b. Excavate for concrete slab.
6. Concreter
   a. Erect formwork:
      i. Plumber to do internal drainage.
      ii. Termite protection around drainage pipes.
   b. Install reinforcement.
   c. Organise steel inspection.
   d. Pour concrete.
7. Drainer
   a. Lay and connect sewer pipes.
   b. Lay and connect stormwater pipes.
8. Carpenter
   a. Erect wall frames.
   b. Erect roof frames.
9. Metal worker—Install fascia/gutter.
10. Roofer—Install roof covering.
11. Carpenter—Install windows and external door frames.
12. Bricklayer—Lay perimeter course.
13. Termite protection
    a. Install perimeter termite protection.
    b. Organise termite protection inspection.
14. Bricklayer
    a. Lay bricks from DPC level to underside of eaves.
    b. Clean brickwork.
15. Electrician—Internal rough-in.
16. Plumber—Internal rough-in.
17. Carpenter
    a. Install external eaves lining and mouldings.
    b. Hang external doors to lock up.
    c. Organise frame inspection.
18. Insulation—Install wall and ceiling insulation.
19. Plasterer—Install wall and ceiling plasterboard.
20. Waterproofing
    a. Install wet area waterproofing.
    b. Organise waterproofing inspection.
21. Carpenter
    a. Install internal joinery.
    b. Install internal cupboards, vanities, etc.
    c. Install kitchen cupboards.
22. Tiler—Fix floor and wall tiles.
23. Painter—Internal and external painting.
24. External works
    a. Concrete footpaths, driveways, etc.
    b. External landscaping.

25 Final stages
  a Plumbing fit out.
  b Electrical fit out.
  c Floor coverings.
  d Window dressings.
  e Final internal and external clean.
  f Organise final inspection for Occupation Certificate.
  g Hand over to client.

Equipment, tools and site security

Tools and equipment make up a large part of a builder's overheads, so loss or theft will add a considerable cost to future quoting to cover their replacement. Costs are also incurred due to rises in insurance premiums and the payment of an excess on premiums when a claim is made. Therefore, a system or **checklist** needs to be followed so that tools and equipment are accounted for and locked up or secured at the end of each day.

It is also important to cover materials or restack them at the end of the day to avoid spoiling or loss. Providing a security compound or a chainwire fence around the job with lockable sheds and containers is advisable to secure equipment that cannot be taken home each day.

Procedures for leaving the worksite clear and safe at the completion of each day's activities could be:
- clearing the site of debris using bins;
- keeping all work areas clean and tidy;
- cleaning tools and equipment;
- storing tools in sheds;
- rolling up electric leads;
- locking up sheds and the site;
- turning off power and water;
- putting barricades in place;
- covering materials in case of rain.

Items of plant and equipment commonly at a building site that the builder is responsible for are as follows:
- crane/forklift for hoisting and loading;
- air compressor for a variety of pneumatic tools;
- bobcat, with a variety of other attachments, such as a 4-in-1 bucket, auger bit and backhoe, etc., for minor excavation tasks, piers, trenching, site cleaning, etc.;
- assorted power tools, chainsaws, hand tools, etc., for a variety of construction tasks;
- generator for power supply;
- assorted shovels, rakes, brooms, etc., for site work and cleaning;
- concrete mixer for mixing mortar;
- wheelbarrow for transporting;
- elevator/conveyor belt for transporting materials.

BASIC QUALITY CONCEPTS

Any product that is manufactured or processed has to maintain a quality or tolerance that is consistent over time. Inspections and testing are carried out to ensure that this happens.

Quality assurance has been included in building contracts to ensure that the clients actually get what they pay for. The contractor takes more responsibility for the end product, and this results in quality control for the building industry.

Terminology

Quality can be expressed as the level of excellence that goes into a product or service. It is the skill and commitment that you and everyone you work with bring to your job, each time, all the time. This will help bring about an error-free performance on every job you do.

Quality assurance and quality management is addressed in AS/NZS ISO 9000.2:1998, 'Quality management and quality assurance Standards— Generic guidelines for the application of ISO 9001, ISO 9002 and ISO 9003'.

Quality

Contracts now include quality assurance, and this puts the responsibility back on the supplier in much the same way as a guarantee or warranty would apply if you bought a new car or computer. It puts more meaning into 'doing the job right the first time' and 'using the right tool for the job'.

Quality is of benefit to everyone: from the consumers, who expect and demand quality in the goods and services they receive; to management, who know it will increase profits; to the employee, who benefits from job security and better pay; and of course to the whole country, which will be more competitive in the world market.

Licensing

In order to carry out residential building work, where the total cost of labour and materials combined is more than $1000, you must either be the holder of a contractor licence, or employ someone who is the holder of a qualified supervisor's certificate. Issuing a licence ensures that all qualified builders and contractors are registered with their relevant state or territory's licensing regulator. This gives the client the ability to check on the builder or contractor's credibility. It also provides the client with some protection against faulty production and workmanship.

In February 2007, new legislation was passed which allowed licences granted in a particular state or territory to be mutually recognised nationally. This meant that a person who was licensed to practise a particular occupation, such as a bricklayer, in a certain state or territory, would also be allowed to practise the equivalent occupation in another state or territory, provided they notify the local Regulating Authority. It is also the licence holder's responsibility to undertake work *only* for the purpose for which the licence is held, and they must also show their respective licence numbers on all advertising, stationery and **signage**. Should any of these conditions not be adhered to, it could be deemed a direct infringement of the legislation, whereby subsequent fines may be issued to the respective licence holders.

Note: It is recommended that readers become familiar with their relevant state or territory licensing requirements, by either getting in contact or visiting their appropriate Regulating Authority's website, as detailed in Table 2.3.

Continuing professional development

In 2004, various states across Australia introduced a new system which affected building contractor licence renewals. The system is known as 'Continuing Professional Development' (CPD), and requires building contractors to undergo an upgrading process by taking part in an assortment of designated learning activities. Each of these activities equates to a certain amount of nominated credit points. The minimum number of credit points a building contractor must accumulate is 12 points per annum (or 36 points when licences are renewed for a three-year period). The types of activities that generate credit points include, although are not limited to:

- employing an apprentice;
- attending industry forums;
- gaining additional licences;
- undertaking academic training related to building activities;
- completing certificate training;
- subscribing to industry magazines.

CPD must be relevant to the building contractor's area of practice, and that which ultimately enables the building contractor to:

- extend or update their knowledge, skill or judgment;
- become more productive;
- understand and apply advances in technology;
- face changes in the industry;
- improve their individual career paths and opportunities for advancement;
- better serve the community.

Whilst the CPD program is currently administered in Victoria, NSW and Tasmania (with other states currently considering its introduction), there are significant differences between how each state has decided to regulate its own particular

Table 2.3 Regulators for carpenters, joiners, builders and bricklayers (as at February 2009)

State / territory	Regulator	Location	Contact details	Web link
WA	Builders Registration Board of Western Australia	Suite 10 18 Harvest Terrace West Perth WA 6005	(08) 9476 1200	<www.builders.wa.gov.au>
Vic	Victoria Building Commission	PO Box 536E Melbourne VIC 3001	(03) 9285 6400	<www.buildingcommission.com.au/www/html/7-home-page.asp>
Qld	Building Services Authority	PMB 84 Cooparoo DC QLD 4151	1300 272 272	<www.bsa.qld.gov.au/Pages/default.aspx>
NSW	Office of Fair Trading	NSW Office of Fair Trading PO Box 972 Parramatta NSW 2124	13 32 20	<www.fairtrading.nsw.gov.au/building.html>
SA	Office of Consumer and Business Affairs	GPO Box 1719 Adelaide SA 5001 DX225	(08) 8204 9696	<www.ocba.sa.gov.au/licensing>
Tas	Workplace Standards Tasmania	PO Box 56 Rosny Park TAS 7018	(03) 6233 7657	<www.wst.tas.gov.au/industries/building/bpa>
ACT	ACT Planning and Land Authority	GPO Box 1908 Canberra ACT 2601	(02) 6207 1923	<www.actpla.act.gov.au/customer_information/industry>
NT	NT Building Practitioners Board	GPO Box 1680 Darwin NT 0801	(08) 8999 8964	<www.nt.gov.au/bpb/index.shtml>

version; for example, in Victoria the program is voluntary, whereas in NSW and Tasmania it is compulsory.

Note: It is recommended that readers become familiar with their relevant state or territory CPD system, by either contacting their appropriate Regulating Authority, or accessing its website, as detailed in Table 2.3.

Cost savings

Cost savings will be achieved if staff are trained to be more skilful and more productive. Each member of staff will have job security and will know that she or he can do the job better, thus being a valuable human resource for the company.

Quality, when applied to all the materials supplied to a building, results in less **waste** and a better-quality end product. When quality standards are specified, suppliers are responsible for orders placed. This guarantees the quality of the materials. Having quality assurance in place improves the standard of work, which is required of the end product. The benefit of quality assurance will be a saving in cost, as the material supply and the quality of the work will be of a good standard the first time.

Personal responsibility

If the employer knows as much as possible about the product or service (this might involve working to tolerances), and is confident that each employee pays attention to detail in every area of job performance, then personal responsibility can be an added bonus for the employer.

Having the right attitude certainly affects the world around you. Problems can be prevented by maintaining standards and building quality into procedures, products and design at the start of a job. Rather than fixing mistakes, one should take the time to analyse failures and succeed the next time.

Take the initiative when it comes to accepting new challenges, and stick with them. Set goals to do your job better and aim to do it quickly and accurately. Share any ideas you might have to improve procedures, save money or increase **efficiency** or productivity with your employer. All of these personal responsibilities will lead to a quality improvement for everyone.

Quality standards

Building contractors are required to plan, implement and maintain a quality system for all projects undertaken in line with the requirements of AS/NZS ISO 9000.2:1998 and apply the requirements of AS/NZS ISO 9001, 9002 and 9003.

Refer to section 4 of ISO 9000.2, 'Quality system requirements for quality systems, quality procedures, quality planning and quality design'.

Quality principles

The eight key principles of quality control are:
- All systems exhibit variation.
- High quality does not cost—it pays (in the long run).
- People work within systems.
- Everyone serves a customer.
- Improvements should be plan-driven, not event-driven.
- Improvement should be a way of life.
- Management should be controlled by facts and data.
- Control the process, not the output.

Terms

The six key terms you need to know:
- **Quality**—fitness for purpose.
- **Reliability**—the system or product will continue to work for its guaranteed life.
- **Maintainability**—parts and service are readily available and the system or product can be repaired if necessary.
- **Availability**—the system/goods work instantly when required.
- **Supplier**—a person giving customers goods and materials with which to work.
- **Customer**
  - internal—the person who receives your work next;
  - external—the person who buys goods and services.

Standards

Standards ensure that goods, services and products are accurate and safe. They are put together by a panel of experts in the specific area of the scope of the standard. They may be representatives of safety committees, manufacturing industries, suppliers, customers and government departments. Standards Australia was founded in 1922 as the Standards Association of Australia, an independent, non-profit body incorporated by Royal Charter. Its work is conducted solely in the national interest and its principal functions are to prepare and publish Australian standards and to promote their adoption.

Examples are:
- AS 1478.1-2000, 'Chemical admixtures for concrete, mortar and grout—Admixtures for concrete';
- AS/NZS 2311:2009,[1] 'Guide to the painting of buildings';
- AS 3958.1-2007, 'Ceramic tiles—Guide to the installation of ceramic tiles';
- HB 161-2005,[2] Guide to plastering';
- AS 3740-2004, 'Waterproofing of wet areas within residential buildings'.

---

1. Note that some 'Australian' standards are also effective in New Zealand.
2. HB stands for 'Handbook' which is a guide rather than a fixed set of standards.

WORK IN A TEAM

Value of teamwork

Teams of workers, which may consist of two, three, four or more, are formed when building site tasks require rapid completion, such as bricking up a new home (Figure 2.19), or when they require completion within a given timeframe, such as tiling a large entertainment area. Teams are also needed when tasks consist of a number of related parts or elements, such as plastering out a house, from sheeting to stopping off.

Once the team members understand the role they play in the overall construction of the building, they will be able to contribute more effectively to meet the 'site goals' and to work more efficiently with other teams or individuals.

Individuals within the team

To allow smooth operation and cohesiveness within a team, each person should be allocated a job or function, which will complement that of other team members and not duplicate their effort. Therefore, a simple meeting is required to establish the role of each member within the team. Ideally, one person in the team will be nominated and appointed as the 'team leader' to ensure consistency of effort and communication.

The old adage of 'many hands make light work' applies to a team effort; however, 'too many cooks spoil the broth' also applies if there is more than one team leader.

Regular rotation of roles within the team is also important, so that there will not be just one or two members doing all the work. This rotation of roles and responsibilities also reduces the risk of individual egos getting in the way of a productive team effort.

Individual strengths and weaknesses

Rotation of roles and responsibilities is important to effective function, but there is also room for specialisation within the team. If individuals have special skills or talents they should be able to maintain a particular role or position in the team and allow the other team members to assist them. This may occur where only one member of the team holds a qualification or has a licence to do specific work.

Where a member, or members, within the team lacks skill or knowledge to carry out a task on his or her own, then this person may 'buddy-up' with another member who can supervise and support him/her. This may occur where one or more members of the team are labourers and the others are tradespersons.

Enhancement of roles

Team roles may be enhanced by giving individuals more responsibility than they would normally accept, or by encouraging individuals to undertake further structured training to gain a qualification. This may mean, for example, enrolling in a trade course or

Figure 2.19 A bricklaying team in action.

post-trade course, or attending a skills recognition assessment program.

Team rules

Each team should discuss and set ground rules by which all members must abide. This is important to maintain harmony and allow the team to function effectively. Any changes to these rules, and/or improvements made, must be discussed and agreed on by all the members. All members should have equal rights and responsibilities and be able to suggest changes and improvements without fear of being ridiculed or vilified by the others.

Disharmony

Where conflict within a team arises, it should be dealt with immediately and not allowed to fester and grow into a full-blown confrontation. Ideally, all conflicts should be resolved within the team so that other unwanted persons are not involved, as this may lead to individuals left feeling resentful. If a resolution cannot be found, another person from outside the team should be brought in to arbitrate the situation. This person should be someone whom all the team members know and respect, otherwise an amicable resolution may not be found or bitter resentment may result.

The key to successful **teamwork** lies in the selection of members with suitable skills, who are willing to do their bit and abide by the team rules, have an equal say in the running and organising of the team, and hold skills that are valued and complement the other team members, as well as putting a team structure in place that encourages participation and appreciation of each member.

PERSONAL DEVELOPMENT NEEDS

Workplace competencies

Before a person is able to progress beyond any given academic level in the building industry, a list of required competencies must be identified. In most trade qualifications, the individual is required to attempt and successfully complete a range of 'core' or compulsory competencies plus a minimum number of elective competencies. Several of these core competencies may be 'common units' to a wide range of trade and non-trade areas. These core and elective competencies are grouped together to form the structure of the 'qualification'.

Identifying own learning needs

The nature or kind of work has changed in Australia during recent times, which reflects the changing needs of business and the growth of the global economy. Many industries are meeting these changes by varying employment patterns, such as changing from the traditional full-time 35-hour week to offering part-time, casual, contract, outsourcing and labour hire as alternatives.

Career pathways for workers are also changing and in some industries are breaking down, leaving little or no option of advancement. Some employers expect their employees to contribute new knowledge and innovation to their organisation to assist the growth, effectiveness and competitiveness of the business. Therefore, individuals may need to seek advice from a range of personnel to find out what they need to do to move in another direction. Good advice is usually gained from the most experienced persons with a good understanding of the trade or industry, which may include:
- the foreperson on the job;
- architects and/or engineers;
- progressive builders;
- teachers/lecturers from vocational education organisations;
- environmental experts;
- designers;
- building material manufacturers.

WORKPLACE MEETINGS

There are two main methods used to raise issues or discuss problems on a building site: these are the formation of committees, and a more formal site **meeting**. A committee has a limited membership that has been elected or appointed to perform specified functions, such as a site occupational health and safety committee. A formal meeting consists potentially of all members of an association or their representatives. Special meetings may be held on-site, involving the builder, client, architect and subcontractors, to consider important matters relating to the project. Where deals involving variations are made, all affected parties may sign a statement on the spot, confirming changes that have been agreed to. Copies of the report of the meeting should be made available to all parties who were present.

Formal **minutes** should be kept of all site meetings arranged and conducted by the project manager or general foreperson. Copies of these minutes, which represent a report of proceedings, can be distributed to all persons present at the meeting and any other people who need to be informed.

Trade unions also have meetings, such as annual, delegate and branch, and committee meetings. The branch meeting may take place once a month, but the executive committee may meet once a week.

Participation and representation

These two roles are not the same. Voting for a representative to do what he or she thinks would be best for you is different from being there yourself to make your own alliances and say your piece. Sending a delegate instructed to vote in a specific way is different again.

Rights

In a democratic society we have the right to be represented in our national government, to be consulted in this decision or that decision, and to vote, among others. These are rights that people have argued and fought for, which we continue to assert, which are accepted by our society, and which are continually subject to readjustment as our society changes. Therefore, each person or representative of a committee or part of a meeting has the right to voice an opinion on the subjects raised. You should not sit and say nothing if you disagree strongly, as you may have to live with the decision. Your say may sway the argument in your favour by convincing others who were undecided.

Communication and consultation

Representation or participation is seen as a means of finding out what people want, and of communicating information and decisions to them. This aids in understanding, freeing processes of decision making, and action. It may be a matter of persuasion as much as sharing in making decisions.

Efficacy

People are more likely to implement a decision effectively if they have shared in making it, possibly because it is felt that their contribution makes it a better decision, they feel responsible for it, and they trust and understand it.

Group dynamics

To gain the greatest input of people at the meeting, it is important to have a room layout where everybody can see and be seen easily. This will promote a more inclusive environment than having members pushed into a corner, making them feel insignificant. Two of the most important requirements for effective meetings are that **communication** takes place in an orderly manner, and that content is relevant. Too often it is assumed that both of these requirements are the responsibility of the chairperson, who achieves them by controlling the meeting. There are obviously instances where such control is necessary, but responsibility for orderly communication and relevance lies with every individual member. The chairperson does not need to intervene if every member adopts the self-discipline needed to achieve an effective meeting. It is easy to get off the track when many people have input, so it is necessary to outline a plan at the beginning of the meeting to which everybody agrees. Everyone should have the opportunity to contribute to the discussion, which may mean that some people will have to be restrained while others may need to be encouraged.

Effective participation

If the meeting or committee is to function effectively, the following areas will need to be addressed.
- The aims and objectives should be clearly formulated and understood. They must also be accepted by all members.
- Discussion should be relevant, with virtually everybody contributing.
- Everyone should be prepared to listen to each other and consider the points made. Members should not be afraid to put forward views and ideas.
- The atmosphere should be informed and relaxed, with all members of the group being involved and interested.
- Only constructive criticism should be offered and accepted.
- Disagreements should be dealt with openly and examined to find a compromise.
- Members should feel free to express their own feelings and attitudes towards a problem.
- Decisions should be arrived at by consensus; individuals should not be afraid to disagree and should be given fair consideration.
- Decisions and follow-up action should be determined, with everyone fully aware of them. Jobs should be allocated clearly and appropriately.
- The chairperson should not dominate; the members should not be subservient to that person.
- The group should be aware of and be prepared to discuss its own deficiencies.

Limiting factors

Things that may limit a person's performance at a meeting or within a committee are:
- experience;
- knowledge;
- fear;
- behaviour;
- dominance.

People's lack of experience and knowledge of how meetings and committees operate may inhibit their performance. In addition, fear of **conflict**, hidden hostilities and underlying personal factors are difficult to overcome.

Another inhibiting factor in the success and effectiveness of a meeting is the misapprehension that the effectiveness of a group rests solely with the leader. Research indicates that skilful membership behaviour is the operative factor. The greatest danger of all is that the person with the loudest voice or strongest personality will become the chairperson and totally dominate the meeting, so that those who have relevant knowledge and opinions are not heard.

Purpose of meetings

It is important to be quite clear about the purpose of the meeting or committee discussion so that each member is fully aware of the reason for his/her participation. Common reasons for calling a meeting would be to:
- pass on information;
- seek ideas to solve a problem;
- coordinate information gathered over a period of time;
- negotiate changes to contracts;
- resolve disputes;
- form new policies or revise old ones to improve the running of the site;
- plan for future work;
- decide on proposals or outstanding issues;
- act on decisions made as quickly as possible to ensure that enthusiasm and credibility are ongoing.

Use of committees

There are three main ways of dealing with a situation or settling a problem:
- by imposing a solution either from above or below;
- by accepting a compromise reached by bargaining;
- by agreeing to the best answer arrived at after joint discussion.

The first method implies the use of force, which emphasises a permanent division, as it represents either a victory to be maintained or a defeat to be avenged at the next opportunity. Equally, the second method is an uneasy armistice decided by relative powers, rather than on true assessment of the facts.

By contrast, the employment of discussion ensures a genuine solution based on conviction, with the maximum chance of acceptable and harmonious action. Thus committees or informal group meetings can play an important part in the organisational structure. Committees can be used for different purposes, with appropriately varied membership as follows:
- as an advisory body, such as a site occupational health and safety committee with elected representatives from each activity, to deliberate on particular problems and so assist an executive by combining the total knowledge and experience of its members;
- as a means of consultation, e.g. a planning meeting, where members are appointed by reason of their individual functions and contribute their different viewpoints, ensuring that all aspects are properly considered;
- as a channel for information and a method of communication, usually between a supervisor and his or her managers, where reports are tabled and resources are pooled to find a suitable solution to a problem;
- as a process of coordination, usually between a specialist and several production units, such as a transport manager allocating his or her daily vehicle tasks to meet the requirements of the respective general foreperson.

All these elements may go into forming an effective working committee to be able to resolve problems and make judgments.

SUSTAINABILITY— RESOURCE EFFICIENCY AND WASTE MINIMISATION

Estimates suggest that when renovating, demolishing or constructing a house from new, up to 200 tonnes of waste could be generated. Furthermore, approximately 80% of this waste material could be reused by means of adopting thoughtful recycling processes and procedures. In doing so, vast quantities of energy, water, resources and, of course, money would be saved.

Averaged nationally, per annum each Australian produces in excess of one and a half tonnes of so-called 'waste', with approximately 40% of these waste products resulting from the undertaking of renovation, demolition and/or new construction activities. What's more, it would be safe to suggest that a major percentage of this specifically discarded construction waste, if prudently managed, could be segregated for future recycling and reuse.

By recycling we are saving the earth's resources, i.e. the raw materials of all the products we buy—the minerals, oil, petroleum, plants, soil and water. We also reduce our consumption of energy, limit pollution and lessen global warming by cutting down on the harvesting, construction, transportation and distribution of new products.

For example, let's look at aluminium. Recycling one tonne of aluminium saves four tonnes of bauxite and 700 kilograms of petroleum. It also prevents the associated emissions (which would include 35 kg of the toxic air pollutant aluminium fluoride) from entering our air. As an added bonus, it would reduce the load on our already stretched landfill sites.

Remember, 'one person's junk is another person's treasure', so skip the skip – and start recycling and re-using your building materials.

How to do it now!

Follow the waste minimisation hierarchy—avoid, reduce, reuse and recycle—in that order.

Avoid

Building and construction waste often enters our waterways through stormwater drains, a major cause of water pollution. A typical instance of this could be a stockpile of sand: if not properly stored or sensibly located, a sudden downpour of rain could see the sand washed away into the stormwater system. Ensure that you do not contribute to the pollution of our waterways by adopting a careless attitude when undertaking the simple task of storing building products on a construction site. The two obvious advantages gained would be:
- minimal affects to our water system as a result of unnecessary contamination;
- minimal replacement of building materials already purchased.

Reduce

The general trend these days is to build from boundary to boundary. Over time the typical house in Australia has evolved from having three bedrooms, one bathroom and separate living areas, into a more open plan, including extra bedrooms, ensuites, rumpus, walk-in wardrobes and pantries, studies, specific movie rooms and the like. All this has had obvious consequences, notably the overall increase in size of the modern home. In fact, according to the Bureau of Statistics, nationally, since the mid-1980s, residential houses have grown in size by approximately 40%, (from 162.2 m^2 to 227.6 m^2). This simply equates to the fact that 40% more energy, resources and materials, across the board, are required to construct a house today than were required in the mid-1980s.

You can considerably reduce waste by planning carefully and sensibly. Determine exactly what it is that you require within your own residential home. Don't over-indulge, design to accommodate standard sizes of materials, utilise prefabricated frames and trusses, as this is a proven method of reducing waste, and don't fall into the trap of 'keeping up with the Jones's'. Obviously, reducing the size of your construction will see a reduction in the use of energy, resources and materials and in the generation of waste.

Reuse

Undertake a commitment to use, where ever possible, recycled materials. The following items are the most common recyclable materials and are therefore generally easy to locate:
- steel;
- recycled or plantation timber;
- recycled concrete;
- second-hand bricks;
- soil and fill.

Remember that there are many fittings and fixtures (such as doors and windows) that are also available second hand. A list of recycled building products and/or suppliers can be sourced by visiting any of the following online services:
- *Yellow Pages*—at <www.yellowpages.com.au>, under 'Building Materials–Second Hand';
- Eco-Buy Online—<www.ecobuy.org.au/index.cfm>. The site lists suppliers of second-hand construction and building materials and other green alternatives;
- Construction Connect—at <www.arrnetwork.com.au>. The site's aim is to help improve the environment by reusing building materials.

Buying recycled products increases the market for them, making it more viable for businesses to supply them.

Recycle

Some materials can be recycled directly into the same product for reuse. Others can be reconstituted into other usable products. Unfortunately, recycling that requires reprocessing is not usually economically viable unless a facility using recycled resources is located near the material source. Many construction waste materials that are still usable can be donated to non-profit organisations. This keeps the material out of landfill and supports a good cause.

The most important step in recycling construction waste is on-site separation. Initially, this will take some extra effort and the training of construction personnel. Once separation habits are established, on-site separation can be done at little or no additional cost.

The initial step in a construction waste reduction strategy is good planning. Design should be based on standard sizes and materials should be ordered accurately. Additionally, using high-quality materials such as engineered products reduces the number of rejects. This approach can reduce the amount of material needing to be recycled and bolster profitability and economy for the builder and customer.

The following list details the most common building products that can be recycled and suggests particular pathways for which they could be re-used:

- Concrete
  - Un-set: washed and used on future projects.
  - Set: crushed and used for future concrete works, or as crushed and used as road base or fill.
- Bricks and tiles
  - Cleaned and used on future projects.
  - Cleaned and sold on.
  - Crushed and used as backfill or gravel.
- Paints
  - Retained for use on future projects.
  - Sold on as a second-hand product.
  - Recycled into new products.
- Aluminium products
  - Retained for use on future projects.
  - Sold on as a second-hand product.
  - Recycled into new products.
- Gypsum plasterboard
  - Large sheets retained for use on future projects.
  - Recycled into a new plasterboard/gyprock product.
  - Used as a soil conditioner or for composting.
- Timber/green waste
  - Large beams/sheets/etc. retained for use on future projects.
  - Reprocessed into other timber building and landscaping products.
  - Untreated materials can be used as firewood.
  - Chipped and used as mulch either on-site or at other projects.
- Plastics—Recycled into new products.
- Clean fill/soil
  - Used in on-site landscaping.
  - Stockpiled for use on future projects.
  - Sold on as a landscaping material.
- Glass
  - Large sheets retained for use on future projects.
  - Crushed and used as aggregate in concrete.
  - Recycled into new products.
- Carpet
  - Large pieces retained for use on future projects.
  - Sold on as a second-hand product.
  - Used on-site to prevent erosion, dust mobilisation and weed invasion.
  - Natural fibre carpets can be shredded and used as fill in garden beds or composted.

Also consider the food and drink containers used by on-site workers; after all, one recycled aluminium can saves half a can of petroleum, and 20 litres of water.

You can also take the materials yourself to your local recycling centres or transfer stations. Your local council will be able to provide details of your nearest station. Your local waste facility or landfill operator might also handle some recycled products—it might be worthwhile giving them a call.

Why is this action important?

It is possible to recycle and reuse up to 80% of demolition and construction materials. This would greatly alleviate the huge and growing pressure on the earth's resources, on our forests, our land and our human effort. In addition, instead of carting hundreds of tonnes of material from mine to house to landfill we would simple use what is at hand, saving transport and carbon emissions and their contribution to climate change.

Towards resource efficiency and waste minimisation

The following steps will assist in the initiation of a resource recovery program on a typical construction site:

1. Commit to responsible waste management.
    a. Develop and implement a business waste minimisation policy.
    b. Involve all staff members in this process.
    c. Incorporate waste minimisation into position descriptions.
    d. Request that subcontractors sign project waste minimisation plans.
    e. Incorporate waste minimisation into site induction programs.
    f. Provide positive feedback to staff successfully minimising waste.
2. Identify resource pathways
    a. Review materials utilised during construction and demolition activities.
    b. Assess volumes of resources currently going to waste.
    c. Assess material avoidance, reduction, reuse and recycling options.
    d. Determine avoidance, reduction, reuse and recycling methods of each material.
3. Develop a project waste minimisation plan
    There are several keys to implementing an achievable project waste minimisation plan, including:
    a. Committed key field staff
        i. Ensure that all staff understand what is to be achieved.
        ii. Keep staff informed of progress.
        iii. Ensure provision of up-to-date training.
    b. Project specific planning
        i. Conduct a site assessment.
        ii. Prepare a project waste minimisation plan.
        iii. Determine resources for the project, pathways for excess and/or waste materials.
        iv. Determine the location of both waste and materials recovery stations.
        v. Set targets and objectives for each project.
        vi. Require subcontractors to recycle and dispose of their own waste.
    c. Understand options and limitations
        i. Identify collection, sorting and resource utilisation options.
        ii. Determine the suitability of each option for each job.
    d. Establish a monitoring and reporting program to:
        i. Quantify results and identify shortfalls.
        ii. Provide a record for comparison across work sites and methods.
        iii. Set targets and objectives.
        iv. Determine financial outcomes.
    e. Focus on high potential materials and practices that:
        i. Are high volume, can be readily separated and collected.
        ii. Have a viable economic value for recovery.
4. Educate staff about waste minimisation plan
    a. Communicate with staff.
    b. Inform staff of the methods and objectives of maximising resource recovery.
    c. Provide copies of the no-waste project plan.
    d. Involve staff during development and review of the project waste minimisation plan.
5. Implement waste avoidance strategies
    a. Limit the types of resources being consumed on the work site.
    b. Design works to avoid waste generation.
    c. Use modular/prefabricated frames and fit-outs when possible.
    d. Request minimal packaging from material suppliers.
    e. Ensure that materials that will generate minimal waste are used.

Note: It is recommended that readers become familiar with their relevant state or territory Environmental Protection Authority (EPA), by either contacting them or by visiting their respective website, as detailed in Table 2.4.

Table 2.4 Current state and territory Regulating Environmental Protection Authorities

State / Territory	Regulating Authority	Current EPA Act	Web link & contact numbers
WA	Environmental Protection Authority	*Environmental Protection Act 1986*	<www.epa.wa.gov.au/contacts.asp> Contact: 08 6467 5000
Vic	Environmental Protection Authority	*Environmental Protection Act 1970*	<www.epa.vic.gov.au> Contact: 03 9695 2722
Qld	Environmental Protection Agency	*Environmental Protection Act 1994*	<www.epa.qld.gov.au/about_the_epa/contact_us> Contact: 1300 130 372
NSW	Department of Environment and Climate Change	*Protection of the Environment Administration Act 1991*	<www.environment.nsw.gov.au> Contact: 131 555 or 02 9995 5000
SA	Environmental Protection Authority	*Environmental Protection Act 1993*	<www.epa.sa.gov.au> Contact: 1800 623 445 or 08 8204 2000
Tas	Environmental Protection Authority	*Environmental Management and Pollution Control Act 2007*	<www.epa.tas.gov.au/index.aspx?base=10> Contact: Via email only
ACT	ACT Commissioner for Sustainability and the Environment	*Environmental Protection Act 1997*	<www.environmentcommissioner.act.gov.au/index> Contact: 02 6207 2626
NT	Environmental Protection Authority	*Environment Protection Authority Act 2007*	<www.epa.nt.gov.au> Contact: 08 8924 4056

Sources: Information from <www.naturalstrategies.com.au>; <www.wyndham.vic.gov.au>; <www.abs.gov.au>; <www.tams.act.gov.au>

SUSTAINABLE HOUSING

Sustainable housing is about designing and building homes that are comfortable and practical to live in, economical to maintain and cause the least possible burden on the **environment**. This balanced approach takes into account the social, economic and environmental aspects of housing development, and ensures that all the key issues are considered together at the design stage.

Economically sustainable homes are cost-efficient over the lifespan of the dwelling. The design balances upfront construction and fit-out costs against ongoing running and maintenance costs. The building may be constructed of low maintenance materials and feature efficient fittings and appliances.

Environmentally sustainable homes are **resource-efficient** in terms of materials, waste, water and energy. They are designed for water efficiency in the house and garden, waste reduction during construction and occupancy, and energy efficiency in terms of orientation and energy consumption. An environmentally sustainable home can reduce household running costs by up to 60%, saving over three tonnes of greenhouse gases and more than 100 000 litres of water a year. Environmentally sustainable elements include water-efficient and energy-efficient appliances, solar hot water, insulation and efficient lighting.

Does a sustainable home cost much extra?

Current estimates suggest that for a new home of average size, an initial outlay of approximately $3000–$4000 will allow a new home builder to install:

- a solar or gas hot water system instead of electric;
- low-flow taps and fittings;

- water tank for the garden;
- dual-flush toilet;
- bulk ceiling and wall insulation.

Besides the eventual financial savings the householder will receive, they can also enjoy greater living comfort with the knowledge that they are helping the environment. At the same time they are adding to the value of their home and improving its future saleability.

Sustainable alternatives

The Environmental Protection Agency in Queensland (see <www.epa.qld.gov.au>) recommends that you consider these design features, product decisions and site management practices when designing or renovating a home.

General
- An open plan and northerly orientation will maximise breezes and avoid the western sun.
- Bathroom, kitchen and laundry should be located close to the hot water system.
- Ensure that living areas are positioned to capture winter sun and summer breezes.
- Plan the window size, style and location to optimise protection against summer sun and access to winter sun.
- Minimise windows on the western side to avoid the afternoon sun.
- Use materials with low long-term maintenance costs.
- Install awnings and eaves to reduce heat.
- Install insulation in the roof, ceiling and walls.
- Consider an insulated skylight to let in natural light and not heat.
- Install compact fluorescent lighting including down-lights with efficient 12-volt task lighting.
- Incandescent lighting can be used for shorter duration lighting in selected areas.
- Paint the exterior of your house and roof in a light colour to help cooling.

Kitchen
- Install double sinks so that you can rinse in a second sink and not under a running tap.
- Install AAA-rated water-efficient taps.
- Provide task lighting over sink, stove and work surfaces.
- Choose water-efficient and energy-efficient white goods: oven, dish washer, refrigerator and freezer. Look for the AAA rating on water products and the highest star rating on energy-efficient appliances.
- Place your fridge in a cool spot away from the stove and direct sunlight.
- Stove range-hoods should be vented to the outside.

Bathroom and laundry
- Use rain water for toilet flushing, hot water, washing and showering.
- Greywater from the laundry and bathrooms may be used for the garden irrigation system (depending on state and territory legislative requirements).
- Use AAA-rated taps and shower rose for water efficiency.
- Install mixer taps in showers to reduce hot water loss while you adjust the temperature.
- Six-litre/three-litre dual flush toilets will reduce water use.
- Choose an AAA water conservation rating and a high star energy-rated front-loading washing machine.
- If you must install a clothes dryer, choose an energy-efficient one, but it's best to use the outdoor clothes line.

Finishes
- Use non-toxic paints, renders and floor finishes with either no or low VOC (volatile organic compound) emissions to give superior air quality compared to a standard house.
- Consider using floor tiles in rooms reached by the winter sun.
- Bamboo flooring is an efficient renewable resource with low VOC (volatile organic compound) emissions.
- Ensure your carpet underlay is fully recyclable and that the carpet has some natural fibre.

Hot water systems and energy supply
- Install a gas, solar or heat-pump hot water system for the greatest energy efficiency.
- A solar photovoltaic (PV) electricity system converts sunlight into electricity. This will eliminate electricity bills for the life of the system, and you can sell any excess electricity. The federal government currently offers a generous rebate.

Garden and outdoor areas
- Position trees to maximise shade on your property.
- Local native plants in well-mulched gardens will minimise the need for external watering.
- An automatic underground irrigation system will also minimise water use.
- Where practical, create porous surfaces outside the house to allow stormwater to soak into the soil.
- Use recycled timber for outside decking.
- Install an external clothes line.
- Compost bins and worm farms encourage the recycling of all food wastes.
- Use pervious materials such as rocks and pebbles for driveways and paths to slow water run-off into gutters and stormwater drains.

Rainwater tanks
- Install a rainwater tank to supply water for purposes such as toilet flushing, hot water, washing and garden irrigation.

During construction
- Use renewable resources and materials with low VOCs.
- Work around established trees rather than cut them down.
- A site management plan will help control stormwater and waste, minimise soil loss, ensure materials are handled efficiently and that the site is clean and safe.
- Recycle construction waste where possible.
- Direct stormwater to a stormwater drain, not to the sewerage system.

Worksheet 1

Student name: _____

Enrolment year: _____

Class code: _____

Competency name/Number: _____

To be completed by teachers:
Student competent ☐
Student not yet competent ☐

Task: Read through the sections beginning at *Historical industry background* and up to and including *Current trends and technological development*, then complete the following.

Q. 1 According to which humanist psychologist is shelter one of civilisations most basic needs?
_____

Q. 2 Name the Danish architect who designed the Sydney Opera House:
_____

Q. 3 What was the main type of construction adopted by the early settlers once tents were discarded?
_____

Q. 4 What type of roof-covering material was used for early colonial buildings of around 1800?
_____
_____

Q. 5 State the approximate year when the 'Federation period' of architecture appeared:
_____

Q. 6 State the name of the architectural style that appeared between 1930 and 1940 and had similar lines to the bungalow style:
_____

Q. 7 Around what year did brick veneer construction become widely accepted by municipal councils and lending institutions?
_____

continued ➤

Q. 8 The building industry is divided into two main groups: the commercial and residential sectors. List seven types of structure that fit into the residential sector:

1. _____
2. _____
3. _____
4. _____
5. _____
6. _____
7. _____

Q. 9 State the approximate period in which 'plasterboard' was introduced for use in Australian residential buildings:
_____

Q. 10 State two concrete products that have been developed as prefabricated elements commonly used in building today:

1. _____
2. _____

Worksheet 2

Student name: _____

Enrolment year: _____

Class code: _____

Competency name/Number: _____

To be completed by teachers:	
Student competent	☐
Student not yet competent	☐

Task: Read through the sections beginning at *The role of employers, employees and workplace committees* and up to and including *Quality standards*, then complete the following.

Q. 1 Industrial relations addresses many areas of employment. List six issues relating to industrial relations:

1. _____
2. _____
3. _____
4. _____
5. _____
6. _____

Q. 2 Apart from setting minimum wages, state three conditions within an 'award':

1. _____
2. _____
3. _____

Q. 3 Disputes and conflict are common in the building industry. A common trigger for these disputes occurs as a result of 'demarcation'. Explain this term:

_____

_____

Q. 4 Give three examples of procedures used to resolve grievance disputes:

1. _____
2. _____
3. _____

continued ➤

Q. 5 What is the name given to union members who promote membership, represent workers in negotiations etc.?

_____

Q. 6 The Australian apprenticeship system offers apprenticeships in many areas. Give three examples of these areas that relate to building:

1. _____

2. _____

3. _____

Q. 7 Briefly describe the 'enterprise bargaining' system:

_____

_____

_____

Q. 8 Apart from the final qualification, state the two main components of a training package:

1. _____

2. _____

Q. 9 What is the main function of workplace consultative committees?

_____

_____

_____

Q. 10 Complete the following statement: 'The construction safety regulations require self-employed individuals to complete _____ before they carry out any construction work.'

Q. 11 What is the name of the OH&S Regulating Authority in your relevant state or territory?

_____

Q. 12 State the five main skill-stream areas that make up the building and construction industry training framework:

1. _____

2. _____

3. _____

4. _____

5. _____

Q. 13 Briefly describe the role of an architect within a building company:

_____
_____
_____
_____
_____
_____

Q. 14 Briefly describe the role of a general foreperson within a building company:

_____
_____
_____

Q. 15 In relation to basic quality concepts, state four benefits of quality work:

1. _____
2. _____
3. _____
4. _____

SUGGESTED ACTIVITY one

Task: After reading through the section on *Workplace structure*, **prepare a simple site hierarchy chart for your workplace.**

Briefly describe the role of each person in the chart.

SUGGESTED ACTIVITY two

Task: After reading through the sections on *Time management* and *Sequencing of major building activities*, **prepare a simple Gantt chart for the preparation and construction of a basic garden shed, garage, or other structure found on a residential site.**

You may require assistance and guidance from your teacher/instructor in relation to timing, trades involved and materials to be used.

SUGGESTED ACTIVITY three

Task: After reading through the section on *Work in a team*, **carry out simple team activities, as directed by your teacher/instructor, and follow the processes outlined in the section. Each person should be allocated a job/role in the group with one person nominated as the team leader.**

Set ground rules for your team to follow and rotate roles as often as you can. These team functions will be used throughout your training and in the workplace.

Worksheet 3

Student name: _____

Enrolment year: _____

Class code: _____

Competency name/Number: _____

To be completed by teachers:
Student competent ☐
Student not yet competent ☐

Task: Read through the section *Workplace meetings* and then complete the following.

Q. 1 List the two main methods used to raise issues and discuss problems on a building site:

1. _____

2. _____

Q. 2 What is the name given to the report that records the proceedings of a meeting?

_____

Q. 3 Circle or underline the type of society in which we have the right to free speech and an individual opinion (like the way a site meeting should be conducted):

a. Autocratic   b. Hippocratic

c. Democratic   d. Hypocritical

Q. 4 Complete the following statement in relation to individual behaviour and group dynamics: 'The _____ does not need to intervene if every member adopts the _____ needed to achieve an effective meeting.'

Q. 5 List nine common reasons for calling a meeting on a building site:

1. _____
2. _____
3. _____
4. _____
5. _____
6. _____
7. _____
8. _____
9. _____

Worksheet 4

Student name: _____

Enrolment year: _____

Class code: _____

Competency name/Number: _____

To be completed by teachers:	
Student competent	☐
Student not yet competent	☐

Task: Read through the section *Sustainability—Resource efficiency and waste minimisation* and up to and including *Sustainable housing* and then complete the following:

Q. 1 How could 80% of construction waste material be reused?
_____

Q. 2 By recycling we are able to save on the demands for the earth's natural resources such as minerals, oil, petroleum, plants, soil and water. What are three other distinct advantages gained by recycling?

1. _____
2. _____
3. _____

Q. 3 What are the four steps to follow in a waste minimisation hierarchy?

1. _____
2. _____
3. _____
4. _____

Q. 4 What are the 10 most common building products that can be recycled?

1. _____
2. _____
3. _____
4. _____
5. _____
6. _____
7. _____
8. _____
9. _____
10. _____

continued ➤

Q. 5 What are the five steps to follow in achieving resource efficiency and waste minimisation?

1. _____
2. _____
3. _____
4. _____
5. _____

Q. 6 Define 'sustainable housing'.

_____
_____
_____

Q. 7 Environmentally sustainable homes are designed to achieve three main goals. What are they?

1. _____
2. _____
3. _____

Q. 8 What are three general design considerations that will assist in achieving a sustainable home?

1. _____
2. _____
3. _____

REFERENCES AND FURTHER READING

Acknowledgment

Reproduction of the following *Resource List* references from *DET, TAFE NSW C&T Division (Karl Dunkel, Program Manager, Housing and Furniture) and the Product Advisory Committee* is acknowledged and appreciated.

Texts

Construction Industry Training Advisory Board (NSW) (1999), *The building and construction industry vocational education and training plan industry extract for the period 2000–2002*, CITAB, Sydney

Construction Industry Training Advisory Board (NSW) (2000), *Industry VET plan 'beyond 2000': The annual plan for vocational education and training for the building and construction industry in NSW*, CITAB, Sydney

Graff, D.M. & Molloy, C.J.S. (1986), *Tapping group power: A practical guide to working with groups in commerce and industry*, Synergy Systems, Dromana, Victoria

National Centre for Vocational Education Research (2001), *Skill trends in the building and construction trades*, National Centre for Vocational Education Research, Melbourne

NSW Department of Education and Training (1999), *Construction industry: Induction & training: Workplace trainers' resources for work activity & site OH&S induction and training*, NSW Department of Education and Training, Sydney

NSW Department of Industrial Relations (1998), *Building and construction industry handbook*, NSW Department of Industrial Relations, Sydney

TAFE Commission/DET (1999/2000), 'Certificate III in General Construction (Carpentry) Housing', course notes (CARP series)

WorkCover Authority of NSW (1996), *Building industry guide: WorkCover NSW*, Sydney

Web-based resources

Regulations/Codes/Laws

<www.industrialrelations.nsw.gov.au/awards/payrates> NSW Department of Commerce—Office of Industrial Relations—Awards/Pay Rates

<www.austlii.edu.au/au/other/nsw-awards> Australasian Legal Information Institute

<http://apprenticeship.det.nsw.edu.au> DET Apprenticeships & Traineeships

Resource tools and VET links

<www.hsc.csu.edu.au/construction> NSW HSC Online—Construction

<www.edna.edu.au/go/browse/12650:resource#resulttab> Education Network Australia

Industry organisations' sites

<www.citb.org.au/publications/resources.asp> SA Construction Industry Training Board

<www.lspc.nsw.gov.au> Long Service Building and Construction

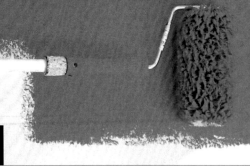

CHAPTER 3
PLANNING AND ORGANISING WORK

This chapter addresses the skills required to plan allotted tasks to maximise personal productivity on a general construction building site. Areas addressed include pre-work planning and safe, efficient work sequencing.

Areas addressed from the unit of competency include:
- planning and preparing work;
- sequencing work safely; and
- cleaning up.

PLAN AND ORGANISE YOUR WORK

Purpose of planning

The ability to plan and organise is an extremely important life skill. Whether you are planning and organising a holiday, a party, a shopping trip or a task at work, your life will probably be more satisfying, more fulfilling, more successful and far less stressful if things turn out the way you want them.

Of course, in your private life, if you are not a very good planner and organiser, it probably does not matter much if you take the wrong clothes on holiday, or that you forgot to buy food at the supermarket. If you are a poor planner and organiser in your private life you mainly cause problems for yourself.

But if you are a poor planner and organiser in your working life you cause problems for others (Figure 3.1). If you cause enough problems for others at work they will not want you around, and you are likely to find it difficult to get work or keep a job.

Aim to work efficiently and safely

In the construction industry it is important that everybody is a good work planner and organiser, because poor work organisation can not only waste time and materials—it can also cause accidents and injuries. You must be able to work efficiently and safely.

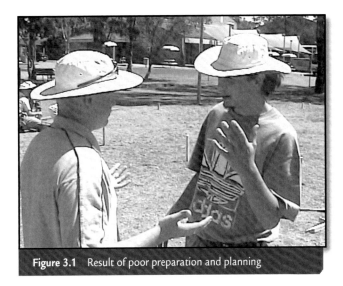

Figure 3.1 Result of poor preparation and planning

To work efficiently means not wasting your own or other people's energy by creating unnecessary work. Work in the building industry can be very physically demanding. Wasted energy means less work completed. Less work completed means less money earned. Building workers do not like having to do unnecessary work because someone else has not done their job properly.

It also means not wasting materials, tools or equipment. Wasted materials, tools and equipment are a cost that must come out of the price of the job. Money spent on replacing materials, and replacing or repairing lost or damaged tools and equipment, means either less money for wages or higher prices quoted to do jobs. The higher the prices quoted to do jobs, the fewer jobs won and the less work.

This does not mean that working efficiently means working unsafely. On the contrary, work injuries lead to lost work time, higher costs, and higher prices quoted to do jobs. Furthermore, working unsafely for an imaginary rise in your pay packet will never make up for your or another worker's damaged back, lost sight or hearing, damaged lungs or hands. More money in your pay packet cannot make up for you having to leave the building industry to work in a lower-paid job because your injuries make you unfit for building work.

Working *efficiently* means working skilfully and productively so that you produce a high-quality product that is good value for money and that will ensure the success of the company and a continuity of work. Working *safely* means to work in such a way that you create a healthier and safer workplace for everybody. Planning and organising your work clearly has general benefits for everybody on a construction site by increasing efficiency and safety, but planning and organising your work can also have specific benefits just for you.

When you plan and organise a task you have been given on a construction site, you can:

- reduce later problems by making important decisions beforehand;
- get the cooperation of other workers when you inform them about the task;
- have materials, tools and equipment ready when and where you need them;
- save your time, materials and tools;
- eliminate or reduce hazards to your safety;
- easily adapt to overcome problems you did not foresee.

On a busy construction site these are important benefits that can improve the quality of your day. Planning your tasks can make your work more satisfying, more fulfilling, more successful, and definitely less stressful.

Make sure you know what you have to do

Before you begin any task, it is essential that you make sure that you know exactly what you are required to do. Your supervisor may give you spoken **instruction**, written instructions, drawings, or a combination of all three.

While your supervisor is doing this, it is important for you to be patient and to check that you have clearly understood exactly what is required. Utilise your communication skills to check that you know, not just think you know, what the task is that you must carry out. Always seek and offer feedback.

It is very easy, through eagerness, enthusiasm or impatience, to get on with the task, to not listen properly, to ignore what is being said, or to dismiss information that does not seem important to you. Many a task has been botched right from the start because crucial information has not been paid attention to. Because it is difficult to remember a lot of spoken instructions, a number of different sizes or dimensions, or a number of different steps in a procedure, always be prepared to write down what is required.

For simple tasks, it is a good idea to jot down a small reminder on a piece of board or piece of sheet material to keep you focused on what is required and stop you making foolish mistakes (Figure 3.2). For other tasks, a safe work method statement will be required.

Before you begin any tasks, be clear in your mind about what it is you are trying to achieve.

Work out how you are going to do the task

When you have a clear idea of exactly what the task is that you are required to complete, the next step is to decide the sequence of actions that you will follow. This sequence may be best set out in a safe work method statement.

In the building industry, construction follows a logical order. For example, the construction of an entire building can be broken down into stages. Each of these stages can be broken down into a series of steps. Then each of these steps can be broken down into a series of tasks. Then each of these tasks can be broken down into a series of actions.

Some stages must be completed in preparation for the next stage, and some steps must be completed in preparation for the next step. Similarly, some tasks must be completed in preparation for the next task.

Some tasks have to be done at a specific point, because that is the only possible time. Tasks that are not done at the correct time may never be able to be done, which will have serious consequences, whereas tasks that are done too soon will have to be done again.

Example 1

Bathrooms require the tile layer to follow a series of steps, such as waterproofing; applying and grouting the floor and wall tiles; then applying sealant to all joints liable to movement. However, each step may be broken down to a number of tasks composed of numerous actions. Waterproofing, for example, requires identifying the position and nature of each joint prior to applying backing rods, bond breaking and reinforcement tapes, and ultimately some form of liquid flash (Figures 3.3–3.4). If any of these are not done exactly right then there may be problems when laying the tiles, or, worse, the flashing will fail and you will be liable for repairs in the future.

An example of a complete sequence is as follows:
1. Identify the characteristics of each joint.
2. Apply backing rods.
3. Apply bond breaking tapes.
4. Apply reinforcing tapes.
5. Apply the first coat of liquid flash.
6. Apply the second coat of liquid flash.

Figure 3.2 Prepare a checklist

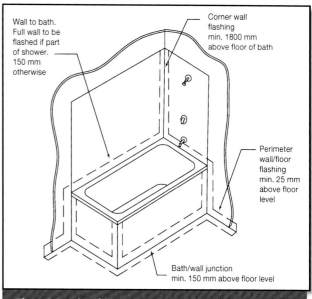

Figure 3.3 Identifying joint type and characteristics

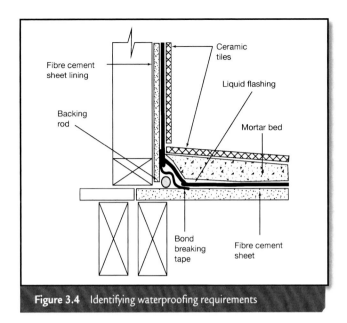

Figure 3.4 Identifying waterproofing requirements

An example of an incomplete sequence is:
1 Apply backing rods.
2 Apply reinforcing tapes.
3 Apply the first coat of liquid flash.
4 Apply the second coat of liquid flash.

Example 2
Similarly, bricklayers must install damp-proof courses, flashings and brick ties as they build a wall, because these things are impossible to install after the wall is finished (Figure 3.5). If this is not done, the wall will have to be demolished and built again.

Figure 3.5 Preparing the wall frame ready for the brick veneer

Example 3
When a house is being built, the plumber and electrician must run their pipes and cables through the walls before the plasterboard is put on (Figure 3.6). If the plasterboard goes on first it will have to be removed, the pipes and cables installed, and then the plasterboard replaced.

Figure 3.6 Rough-in the services before fitting linings

Conclusion
Obviously, mistakes can not only be very expensive but can delay a job and make it unprofitable, which could send the builder broke. If the builder goes broke, many people will be out of a job.

When you are carrying out a task, make sure that you are clear in your mind about the sequence of actions you will follow.

PREPARE A SAFE SITE

Barricades
Barricades include any physical barriers placed to prevent entry and signify that a danger exists. They may be as strong as a timber or steel hoarding or as lightweight as a coloured plastic strip or tape.

Hoardings
There are two main types: type A (fence-type hoardings) and type B (overhead-type hoardings).

Fence-type hoardings are used for residential, commercial or industrial sites where there is a minimal risk of objects falling on the heads of passers-by.

Overhead-type hoardings are primarily used on commercial or multistorey construction sites where there is a high risk of objects falling on passers-by.

Fence-type hoardings may be constructed of timber and sheeted with plywood or galvanised steel posts and chainwire (Figure 3.7). They range in height from 1.8 m for most residential sites up to 2.1 m for commercial or industrial sites, and may be described as follows:

A high temporary fence or structure enclosing a demolition site, or a building site during construction or alterations, to restrict access, prevent theft, and provide side but not overhead protection to passersby.

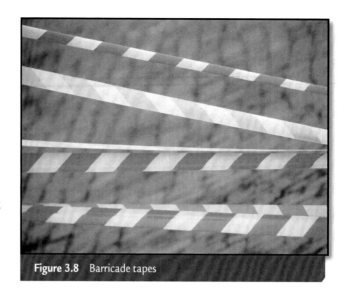

Figure 3.8 Barricade tapes

Figure 3.7 Typical chainwire fence type

Other barricades

These include any form of physical barrier used to draw attention to the existence of a hazard or danger zone. They may be in the following forms:
- *Barricade tapes*—similar to those used by police to close off a crime scene (Figure 3.8);
- *Warning tapes*—placed on top of underground wiring or pipes to indicate their presence (remember, always dial before you dig—1100);
- *Tube and fittings*—these may be placed around open trenches or pits to prevent people from falling in;
- *Plastic water-filled units*—hollow plastic interlocking units used to create a lineal barrier to prevent the entry of vehicles;
- *Board and trestles*—similar to those found on the side of the road where roadworks are underway; that is, a horizontal timber board painted with black and yellow angled stripes held up at either end by a metal 'A'-frame trestle;
- *Bollards*—solid vertical barrier of less than 1 m high to visually or physically deter the entry of vehicles into an area of free pedestrian movement (similar to ram-raid posts).

DETERMINE TOOLS AND MATERIALS

After you have worked out the sequence in which you will do the task, you will have an idea of the materials you will require. You must list the materials necessary to do the task and in the order in which you will use them.

A good motto to remember is 'just in time'. All materials need to be on the site 'just in time' to be incorporated into the building. The materials have to be on the site when you need them.

While it is important to have materials delivered just before they are needed, it is just as important to not have materials on the job that are not needed at the time. This is because the site can become congested with materials that are not needed and therefore get in the way. The materials may be damaged, lost, stolen or vandalised, or simply slow the entire job down because they have to be moved all the time so that other people can get on with their work.

Therefore, you must have your materials on the site, ready to be used, in the sequence in which you will need to use them, *just in time* for you to use them.

The following is a material list for an ensuite (Figures 3.9, 3.10 and 3.11). Materials are required in the following order:
- wall and ceiling sheets;
- stopping-off compounds;
- approved primer (for waterproof and adhesive compounds);
- tile to carpet metal angle;
- waterproofing and flashing materials;
- floor tiles and appropriate adhesive;
- floor grout;
- coloured mastic/silicone to suit floor edges;
- wall tiles and appropriate adhesive;
- chromed metal edging;
- wall grout;
- coloured silicone for wall/floor junctions etc.;
- primer/sealer paint;
- topcoat paints.

Decide where you will store the materials

The next step is to make sure that the materials you will need are where you will need them. Do not make the mistake of simply picking up what you need and starting off to carry it to where you will be doing the task.

If the material you need is large, heavy or bulky, check the route you will be following through the job site. Carrying large and cumbersome items of building materials can be difficult. If you are going up stairs, along corridors, through doorways, up

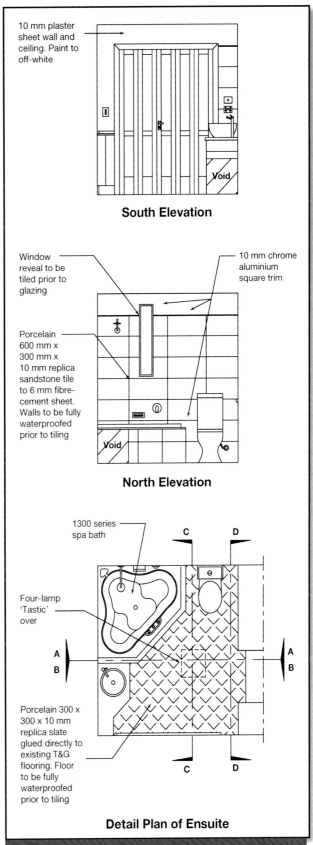

Figure 3.9 Plan and elevations of an ensuite bathroom

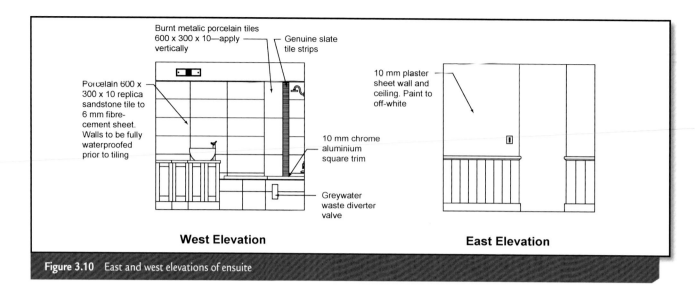

Figure 3.10 East and west elevations of ensuite

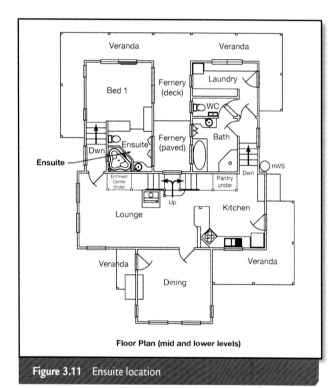

Figure 3.11 Ensuite location

ramps, along planks, along scaffolding, around corners, through tight spaces, over trenches, through mud, you can place an unnecessary strain on your body by getting into trouble while you are under load. It is important to inspect your route beforehand to make sure that all is clear and that you are not going to injure yourself or others because you have got yourself into a difficult situation.

When you reach the place where you will be carrying out your task, make sure you locate your materials in a safe, secure position out of harm's way.

There are many types of building materials, and each type must be stacked or stored in a specific way. Stack or store your materials in the approved way for each material. To find out the best way to stack each type of material, pay attention to the way each one is delivered from the manufacturer. The manufacturer's stacking or storage method will be intended to maintain the material in the best possible condition in the safest possible way.

When you stack or store your materials, simply repeat the manufacturer's stacking or storage methods.

Determine the tools and equipment required

After you have worked out the sequence in which you will do the task, and worked out the materials you will require, you will have an idea of the tools, equipment and personal protective equipment you will require (Figures 3.12 and 3.13). You must list the tools, equipment and personal protective equipment necessary to do the task, in the order in which you will use them (Table 3.1).

If you need special equipment, keep in mind that much special equipment is hired. Therefore, you must remember to order special equipment in time for it to be delivered to the site so that it is there when you need it.

Figure 3.12 Determine tools required

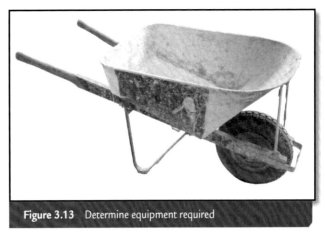

Figure 3.13 Determine equipment required

Table 3.1 Checklist for slab placement

\multicolumn{2}{c}{TOOL CHECKLIST}
\multicolumn{2}{c}{Project: Construct a concrete slab for a small shed}
Check	Prepare the slab
✓	Shovels, pick, mattock, spud bar, spade, wheelbarrow, tape, dumpy or laser level
	Form up for the slab
✓	Hammer, nail bag, tape, rule, pencil, string lines, hand saw, power saw, lump hammer, sledge hammer, nails, dumpy or laser level
	Place the reinforcement
✓	Bolt cutters, wire snips
	Pour and finish the concrete
✓	Short-handled square mouth shovel, immersion vibrator, aluminium screed, wheelbarrow, edging tool, steel trowel, woof float, hose, cleaning brush

After you have prepared your list of tools and equipment, you must plan how you are going to get them to the place where you are going to do the task.

You will probably have to carry them there from the truck, from the store, or from another part of the site. In some situations you may even have to bring them from another site or from home. Most tradespeople store their equipment at home in a shed or garage and take what is required to the job only when it is required. It would be impractical for most tradespeople to carry every tool they own to every job they attend. They would need a huge truck to cart around tools and equipment they did not need at the time, which would be not only a waste of time and effort but also a waste of money.

Therefore, it is important to plan what equipment you will need each day. If you do not take a tool to the job when you need it, you will either have to go and buy it again or set the job back by having to do another task that you do have the tools for.

Once you have the correct tools and equipment on the site, you must make sure that you have the tools you need with you right where you are doing the task. This means you must have a tool carry-all, toolbox or wheelbarrow to transport the tools from where they are stored to where you will use them.

Check tools and equipment

You cannot afford to waste time and energy going backward and forward to the truck, or store, all day because you are not mentally organised enough to plan what you need and to take what you need to where you will be working. As with the materials, do not make the mistake of simply picking up the tools, equipment and PPE you need and carrying them to where you will be doing the task. Check that the tools are working properly. Check that the equipment will do what you want. Make sure that the PPE is correct for the task at hand. Make sure that you have the means of carrying them to where you will be doing the task, such as a toolbox or wheelbarrow.

If some time has elapsed since you delivered the materials to the place where you will be working, make sure that the path is still clear for you to carry the tools, equipment and PPE safely. Building sites are busy places: other workers are continually shifting gear and moving things around out of their way. In doing this they may unwittingly put things in your way. Always check that your task has not been made more difficult in such a situation.

CARRY OUT THE TASK CORRECTLY

Now that you have your materials, tools, equipment and PPE on the spot where you will be working, it is time to take care to follow your instructions.

To do the task properly you must follow a safe and efficient sequence of work. Make sure that you have your safe work method statement (SWMS) if one is required for the task.

If your instructions were spoken, think back to when you received them and be clear about what it is you were instructed to do.

If your instructions were written, or in the form of a drawing, make sure you read them again. Do not make the mistake of relying on your memory.

Expensive errors are made by people who think they can remember the written instructions or what was on the drawing. Never be too lazy to double-check (Figure 3.14). Double-checking may help you to keep your job, or save a lot of money.

Finally, it is time to do the task. Make sure you follow the logical sequence of work. Check your safe work method statement for the steps in the procedure. It is important that you do everything at the correct time. Do not rush the task. If you rush the task you are likely to botch it. Speed is good in building only if the task is done correctly the first time. Having to go back and redo things, or demolish them, costs much more than some imagined small saving in time gained by rushing. A good rule to remember when setting out materials for cutting is: 'Measure twice, cut once'.

Regularly take the time to make sure that you keep your work area clean and clear of debris. A few moments taken to get rid of waste and to sweep up will save you having to stumble over waste, will save you lost time, and may even save you or another worker from having an accident.

While you are working, you will not be using all your tools and equipment all the time. Therefore:

- Make sure that the tools and equipment not in immediate use are safely located.
- Do not block walkways.
- Never stand up, stack or lean tools or equipment in such a way that they can slip, slide or fall over and injure another worker, or be damaged.

Solve problems as they arise

Working on a construction site is not like working in a factory. Construction workers are not just assembling parts that have been manufactured somewhere else. A building is like a huge jigsaw puzzle, in which the people who are trying to put the puzzle together are also making the parts as they go along (Figure 3.15).

To an observer it may look as though all the workers are part of a mass production process, but this is not the case. The building that the observer is watching being constructed is probably of an

Figure 3.14 Double-check details before you start

Figure 3.15 A building is like a huge jigsaw puzzle

original design, which has never been constructed before and is being erected on that particular site for the first time.

Therefore, although the workers have a great deal of experience and may be repeating tasks they have done many times before, they are doing that particular task on that particular spot on that particular site for the first time. In doing so they are dealing with different people and different materials in different situations. This means that problems and difficulties can arise. How the workers deal with these problems and difficulties will depend on their individual talent, training, knowledge and experience. In general, the most successful construction workers have highly developed problem-solving skills.

In the construction process, it is important to solve each construction problem as it arises and not to allow problems to accumulate at the end of the job when it is too late to do anything about them. The accumulation of errors on a job can lead to structural failure, faulty finishes on the building, and accidents and injuries to other workers.

Expect problems to arise and be prepared to overcome them. Problems can arise with the procedure you had intended to follow. Because of work going on nearby you may not be able to follow the procedure you had initially planned. You will have to change your sequence of work to accommodate what is going on near you or around you.

Problems can also arise with the materials you had intended to use. The materials may not be available in the order you need or at the time you need them. On the other hand, some of the materials may be incorrect or faulty and you will have to arrange for replacement material but still keep the work progressing. This also will require you to change your sequence of work. Weather may affect your task sequence and your materials. This would require a radical replanning and reorganisation of the sequence of the task that you are doing (Figure 3.16).

Construction processes rarely go according to plan. This does not mean that a plan is therefore unnecessary. A plan is important because it tells when we must start and when we must finish a task. It also identifies key points along the way. It tells us when various important sections must be finished.

Figure 3.16 Discuss and solve problems as they occur

But along the way we must be flexible. What we lose here we can gain there. What we gained there we lost here. The important thing to remember is that the task must be completed safely, efficiently, to the required quality standard, in an appropriate time.

Clean up when work is complete

Cleaning up correctly is just as important a part of doing the task as planning and organising, but cleaning up does not just mean sweeping a floor. Cleaning up means that debris and waste materials must be collected and removed to the waste storage area and sorted for recycling. Unused materials must be returned to the storage area and stacked or stored neatly. The work area must be left clean, safe and secure on completion of the task (Figure 3.17), which includes the following operations:

- Tools and equipment must be cleaned, maintained and stored correctly (Figure 3.18).
- Hire equipment items must be counted, accessories noted, and all must be located where they can be picked up by the hire company, with any faulty parts noted.
- Company-owned tools and equipment must be checked and maintained.
- Timber handles on tools can be wiped with raw linseed oil.
- Metal **blades** on tools can be wiped with mineral oil.

Figure 3.17 Clean work area daily

Figure 3.18 Cleaning tools using a bucket

- Equipment with its own storage case must be returned to its case together with blades, attachments and accessories. Any damaged or faulty equipment must be reported so that it can be repaired.
- Safety signs and barricades must be removed and stored for later use.

Report on completed work

After you have carried out all your responsibilities, it is important to report to your supervisor on the completion of the task (Table 3.2). Make sure that you tell your supervisor about any problems or difficulties you experienced. It is also important to report any damaged or faulty equipment so that it can be repaired before it is required again. It is very frustrating when you are going to do a job to find that the tools or equipment you need are

Table 3.2 Create a checklist for reporting	
✓	Make a report for the supervisor.
✓	List any problems or difficulties you experienced.
✓	Report any damaged or faulty equipment.
✓	Report any unsafe or dangerous behaviour.
✓	Report any uncooperative workers, which may have hindered your progress.
✓	Any other matters you feel are relevant.

not working or not working properly. Advise your supervisor of any unsafe or dangerous behaviour or procedure you have experienced or witnessed. Any unsafe or dangerous behaviour or procedures can risk everybody's health, safety and welfare. Your supervisor needs to know about OH&S breaches because the unsafe worker, supervisor, manager and company can all be heavily fined for failing to follow the OH&S regulations.

Finally, tell your supervisor about any lack of cooperation or assistance from other workers. On building sites, many tasks are carried out by subcontractors who are paid to come onto the site, do a task and then leave. Some subcontractors feel pressured to get the task done as quickly as possible so that they can get on with another task somewhere else. This can lead to a lack of consideration for, cooperation with and assistance of other workers on the site, which may generate friction and lead to arguments and fights. These actions can disrupt the smooth running of a job, slow the job down, lead to faulty and/or unsafe work and waste time, materials and money. Your supervisor needs to know anything that can affect the efficient running of the job.

ENVIRONMENTAL PROTECTION REQUIREMENTS

Environmental Protection Authority and waste

According to the NSW TAFE/EPA (1997), *Minimising Construction and Demolition Waste*, waste may be defined as 'anything unused, unproductive, or not properly utilised' (*Macquarie Dictionary*).

The NSW *Waste Minimisation and Management Act 1995* gives a legal definition of waste, which is very broad:

Waste includes:
(a) Any substance (whether solid, liquid or gaseous) that is discharged, emitted or deposited in the environment in such volume, constituency or manner as to cause an alteration in the environment, or
(b) Any discarded, rejected, unwanted, surplus or abandoned substance, or
(c) Any otherwise discarded, rejected, unwanted, surplus or abandoned substance intended for sale or for recycling, reprocessing, recovery or purification by a separate operation that produced the substance, or
(d) Any substance prescribed by the regulations to be waste for the purposes of this Act.
A substance is not precluded from being waste for the purposes of this Act merely because it can be reprocessed, reused or recycled.

Most wasted

Estimates suggest that about 20 cubic metres of waste are produced in constructing a typical residential house in Australia. The different nature and types of construction plus the different stages of work each make a significant contribution to the waste stream.

Whichever measure is used, timber, followed by plasterboard, accounts for the majority of waste volume in residential construction—whether in a single dwelling, multi-unit housing or low-rise apartments. The greatest weight of waste is caused by bricks and rubble. Packaging generally has been identified as a significant contributor to the waste stream, especially for residential construction, commonly constituting more than 5% or more of total construction waste in Australia.

Therefore, it is critical that builders develop a **waste management** plan to help reduce the amount of waste by separating waste materials and reusing or recycling waste materials as much as possible. This will not only reduce the amount of waste that is simply dumped, but will reduce building costs overall. Note: More information may be obtained from the publication by NSW TAFE/EPA (1997), *Minimising Construction and Demolition Waste*, TAFE Construction & Transport Division, Castle Hill, Sydney.

Non-toxic waste disposal

By the nature of the activity, building operations generate a large amount of waste. Good building practice also includes good waste management practice. This means that all waste generated on a building site should be sorted and stored in separate piles so it may be reused, recycled or removed.

It is accepted general practice on building sites to make each subcontract team responsible for the waste it creates and its clean-up and tidy storage on-site. This includes the proper disposal of lunch wrappers, drink bottles, cans, fruit scraps etc., because if this refuse is left lying around it will attract pest insects and rodents, which are capable of spreading disease on the job site.

Unsorted waste adds a great cost to the overall job total, as mixed or unsorted general waste costs a lot more per tonne to dispose of than materials such as sorted masonry, timber, paper products, glass, plasterboard and metal. Many builders include a waste management statement in the requirements for engagement of subcontractors and usually back-charge those who don't tow the line. Many of these waste products may be collected by specialist groups who recycle the base materials; concrete, bricks and tiles, for example, may be crushed at the tip site and sold off for such uses as clean filling, roadbase or all-weather driveways for construction sites.

Therefore, it is every worker's responsibility to clean up as they go to prevent accidents and hazardous situations from occurring, to provide a safe, hygienic site for all and to minimise the amount of waste taken to the tip as general fill.

Worksheet 1

Student name: _____

Enrolment year: _____

Class code: _____

Competency name/Number: _____

To be completed by teachers:
Student competent ☐
Student not yet competent ☐

Task: Read through the sections beginning at *Plan and organise your work* and up to and including *Aim to work efficiently and safely*, then complete the following.

Q. 1 Why is it important that everybody is a good work planner and organiser?

_____
_____
_____
_____
_____

Q. 2 List the meanings of the term 'working efficiently':

_____
_____
_____

Q. 3 What does 'working safely' mean?

_____
_____

Q. 4 List the specific benefits you can get from planning and organising your work:

_____
_____
_____
_____
_____
_____
_____
_____

Worksheet 2

Student name: _____

Enrolment year: _____

Class code: _____

Competency name/Number: _____

To be completed by teachers:
Student competent ☐
Student not yet competent ☐

Task: Read through the sections beginning at *Purpose of planning* and up to and including *Conclusion*, then complete the following.

Q. 1 What are six possible advantages gained from planning and organising a task that you have been given on a construction site?

1. _____
2. _____
3. _____
4. _____
5. _____
6. _____

Q. 2 On a construction site, certain tasks have to be carried out in a logical order or at specific points in time. What are the possible consequences when this is not adhered to?

_____
_____
_____

Q. 3 If mistakes are made, what can happen?

_____
_____
_____

Q. 4 What must you be clear about in your mind before you start work?

_____
_____

Q. 5 Briefly describe a type A hoarding used for residential construction:
_____
_____

Q. 6 Prior to excavating or digging in a new or unidentified area, what must be done to avoid possible damage to underground cables and pipes?
_____
_____

Q. 7 Briefly describe the purpose of barricades known as 'bollards':
_____
_____

Worksheet 3

Student name: _____

Enrolment year: _____

Class code: _____

Competency name/Number: _____

To be completed by teachers:

Student competent ☐

Student not yet competent ☐

Task: Read through the sections beginning at *Determine tools and materials* and up to and including *Check tools and equipment*, then complete the following.

Q. 1 What must you do after you have worked out which materials you need?

_____
_____

Q. 2 What are the possible consequences of having materials delivered to the construction site earlier than they are required?

_____
_____
_____

Q. 3 If you need special equipment, what must you remember to do?

_____
_____
_____

Q. 4 Why is it important to plan what tools and equipment you will need before you head off to begin a new day's work?

_____
_____
_____

Q. 5 What are suitable means of transporting tools and equipment to the location at which you will be undertaking a particular task?

_____
_____

CHAPTER 3 PLANNING AND ORGANISING WORK

Q. 6 What checks should you carry out on the tools, equipment and PPE you require, prior to transporting them to where you will be doing a particular task?

_____

_____

Q. 7 What other check should be carried out before transporting tools, equipment and PPE to the place where you will be working?

_____

Worksheet 4

Student name: _____

Enrolment year: _____

Class code: _____

Competency name/Number: _____

To be completed by teachers:
Student competent ☐
Student not yet competent ☐

Task: Read through the section on *Carry out the task correctly*, then complete the following.

Q. 1 After you have worked out which tools, plant and PPE you need, what should you do next?

_____

Q. 2 To do the task properly, what must you do?

_____
_____
_____

Q. 3 What should you not make the mistake of relying on, and what should you do?

_____
_____
_____
_____

Q. 4 If you rush a task, what is likely to happen?

_____

Q. 5 When you are working, what must you regularly take time to do?

_____
_____
_____

Q. 6 What must you do with tools and equipment not in immediate use?

_____
_____
_____
_____
_____

Q. 7 In the construction process, what is it important to do and why is it important to do it?
_____
_____
_____

Q. 8 What can cause problems with the procedure you had intended to follow, and what will you have to do?
_____
_____
_____
_____

Q. 9 What can cause problems with the materials you had intended to use, and what will you have to do?
_____
_____
_____
_____
_____

Q. 10 If weather affects your task sequence, what might you have to do?
_____
_____

Q. 11 Why is a plan important?
_____
_____
_____

Q. 12 What is the important thing to remember?
_____
_____

Worksheet 5

Student name: _____

Enrolment year: _____

Class code: _____

Competency name/Number: _____

To be completed by teachers:
Student competent ☐
Student not yet competent ☐

Task: Read through the section on *Environmental protection requirements*, then complete the following.

Q. 1 State whether reprocessed, reused or recycled materials are considered to be waste or not:

_____

_____

Q. 2 List the materials that are considered to be the most wasted in the building industry:

_____

_____

Q. 3 State the percentage of packaging waste found in the building industry:

_____

Q. 4 Briefly describe the meaning of 'good waste management practice':

_____

_____

_____

Q. 5 Whose responsibility is it to clean up and reduce hazardous situations from the build-up of waste on-site?

_____

REFERENCES AND FURTHER READING

Acknowledgment

Reproduction of the following *Resource List* references from *DET, TAFE NSW C&T Division (Karl Dunkel, Program Manager, Housing and Furniture) and the Product Advisory Committee* is acknowledged and appreciated.

Texts

Graff, D.M. & Molloy, C.J.S. (1986), *Tapping group power: A practical guide to working with groups in commerce and industry*, Synergy Systems, Dromana, Victoria

National Centre for Vocational Education Research (2001), *Skill trends in the building and construction trades*, National Centre for Vocational Education Research, Melbourne

NSW Department of Education and Training (1999), *Construction industry: Induction & training: Workplace trainers' resources for work activity & site OH&S induction and training*, NSW Department of Education and Training, Sydney

NSW Department of Industrial Relations (1998), *Building and construction industry handbook*, NSW Department of Industrial Relations, Sydney

TAFE Commission/DET (1999/2000), 'Certificate III in General Construction (Carpentry) Housing', course notes (CARP series)

WorkCover Authority of NSW (1996), *Building industry guide: WorkCover* NSW, Sydney

Web-based resources

Resource tools and VET links

<www.resourcegenerator.gov.au> ANTA resource generator

<www.ntis.gov.au> National Training Information Service

<http://hsc.csu.edu.au/construction> NSW HSC Online—Construction

Audio-visual resources

Jolly, R., Baxendale, T., Stubbs, J. (c.1990), *Site Organisation and Planning*, Qi Training, England

CHAPTER 4
WORKPLACE COMMUNICATION

Methods of communication on a building site come in many forms, and can include verbal instruction, written instruction, plans, sketches, two-way radio, signage and site meetings, to mention a few. They are all used to convey or receive messages or information. Whether the information is received and understood depends on how clearly and accurately it is given. Many activities and issues rely on this communication being effectively carried out, not least of which is site safety. Therefore, to avoid misunderstandings, which may lead to costly errors or someone being injured, clear and concise communication is critical.

Areas addressed from the unit of competency include:

- gathering, conveying and receiving information;
- carrying out face-to-face routine communication;
- applying visual communications; and
- participating in simple on-site meeting processes.

WHAT IS COMMUNICATION?

The term 'communication' can be applied in three ways. First, it can be applied to the act of communicating. Talking (Figure 4.1), singing, writing, miming, gesturing, signalling, drawing, sketching, kissing, hugging, hitting, punching or kicking are examples of communication acts. They all carry a message of some kind.

Second, communication can also be applied to the **message** that is being communicated. These messages may be about emotions, feelings, wants, needs, thoughts, ideas, opinions, facts, knowledge, information, warnings, or any one of the many things people need to impart to others. For example, the communication act of talking enthusiastically about a new house that your construction company is going to build, in addition to the information you are conveying,

Figure 4.1 Communicating by talking

imparts a message that you are excited about the upcoming project. The communication act of hugging someone imparts the message that you like that person. The communication act of chastising someone imparts the message that you are upset. The communication act of extending your arm forward with your hand flexed upward and palm toward the viewer imparts the message that you want that person to halt.

Third, communication can be applied to the means of communicating. For example, an audio tape on which you have recorded yourself talking about your new building project, and which you will send to an interstate friend, is a means of communicating. The accompanying letter that you have written is also a means of communicating. So, too, is the poster that you have developed to advertise the new house to prospective buyers.

Communication, then, can be understood as the act, the message, or the means of communicating with another person or group of people.

Communication has been successful only when the person or group towards whom the message has been directed (often called the receiver) has understood the message exactly as the person who imparted the message (often called the sender) intended it to be understood. This does not mean that successful communication is the sole responsibility of the originator of the message. On the contrary, successful communicating is also the responsibility of the receiver of the message. Sender and receiver share the responsibility for successful communication.

All people have different life experiences. These different life experiences shape the character, personality and view of the world of each individual. This will affect the way an individual understands a message from someone else.

To be sure the correct message is received, the sender will require 'feedback'. For the sender, feedback may range from a simple request for the receiver to repeat what was said, to an invitation for the receiver to discuss the message. For the receiver, restating the message for the sender is a good check. Communication is a circular procedure that requires checking and rechecking that the message received matches the message sent. Without feedback, failure to communicate is likely.

Unfortunately, successful communication is not as simple as Figure 4.2 implies. In reality, there are many impediments to effective communication. These impediments, sometimes called barriers or blocks, can exist in the environment in which the people are communicating; in the system by which they communicate; or in the people who are trying to communicate.

Barriers to communication can be conveniently categorised as physical, emotional, psychological or intellectual.

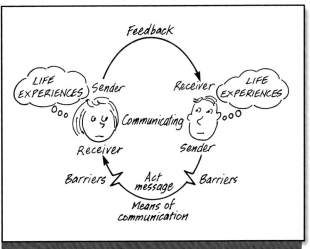

Figure 4.2 Communicating = creating understanding

Physical barriers in the environment or the workplace, such as loud machinery noise or obscured safety signage, can affect communication. In a telephone or two-way radio communication system, interference can interrupt contact. In people, disabilities such as deafness or failing hearing, and blindness or failing eyesight, can create a barrier to communication.

Emotional barriers, of course, occur in people. Emotions such as anger, resentment, frustration, dislike and hatred can get in the way of good communication. The presence of these emotions in the sender can override the content of the message. The presence of these emotions in the receiver can distort the perception of the message.

Psychological barriers also occur in people. People's psychological characteristics can influence the communication process. For example, people who as senders are continually aggressive, arrogant, judgmental, manipulative, confrontational or guilt-inducing in their communication behaviour are unlikely to encourage the feedback that is essential to successful communication. Similarly, people who as receivers avoid, make light of, refuse to acknowledge or divert communication messages are avoiding communicating frankly about an issue.

People also have *intellectual barriers* to good communication. A person's lack of knowledge can limit his or her effective use of complicated communication systems such as computers. Not understanding the language of signs or instructions might jeopardise a person's safety. Similarly, the lack of a shared language could cause serious misunderstanding between a sender and a receiver. In the workplace, the use of language that is too technical for beginners, or the use of jargon with the uninitiated, can lead to problems. Also, the inability to read technical **plans** correctly can lead to expensive mistakes.

How do we overcome these barriers? First, we must not make the assumption that it is the receiver's responsibility to understand the message we are trying to impart. If it's our message, it is our responsibility to get it across. Second, we must anticipate the possibility of the operation of physical, emotional, psychological or intellectual barriers to communication, and, by imagining ourselves in the shoes of the receiver, try to make sure that our communication avoids these barriers.

THE COMMUNICATION PROCESS

The ability to communicate well is an important skill in the building industry. On building sites, poor communication is responsible for injuries, faulty work, and wasting of time and materials. Nobody on a building site wants to be injured, to have to go back and fix a bad job or to waste their time or expensive materials. Good communication benefits everyone. Nearly everyone acknowledges the need for good communication, but few people ever think of themselves as being poor communicators.

So, if people think they are communicating properly, why are people being injured, and why is faulty work being done, and time and materials wasted? This is often because people think that if they can talk, read and write they can communicate. Unfortunately, this is not so.

These skills are an important part of being a good communicator, but are not sufficient on their own. Some people who are not expert talkers, readers and writers are excellent communicators. Why? Because they understand that to communicate they must create understanding in the mind of the person with whom they are communicating. No matter how good a talker, or reader or writer you are, you cannot get your message across if you are not communicating.

Good communicators create understanding. When we are communicating we are creating understanding. If we are not creating understanding we are not communicating. When we communicate we send a message to someone. We are the sender, they are the receiver. If the receiver does not understand the message the way the sender intended it to be understood, the communication has failed.

Both sender and receiver have the responsibility for successful communication. They work together to create understanding. They do this by checking that the message received is the same as the message sent. They check by seeking and giving feedback (Figure 4.3).

Figure 4.4 Communicating through plans and drawings

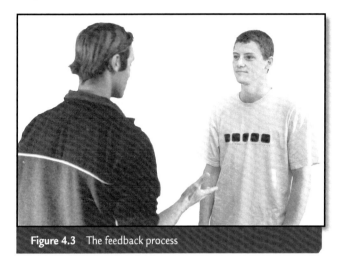

Figure 4.3 The feedback process

The sender checks by:
- asking for the message to be repeated;
- asking questions; or
- asking for the message to be restated.

The receiver checks by:
- repeating the message;
- asking questions; or
- restating the message.

To make sure that you have overcome the barriers to communication, always seek and offer feedback.

Communication on-site

On a building site, face-to-face communication is not the only form of communication. Information comes in different ways (Figure 4.4):
- spoken language;
- written language;
- body language;
- diagrams;
- sounds;
- lights;
- touch sensations;
- odours.

All of these channels of information are important.

Spoken language may be:
- face-to-face;
- via a telephone;
- via a two-way radio.

Written language may be in:
- letters;
- memos;
- instructions for use;
- safe work method statements;
- product directions for use;
- 'must not do' safety signs (prohibition);
- 'must do' safety signs (mandatory);
- 'must be careful' safety signs (caution);
- safety help signs (emergency);
- firefighting equipment signs;
- notices;
- orders;
- contracts;
- drawings;
- specifications;
- job schedules;
- emails;
- text messages;
- faxes;
- door, window or steel schedules;
- delivery dockets;
- material safety data sheets (MSDS);
- safety tags.

Body language may be in:
- hand signals; (Figure 4.5.)
- gestures;
- facial expressions;
- movements;
- posture, colour, breathing.

Figure 4.5 Communicating using hand gestures and facial expression

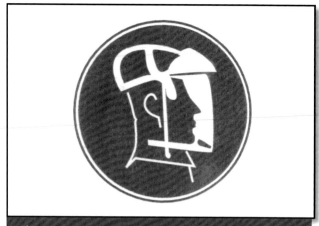

Figure 4.6 Communication using pictograms

- stressed or straining machinery;
- failing machines or tools;
- collapsing formwork/falsework;
- collapsing trenches;
- collapsing scaffolding;
- unusual occurrences.

Diagrams may appear in:
- 'must not do' safety signs (prohibition);
- 'must do' safety signs (mandatory); (Figure 4.6.)
- 'must be careful' safety signs (caution);
- safety help signs (emergency);
- firefighting equipment signs;
- dangerous goods labels;
- barricade tapes;
- underground tapes;
- safety tags;
- instructions for use;
- drawings of buildings;
- drawings of components.

Sounds may come from:
- whistles;
- sirens; (Figure 4.7.)
- bells;
- buzzers;
- horns;
- other workers;
- approaching machinery;
- reversing machinery;
- overhead cranes;

Figure 4.7 Communication through warning siren or bell

Lights may come from:
- rotating amber beacons; (Figure 4.8.)
- rotating red beacons;
- traffic control lights;
- operating machinery;
- approaching machinery;
- reversing machinery;
- overhead cranes.

Note: Rotating amber beacons are used to warn you to be cautious of working or moving machines such as excavators, backhoes, bobcats, forklifts, trucks or cranes. Rotating red beacons may be used to warn you of danger in situations of fire, emergency evacuation, blasting about to begin.

Figure 4.8 Communication using rotating amber beacons on machinery

Touch sensations may come from:
- other workers; (Figure 4.9.)
- faulty machinery;
- stressed or straining machinery;
- unstable structures;
- unstable walls;
- unstable scaffolding;
- unstable formwork/falsework;
- unusual occurrences.

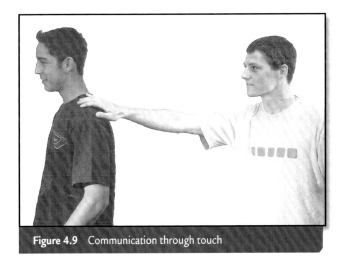

Figure 4.9 Communication through touch

Odours may come from:
- chemicals;
- glues; (Figure 4.10.)
- paints;
- sealants;
- fuels;
- solvents;
- engine exhausts;
- fires.

Figure 4.10 Communication through smell

At all times on a building site, be aware that messages can come through any of these channels that are vital not only for your safety and the safety of others but also for the prevention of faulty work, waste of time and waste of materials.

ENSURE YOU GET THE MESSAGE

Spoken language

When you are being spoken to face-to-face, always listen actively by:
- looking at the person talking to you;
- maintaining an attentive posture;
- showing interest;
- encouraging the speaker;
- asking questions;
- summarising to check your understanding.

Be aware of the barriers to communication that cause misunderstanding and make sure they do not interfere with the message. Always speak with a civil tone and treat people with politeness and respect.

When you answer a business telephone:
- give a polite greeting;
- give your company name;
- give your name;
- offer assistance.

If you are not able to help the caller:
- write down the caller's name;
- write down where the caller is from;
- write down the caller's number;
- write down the caller's message;
- repeat the message to the caller;
- write down the day, date and time;
- give a polite farewell;
- write down your name;
- deliver the message (see Figure 4.11).

Remember to speak at a moderate pace. The caller may never have spoken to you before, and will need time to tune in to your way of speaking.

When you are using a two-way radio:
- use an individual call sign to identify yourself;
- say 'Over' to indicate you have finished speaking so the other person can reply;
- turn your microphone off after saying 'Over' or you cannot hear the other person;
- spell out important words using the international alphabet, e.g. (a)lpha, (b)ravo, (c)harlie, (d)elta.

Note: Speak clearly and at a moderate pace.
Remember that radio frequencies are public, so be careful about what you say.

Written language

When you are seeking information from a document, you can:
- predict the content from the title or pictures;
- skim quickly through the document to get a broad idea of the content;
- scan the document to find a specific piece of information;
- read for the main ideas;
- read in detail for deep understanding.

How you read will depend on your purpose. It is not necessary to read in detail every document. Choose a way of reading to suit your purpose and the document you need to read.

Body language

Be alert to the messages conveyed by people's bodies.

Specific hand signals (Figure 4.12) are used to guide:
- crane operators;
- surveyors;
- truck drivers;
- excavators.

Other workers may use hand or body gestures (Figure 4.13) to tell you to:
- watch out for danger;
- stop what you are doing;
- stay where you are;
- get out quickly;
- watch out above you;
- watch out underfoot or below you;
- watch out behind you;

MESSAGE

To:
Mr, Mrs, Ms ....
Date .... Time ....

From:
Mr, Mrs, Ms ....
Of ....
Phone No. ....

☐ Please phone   ☐ Urgent
☐ Wants to see you   ☐ Will phone again
☐ No message   ☐ Returned your call

Message:
....
....
....
....
....
....

Figure 4.11 Record accurate messages

CHAPTER 4 WORKPLACE COMMUNICATION 147

Figure 4.12 Communicate using hand signals

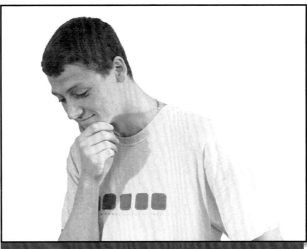

Figure 4.14 Communication using facial expressions

- lift something in unison;
- lower something in unison;
- 'go over there';
- 'come over here';
- 'come and help quickly'.

Other workers' body movements may indicate that they are in trouble or are struggling and need assistance.

Posture, colour and breathing may indicate:
- illness;
- injury;
- poisoning;
- poisonous bite/sting;
- sunburn;
- dehydration;
- asphyxiation;
- intoxication;
- staggering from illness or injury.

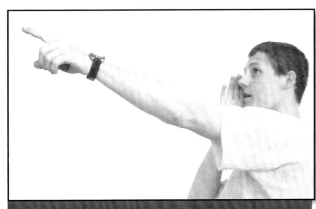

Figure 4.13 Communicate using hand or body gestures

Facial expressions (Figure 4.14) may tell you if another worker is:
- concentrating on the task at hand;
- in difficulty;
- confused;
- happy;
- angry;
- frightened;
- in pain;
- ill.

Figure 4.15 Communication through body posture

WORKPLACE SIGNAGE

Note: It is recommended that readers refer back to 'Safety signs and tags' in Chapter 1, in order to recap on the seven categories of signs available for the construction industry. Refer also to the colour Appendix at the end of this book.

Sign manufacturers have a large range of signs in their catalogues, and new signs can be prepared to suit any situation or circumstance for which an existing sign is not suitable. New signs must always be manufactured according to Australian Standard AS1319-1994, 'Safety signs for the occupational environment'.

The following is a list of many examples of signs available for use in the building industry. Note: The wording may vary from manufacturer to manufacturer, although the intent of the wording must remain the same.

Prohibition (don't do) signs
(Figure 4.16.)

- No smoking (**symbol**)
- No naked flame (symbol)
- No pedestrian thoroughfare (symbol)

Mandatory (must do) signs
(Figure 4.17.)

- Wear eye protection
- Wear face protection
- Wear foot protection
- Wear hand protection
- Wear head protection
- Wear hearing protection
- Wear protective clothing
- Wear protective equipment
- Wear respiratory protection

Figure 4.17 Mandatory (must do) signs

Hazard warning signs (Figure 4.18.)

- Caution—beware of crane
- Caution—beware of hoist
- Caution—beware of traffic
- Biological material hazard (symbol)
- Caution—buried cable
- Corrosive material risk (symbol)
- Caution—crane working overhead
- Caution—do not watch welding arc, protect your eyes
- Electric shock risk (symbol)
- Explosion risk (symbol)
- Caution—explosive powered tool in use
- Fire risk (symbol)
- Caution—flammable liquid
- General risk of danger (symbol)

Figure 4.16 Prohibition (don't do) signs

- Caution—highly flammable
- Ionising radiation risk (symbol)
- Caution—keep clear
- Laser beam in use (symbol)
- Caution—low clearance
- Caution—people working below
- Caution—respirator required
- Caution—slippery under foot
- Toxic material risk (symbol)
- Caution—watch your step

Figure 4.18 Hazard warning signs

Danger hazard signs (Figure 4.19.)
- Danger—asbestos dust
- Danger—asbestos removal
- Danger—authorised personnel only
- Danger—buried cable
- Danger—construction site, unauthorised persons keep out
- Danger—crane working overhead, keep clear
- Danger—deep excavation
- Danger—demolition work in progress
- Danger—explosives
- Danger—explosive powered tool in use, keep clear
- Danger—flammable materials
- Danger—formwork being stripped, keep clear
- Danger—hard hat area
- Danger—high voltage
- Danger—keep hands clear of moving machinery
- Danger—keep out
- Danger—live wires
- Danger—power tools in use
- Danger—scaffolding incomplete, do not use
- Danger—toxic material
- Danger—underground cable
- Water not fit to drink (symbol)

Figure 4.19 Danger hazard signs

Emergency information signs (Figure 4.20.)
- Emergency direction indicating arrows (symbol)
- Emergency exit
- Emergency eye wash (symbol)
- Emergency shower (symbol)
- Emergency use only equipment
- First aid (symbol)
- First aid equipment
- First aid room
- First aid station
- Stretchers

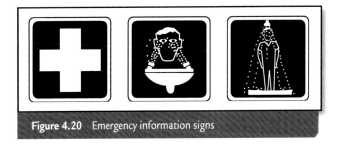

Figure 4.20 Emergency information signs

Fire signs (Figure 4.21.)
- Fire alarm (symbol)
- Fire bucket
- Fire door
- Fire equipment
- Fire escape
- Fire exit
- Fire extinguisher (BCF)
- Fire extinguisher (carbon dioxide)
- Fire extinguisher (dry chemical)
- Fire extinguisher (foam)
- Fire extinguisher (soda acid)
- Fire extinguisher (water)

- Fire hose (symbol)
- Fire hydrant
- Sand for fire only
- Sprinkler valve

Figure 4.21 Fire signs

Placement or erection of signs

It is important to remember not to place or erect a sign in such a way as to create a hazard. For example, an insecurely fixed sign may be knocked or blow down onto someone, causing an injury. A badly placed sign may obscure someone's vision at a critical moment, thereby creating rather than preventing an accident. A projecting sign, erected so that it protrudes into either pedestrian or vehicular traffic, is likely to be run into by someone or something.

Important as it is to place or erect signs correctly, it is equally important to remove signs that are no longer appropriate. As soon as the warning or information set out on a sign is no longer relevant, the sign should be removed. This is particularly important where a sign has been used to warn of a temporary danger—such as demolition work in progress, explosive power tool in use, or people working above—because failure to do so may encourage a tendency to disregard signs.

Generally, signs should be placed or erected just above eye-level in such a way that they are clearly visible and are unlikely to be covered; for example, by stacks of material or machines. (Figure 4.22.) Of course, other heights may be more suitable in some circumstances.

Location of signs

Generally, prohibition (don't do) signs, mandatory (must do) signs and hazard warning signs should be located as close as possible to the hazard that the sign

Figure 4.22 Place signs just above average eye-height

is drawing attention to. However, these signs should not be placed so close to the hazard as to give those seeing the sign insufficient time to react before they are placed in danger.

At no time should signs be located on anything that moves or is likely to be moved. Signs located on doors or windows are likely to be concealed when these are either opened or closed. Signs leant against, or on, moving machinery are likely to fall or be removed from view. Similarly, signs located on stacks or pallets of material will be taken away with the material if it is moved to another position. Care should be taken when locating signs to make sure nothing can happen to them that will make them ineffective.

Barricade tapes

In some situations, where there is a possibility that signs may go either unnoticed or ignored, a physical barrier may be necessary to draw attention to the existence of a hazard or danger zone. A quick and effective way to create a simple barrier is to use barricade tape.

Barricade tape is a roll of coloured polyethylene plastic tape printed with a warning message that can be quickly run out around large areas. The tape can be easily nailed or stapled to, or tied to or wrapped around, posts, poles, stakes, railings or any convenient support, then rolled up again at the end of the job for later reuse. Barricade tape is manufactured in plain yellow, yellow and black diagonal stripes, or red and white diagonal stripes, with a choice of different warning messages in black letters. It comes in 300-metre rolls of either 76 mm or 150 mm width.

Underground warning tapes

Because underground pipes and cables are susceptible to damage during excavation it is important to be able to identify their exact position in the ground. Although above-ground signs can give some idea as to the approximate location of underground utility lines, they do not accurately indicate these along their entire length. On the other hand, an underground warning tape buried above a pipe or cable will give an indication of the installation at every point along its length.

Underground warning tapes are brightly coloured plastic tapes that are usually buried 100–150 mm below the surface of the ground above and along the length of the underground installation. Generally printed with the name of the installation, the warning tape will provide an excavating machine operator with sufficient warning of the existence of the utility line below to avoid expensive damage and service interruption.

Underground warning tapes are available in two types. The first type is a detectable warning tape. This tape is for use with non-metallic installations such as plastic or concrete pipes. It is detectable with a metal detector from above ground. This tape is also brightly coloured for easy visual identification should a metal detector not be utilised.

The second type is a colour-coded plastic tape for visual identification of the installation. The different utilities have the following colours:

Communications	white
Electricity	orange
Firefighting	red
Gas	yellow
Water	green

A close watch should be kept for indications of buried installations at all times during excavation. Damage to underground services not only is likely to be expensive to repair and cause great inconvenience to service recipients but may jeopardise the health and safety of the workers involved if underground electricity, gas or sewer lines are cut.

Note: Always remember to dial 1100 before you dig.

Dangerous goods labels

Many of the materials stored and used on building sites are classified as dangerous goods. There are nine classes of dangerous goods, and each class is identifiable by specific standard labels.

Class 1—explosives. There are five divisions in this class. They range from highly sensitive substances that have a mass explosion potential to very insensitive substances that are unlikely to explode under normal circumstances.

Class 2—gases. There are three divisions in this class. The divisions are flammable gases, non-flammable gases, and poisonous or toxic gases.

Class 3—flammable liquids. There are two divisions: flammable liquids with a flashpoint below 23° C and flammable liquids with a flashpoint above 23° C.

Class 4—flammable substances. The divisions are flammable solids, spontaneous combustible substances, and substances that emit flammable gases when they contact water.

Class 5—oxidising substances. This class contains the divisions oxidising agents and organic peroxides.

Class 6—poisonous or infectious materials. The divisions are toxic substances and infectious substances.

Class 7—radioactive materials. There are three divisions, which correspond with the amount of radiation emitted from the container.

Class 8—corrosive materials. There are no divisions in this class.

Class 9—miscellaneous dangerous materials. There are no divisions in this class. No label is required for this class.

All dangerous goods must be stored, handled and used safely. The information necessary to do this correctly is contained in material safety data sheets (MSDS), which are prepared by product manufacturers. Material safety data sheets give information about the product's constituents, possible health effects, first aid instructions, precautions for use, and safe handling and storage.

By law, employers are required to provide employees with information about the safe use, handling and storage of dangerous substances at work. Having material safety data sheets available for use on the job is an effective way of achieving this.

Sketches and drawings

Instructions for use often contain diagrams. Much technical information is better expressed in pictures. Drawings of buildings and drawings of components contain information that would be nearly impossible to understand if expressed in words.

On-site, supervisors, leading hands and other tradespeople will often do a quick sketch to show what needs to be done because it is quicker, easier, more accurate and more easily understood than a spoken description.

It is neither necessary nor practical for everyone to walk around with a notebook or sketch pad in their pocket. When a sketch is required, a block of timber, an off-cut from a board, a piece of fibre-cement, a piece of plasterboard, a piece of paper torn from a cement or lime bag, or a piece of cardboard torn from a carton will provide a handy drawing medium. Of course, it is necessary to be able to draw and to read drawings. The ability to draw and to read drawings of buildings and components is a special skill that must be learned and developed. Therefore, rather than giving long and involved verbal explanations or instructions, make a quick sketch, which may be more easily understood.

In the building industry the old adage, 'A picture is worth a thousand words' has a much greater meaning. Some things can be described only in diagrams (Figure 4.23).

Figure 4.23 Communicating by sketching a detail

THE ART OF CLEAR COMMUNICATION

Spoken language

When speaking to someone face-to-face, make sure your message is clear:
- Use 'I' messages; talk about your thoughts, your feelings, your needs; do not blame or label others; deal with the issue.
- Focus on an outcome, the situation, the problem, a behaviour; do not focus on the person; be consistent; make sure your body language matches your words; do not send conflicting messages; pick your time and place; make sure the other person is not distracted by something else; and seek and offer feedback to check understanding.
- Always speak with a civil tone and treat people with politeness and respect. If you need to instruct someone, before you begin be certain about what you want the other person to do.

Make sure that you:
- state the overall goal you require;
- describe the main steps in the task in a logical order;
- explain the details of each step slowly;
- emphasise the critical points;

- seek and offer feedback to check the other's understanding;
- summarise the main steps in the task in a logical order.

Be aware that on a building site you may have to combine spoken language, written language and diagrams to make your instructions clear and understandable to the other person.

Written language

When you write, keep in mind that you are writing to be understood. If your readers fail to get your message you have wasted your time and theirs. Your objective must be to get your message across as clearly and quickly as possible. You do this by writing in plain English.

Write your document:
- in a logical order, such as from the least important point to the most important point, or vice versa, according to distance, or according to construction sequence;
- in the active voice (e.g. 'I need more nails'), not in the passive voice, ('More nails are needed'); and
- in short sentences, with one main idea to a paragraph.

Do not obscure your meaning by using:
- ambiguities (e.g. 'I put the saw in the truck that had been damaged');
- clichés (e.g. 'Someone spat the dummy' when you mean 'someone became upset', or 'down the track', when you mean 'in the future');
- tautologies (e.g. 'lineal metres' when you mean 'metres', or 'climb up the ladder' when you mean 'climb the ladder', or 'reverse back the truck' when you mean 'reverse the truck');
- jargon (e.g. 'Optimise output by maximising labour input' when you mean 'Achieve more by working harder');
- slang (e.g. 'mud' when you mean 'mortar', 'helicopter' when you mean 'trowelling machine', 'bubble stick' when you mean 'spirit level', or 'chippy' when you mean 'carpenter');
- technical terms when you are writing for non-technical readers (e.g. 'Place a 100 mm, 20 MPa slab with F72 mesh' when you just mean 'Lay a concrete path').

Note: Jargon, slang and technical terms may also be referred to as 'meta-language'; if used with persons new to the industry or not familiar with them, these terms may be confusing and their meaning may be lost.

Body language

Learn and use the hand signals commonly used on building sites.

Your hand raised with your palm outwards will usually mean 'Stop where you are' (Figure 4.24).

Figure 4.24 Telling others to stop

Your hand and forearm rotated in a circular movement towards your chest will usually mean 'To me or towards me' (Figure 4.25).

A forward pushing movement with your hands, palms outwards in front of your chest, will usually mean 'Move away from me' (Figure 4.26).

Your right arm extended to the right with the hand and forearm swinging in a horizontal arc, gesturing to the right with your index finger pointing to the right, usually means 'Move to the right'. Your left arm extended to the left with the hand and forearm

Figure 4.25 Telling others to move forwards

Figure 4.27 Telling others to move left or right

Figure 4.26 Telling others to move away

Figure 4.28 Telling others to cease what they are doing

swinging in a horizontal arc, gesturing to the left with your left finger pointing to the left, usually means 'Move to the left' (Figure 4.27).

Crossing and re-crossing your right and left hands and forearms horizontally in swinging movements in front of your chest usually means 'Cease what you are doing' (Figure 4.28). Remember, you can use head and body movements to get help, warn others and facilitate teamwork.

Make sure your body language is clear and concise.

It is very easy to confuse others with a sloppy and inaccurate gesture. In your normal interpersonal communication, make sure that your body language and facial expressions are consistent with your spoken language. Do not send conflicting messages.

All the channels of communication are important. We use some more often than others, which makes us more familiar with them. But we cannot afford to ignore the channels we are not familiar with.

If we need to get our message across, we must choose the channel or channels that will give us the greatest chance of success.

ON-SITE MEETING PROCESSES

Types of meetings

There are a number of different types of meetings that may be held on a building site. These are:
- general staff meetings;
- union meetings;
- occupational health and safety committee meetings;
- special-purpose committee meetings;
- team meetings;
- social club meetings;
- special-interest group meetings.

General staff meetings
General staff meetings are used to inform the staff about the company's performance, direction, future, policies and so on. All employees are expected to attend.

Union meetings
These meetings may be called by either union members or union organisers. At these meetings members may discuss pay and conditions, unresolved safety problems, disputes with employers, problems on the site, support for other employees, or support for victims of tragedies. These meetings are usually attended by members of the particular union.

Occupational health and safety committee meetings
Occupational health and safety committees are generally formed only on sites with 20 or more employees. The members of the committee regularly inspect the site and hold meetings where they make policies and recommendations about health and safety on the job. Committee members represent both management and general staff.

Special-purpose committee meetings
Special-purpose committees are formed to undertake particular tasks or responsibilities. Their meetings will usually be conducted according to a program or timetable they have established in order to achieve their objective. When this has been done the committee will generally be disbanded. The committee may be composed of specialists, experts, or simply volunteers who want to get something done.

Team meetings
Team meetings are held by work teams that have been formed to increase efficiency on the job. The teams meet regularly with the objective of planning, organising and carrying out the work as successfully as possible (Figure 4.29).

Figure 4.29 Typical team meeting

Social club meetings
These meetings are usually held by volunteers to arrange social functions after work for interested staff. Activities can range from barbecues to concerts, restaurants or visits to the snow. The success of the social club depends on the enthusiasm and shared interests of the staff.

Special-interest group meetings
These meetings are similar to social club meetings but are focused on specific interests. In a large organisation there may be enough people with a common interest to form a special-interest group and hold meetings to arrange opportunities for them to explore their shared interest.

Conducting meetings

Broadly speaking, there are two ways of conducting these different types of meetings. They may be conducted either formally or informally.

Formal meetings

Formal meetings are run by elected office bearers called chairperson, secretary and treasurer.

Each meeting will follow an **agenda** (Figure 4.30). The agenda is prepared by the secretary and sent to the people attending the meeting before the day of the meeting. It will include the time and location of the meeting, as well as what will be discussed and the order in which it will be discussed.

```
FLYINGFOX CONSTRUCTIONS

PROJECT: Renovation of committee room

PROJECT MEETING

Venue: Site office
Date: Monday 10th October, 2015
Time: 9 am

AGENDA
1. Welcome, introduction & apologies

2. Project update:
   – carpentry work
   – floor tiling and carpet laying
   – painting and decorating

3. Proposed variations to plans

4. General business

Next Meeting:
```

Figure 4.30 Typical example of a meeting agenda

The meeting will be controlled by the chairperson. When each item is raised by the chairperson it will be discussed, and the meeting will follow a set procedure for making decisions using movers, seconders, speakers for and against the motion, then finally a vote by the meeting for or against the motion. During the meeting the secretary will write the minutes of the meeting, which are summaries of the discussions and the decisions agreed on (Figure 4.32). The minutes then become the formal **record** of the meeting.

Note: These procedures are usually followed only when the items to be discussed affect a lot of people, have legal implications, are required by law, or need to be accurately recorded for future reference.

Company meetings, union meetings, occupational health and safety meetings and special-purpose committee meetings are usually conducted as formal meetings.

Informal meetings

Informal meetings, on the other hand, are usually conducted with simple procedures and without the election of office bearers. A group of people will get together to discuss an area of interest (Figure 4.31). There may not be a formal agenda, but everyone will have a general idea of the focus of the meeting.

Figure 4.31 An example of an informal meeting in progress

Someone will probably lead the meeting, and if there are to be a number of meetings group members may take turns at being leader.

There will be no set procedures for conducting discussions and making decisions. The success of the meeting will depend on the reasonableness, fairness, self-discipline and depth of the desire to achieve a

(i) Formal agenda

OCCUPATIONAL HEALTH AND SAFETY COMMITTEE MEETING

Date: Friday 14 October, 2015
Venue: Hut 18
Time: 10 am

Agenda

1. Welcome and apologies
2. Minutes of previous meeting
3. Business arising from minutes
4. General business
   4.1 Non tagging of electrical leads
   4.2 No hand rails on stair wells
   4.3 Bricks falling from fourth floor
5. Other business
6. Next meeting
7. Close

(ii) Minutes of the meeting

OCCUPATIONAL HEALTH AND SAFETY COMMITTEE MEETING

Date: Friday 14 October, 2015
Venue: Hut 18
Time: 10 am

Minutes

Present: A Adams, W Calper (chair), N Couri, G Boon, F Jones, J Melendez, P Nguyen (secretary)

Apologies: B Grune, N Prkic

Minutes of the previous meeting were read and accepted.

2. Business arising from minutes:
   2.1 Missing safety signs have been replaced on entrance gate.

3. General business:
   3.1 Untagged leads. N Couri reported that all untagged leads have been removed from the site.
   3.2 No hand rails on stair wells. J Melendez is to organise a team to install hand rails on all stair wells.
   3.3 Bricks falling from fourth floor. A Adams reported that toe boards had been removed from the bricklayer's scaffold. These have been replaced, which should prevent any further incidents.

4. Other business

5. Next meeting was set down for Friday 16 December at 10 am.

6. The meeting closed at 11 am.

Figure 4.32 Typical examples of meeting documents—(i) Formal agenda, (ii) Minutes of the meeting

---

successful outcome of each of the members of the group.

All members will probably take their own notes about issues that affect them directly. During the meeting, individuals may volunteer or be coaxed into undertaking tasks that have been decided on by the group.

Informal meetings are much more dependent than formal meetings on the ability of the members to work as a team.

Team meetings, social club meetings and special-interest group meeting are most often conducted as informal meetings. Of course, many meetings are neither formal nor informal. In the workplace, depending on the situation, meetings may be hybrids or combinations of both extremes.

Participation in on-site meetings

No matter whether a meeting is formal or informal, it is important to participate. When you attend a meeting, make sure that you:
- understand the purpose of the meeting;
- contribute to relevant discussion;
- are prepared to listen;

- offer only constructive criticism;
- deal with issues or problems, not with people or personalities;
- are prepared to resolve problems;
- are prepared to accept and carry out the decisions reached by the group, because the group's decision is binding on everyone.

During the meeting, it is important to make sure that:
- everyone sits where they can see each other;
- everyone has a place to put their agenda and take notes;
- everyone has an opportunity to make a contribution;
- no-one is allowed to dominate the meeting and control the outcomes.

Because meetings are often called to resolve problems or deal with difficult situations, you are likely to be required to help resolve a conflict. In such a situation, the most common ways of responding are to:
- withdraw from the situation, which allows others to win and, because the conflict is not resolved, may allow it to grow out of control;
- suppress your feelings and refuse to acknowledge the problem, which does not allow others to recognise your feelings and have the opportunity to behave differently;
- compromise, which can lead to dishonesty, which degenerates either into haggling or exaggerated ambit claims;
- confront the other, which can lead to win/lose ego-fired battles of will, which have nothing to do with the pros or cons of the issue at hand.

All of these ways of responding can lead to a win/lose situation. In time, successive win/lose situations can produce a culture of tit-for-tat responses, where it is more important to win and get even than to solve a problem. This leads to a breakdown of harmonious and cooperative work relationships and the creation of an unhappy, unsatisfactory and unproductive workplace.

The ideal response to conflict is the win/win response.

A win/win solution is a solution that meets everyone's needs. Not only do we get what we want but the other people also get what they want. We give up trying to persuade or convince the other that we are right, and we give up trying to destroy the argument that they are using to try to convince us that they are right. We set out to cooperate and to find a solution that will benefit everyone.

Worksheet 1

Student name: _____

Enrolment year: _____

Class code: _____

Competency name/Number: _____

To be completed by teachers:	
Student competent	☐
Student not yet competent	☐

Task: Read through the sections beginning at *What is communication?* and up to and including *The communication process*, then complete the following.

Q. 1 What are we trying to do when we communicate with others?

_____

_____

Q. 2 There are at least two parties in the communication process; what do we call them?

_____

Q. 3 Who has the responsibility for successful communication?

_____

_____

Q. 4 The parties in the communication process can experience barriers. List these barriers:

_____

_____

_____

_____

_____

Q. 5 How does good communication benefit everyone on a building site?

_____

_____

Q. 6 What do we call the procedure for checking whether communication has been successful?

_____

continued ➤

Q. 7 List the procedures for checking whether communication has been successful.

The sender checks by:
_____
_____
_____

The receiver checks by:
_____
_____
_____

CHAPTER 4 WORKPLACE COMMUNICATION 161

Worksheet 2

Student name: _____

Enrolment year: _____

Class code: _____

Competency name/Number: _____

To be completed by teachers:
Student competent ☐
Student not yet competent ☐

Task: Read through the sections beginning at *Communication on-site*, and up to and including *Sketches and drawings*, then complete the following.

Q. 1 List the different forms of communicating information on a building site:

_____
_____
_____
_____
_____
_____

Q. 2 When being spoken to in face-to-face communication, what should you do to make sure you get the message?

_____
_____
_____
_____
_____

Q. 3 When you answer a business phone, what should you do?

_____
_____
_____
_____

continued ➤

PAINTING AND DECORATING, AND MORTAR TRADES

Q. 4 If you cannot help the caller, what must you do?

_____
_____
_____
_____
_____
_____
_____
_____

Q. 5 What is the procedure for using a two-way radio?

_____
_____
_____
_____
_____
_____

Q. 6 List the procedure to follow when reading a document to ensure you get the essential information:

_____
_____
_____
_____
_____

Q. 7 Give four examples of when specific hand signals may be used.

1. _____
2. _____
3. _____
4. _____

Q. 8 Safety signs must be manufactured according to a specific Australian Standard. What is it?
_____
_____

Q. 9 What specific categories of signs do the following types fall under?

1. No smoking (symbol): _____
2. Wear head protection: _____
3. Caution—explosive powered tool in use: _____
4. Fire hose (symbol): _____
5. Toxic material risk (symbol): _____
6. Emergency exit: _____
7. Wear hearing protection: _____
8. First aid equipment: _____
9. No naked flame (symbol): _____
10. Fire extinguisher (water): _____

Q. 10 At no time should signs be located on doors or windows. Explain why:
_____
_____

Q. 11 What is a quick and effective way to create a simple barrier?
_____

Q. 12 Describe underground warning tapes, and why they are used:
_____
_____
_____
_____

continued ▶

Q. 13 List the classes of dangerous goods.

The classes are:

1. _____
2. _____
3. _____
4. _____
5. _____
6. _____
7. _____
8. _____
9. _____

Q. 14 Employers are required to provide employees with information about dangerous goods. State the most effective way of ensuring that the information is made available:

_____
_____

Q. 15 Sketches and drawings are a useful means of communicating information. State six typical materials found on-site that could be used to sketch or draw on:

1. _____
2. _____
3. _____
4. _____
5. _____
6. _____

Worksheet 3

Student name: _____

Enrolment year: _____

Class code: _____

Competency name/Number: _____

To be completed by teachers:	
Student competent	☐
Student not yet competent	☐

Task: Read through the section on *The art of clear communication*, then complete the following.

Q. 1 How do you make sure your message is clear when you are speaking to someone face-to-face?

_____
_____
_____
_____
_____

Q. 2 List the steps you would follow to instruct someone in a practical task:

_____
_____
_____
_____
_____
_____
_____

Q. 3 How must you write a document to make sure you are writing in plain English?

_____
_____
_____
_____
_____

Q. 4 List the types of language you must avoid using if you do not want your meaning to be obscure:

_____
_____

continued ➤

Q. 5 Describe the hand signal that means 'Stop where you are':

Q. 6 Describe the hand signal that means 'Come towards me':

Q. 7 Describe the hand signal that tells others to cease what they are doing:

Q. 8 When using hand signals, how can you ensure that the message will not confuse others?

Worksheet 4

Student name: _____

Enrolment year: _____

Class code: _____

Competency name/Number: _____

To be completed by teachers:	
Student competent	☐
Student not yet competent	☐

Task: Read through the sections beginning at *On-site meeting process* and up to and including *Participation in on-site meetings*, then complete the following.

Q. 1 List the types of meetings that may be held on a building site:

_____
_____
_____
_____
_____
_____
_____

Q. 2 Name and describe the two ways of conducting these different types of meetings:

1. _____
_____
_____

2. _____
_____
_____

Q. 3 List the types of meetings that are often conducted 'formally':

_____
_____
_____

continued ➤

Q. 4 List the types of meetings that are often conducted 'informally':
___
___
___

Q. 5 Describe an agenda for a meeting:
___
___
___
___
___

Q. 6 Describe the minutes of a meeting:
___
___
___

Q. 7 No matter whether a meeting is formal or informal, what must attendees be prepared to do?
___
___

Q. 8 List four important things to make sure of during a meeting:

1. ___
2. ___
3. ___
4. ___

Q. 9 List and describe the four most common ways of responding in a conflict situation:

1. ___
___
2. ___
___
3. ___
___
4. ___
___

Q. 10 Describe the kind of situation these ways of responding can lead to, and the effect this situation can have:

_____
_____
_____

Q. 11 Name and describe the ideal response to a conflict:

_____
_____
_____

SUGGESTED ACTIVITY one

Tasks: Class/Group exercise.

1. Provide students with a timber-cutting list containing typical section sizes, stress grades, species etc. Ask them to contact a timber yard or hardware store to obtain a price for each of the materials. This will allow for practice in telephone technique, and receiving and conveying information.

2. Ask each student to select a basic construction task, preferably one they are familiar with on-site, and create a step-by-step sequence to complete the task from start to finish. Allow students to work in pairs, with one as a sender and one as a receiver. Each one is to communicate the process to the other to see whether the directions/steps are clear and easy to follow.

3. Create a mock site meeting scenario to allow groups of four students to participate in a typical site meeting situation. Rotate the members and vary the scenarios to allow each person to act in a main role, i.e. chairperson.

4. Provide students with a practical workshop task, such as setting out and constructing a carry-all toolbox, setting out and constructing an oilstone case, setting out and constructing a picnic table, or preparing and planning for a small garage or shed. Allow students to work in groups of four to develop a schedule of operations from beginning to end. This will allow for face-to-face communication with others and the planning of activities.

5. Refer students to this book and ask each student to select a section/topic from the book to explain to the class. Each person is to communicate the meaning/purpose/importance/need for the section selected. This will allow students to analyse, clarify and interpret information and learn to communicate it to others.

6. As a class group, view relevant videos and complete prepared questions in relation to the video. Review the answers and provide feedback during class time. Suitable videos available from TAFE libraries may include, but are not limited to, the list overleaf.

continued ➤

Videos:
- Active listening skills
- An easy guide to communication
- Art of taking and making phone calls
- Basic telephone skills
- Best of body language
- Communicating effectively: An essential skill for job success
- Communication: Getting in touch
- Communication: The name of the game
- Constructive communication: Talking your way to success
- Dealing with difficult people
- English at work: Communication change—Part 4, Episode 13
- English at work: Improving communication at work—Part 12, Episode 46
- Failure to communicate
- Meetings
- Tale of 'O': On being different
- Team building: A positive approach
- Team working
- Telephone skills
- Telephone techniques
- Verbal communication: The power of words
- Valuing diversity: Multicultural communication
- What the window cleaner saw!
- When I'm calling you
- You know what I mean
- You're not communicating
- You're not listening

REFERENCES AND FURTHER READING

Acknowledgment

Reproduction of the following *Resource List* references from *DET, TAFE NSW C&T Division (Karl Dunkel, Program Manager, Housing and Furniture) and the Product Advisory Committee* is acknowledged and appreciated.

Texts

Access Series (2001), *Communication for business*, McGraw-Hill, Sydney

Access Series (2001), *Communication for IT*, McGraw-Hill, Sydney

Basic Work Skills Training division, NSW TAFE Commission (1995), *Workplace Communication, NCS001, A Teaching/Learning Resource Package*, Basic Work Skills Training Division, South Western Sydney Institute of TAFE

Eagleson, Robert D. (1990) *Writing in Plain English*, Australian Government Publishing Service, Canberra

Elder, B. (1994), *Communication Skills*, Macmillan Education, Melbourne

Graff, D.M. & Molloy, C.J.S. (1986), *Tapping group power: A practical guide to working with groups in commerce and industry*, Synergy Systems, Dromana, Victoria

National Centre for Vocational Education Research (2001), *Skill trends in the building and construction trades*, National Centre for Vocational Education Research, Melbourne

NSW Department of Education and Training (1999), *Construction industry: Induction & training: Workplace trainers' resources for work activity & site OH&S induction and training*, NSW Department of Education and Training, Sydney

NSW Department of Industrial Relations (1998), *Building and construction industry handbook*, NSW Department of Industrial Relations, Sydney

WorkCover Authority of NSW (1996), *Building industry guide: WorkCover NSW*, Sydney

Web-based resources

Resource tools and VET links

<www.toolbox.flexiblelearning.net.au/search.asp> ANTA

<www.resourcegenerator.gov.au> ANTA resource generator

<www.ntis.gov.au> National Training Information Service

<www.hsc.csu.edu.au/construction> NSW HSC Online—Construction

CHAPTER 5
BASIC MEASURING AND CALCULATIONS

This chapter outlines the competencies required to carry out measurements and perform simple calculations to determine material quantities for tasks or jobs within the general construction work environment.

Areas addressed from the unit of competency include:
- planning and preparing work;
- obtaining measurements;
- performing calculations; and
- estimating approximate quantities.

CORRECT UNITS OF MEASURE

Linear measurements in the building industry are expressed in **millimetres** (mm) and metres (m). **Working drawings** of houses and buildings generally have their **dimensions** shown in millimetres, although some long dimensions on site plans may be in metres. Measuring tools used to construct houses and buildings are marked in graduations of millimetres and metres.

Note: Centimetres are not recognised as true SI units and are therefore not used in the building industry.

Accurate measuring using an appropriate tool is critical (Figure 5.1).

On-site, tradespeople usually speak dimensions in either millimetres or metres and millimetres. For example, a tradesperson may take a measurement of 5432 mm. The tradesperson may say that the measurement is either '5432 millimetres', or '5 metres 432 millimetres'.

Most tradespeople carry two measuring tools on the job. These are the four-fold rule and the retractable metal tape (Figure 5.2). You must be able to use these measuring tools accurately.

CHAPTER 5 BASIC MEASURING AND CALCULATIONS

Figure 5.1 Accurate measuring and marking is critical

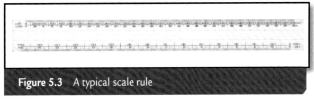

Figure 5.3 A typical scale rule

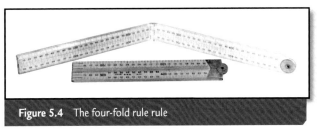

Figure 5.4 The four-fold rule rule

Four-fold rule

The four-fold rule is 1 metre long and is made in four hinged sections so that it can be folded for convenience (Figure 5.4). The rule is best used for measuring and marking out dimensions of less than 1 metre. If the rule is used to measure or mark out a dimension greater than 1 metre an error is likely to accumulate each time the rule is moved.

Depending on whether the rule is made from boxwood or plastic, the rule blades are approximately 4 or 5 mm thick. Because of this thickness, **parallax error** is likely to occur when measuring or marking out with the rule laid flat on a surface. To overcome this problem, the rule should be used on edge so that the graduations marked on the blades of the rule are in contact with the surface of the material being measured or marked out.

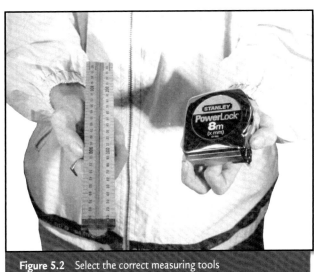

Figure 5.2 Select the correct measuring tools

BASIC MEASURING TOOLS

Scale rule

The scale rule is a plastic rule of 150 or 300 mm in length used to scale off dimensions not given on the drawing (Figure 5.3).

Beware of scaling off a plan. Drawings are not always accurate when no dimension is given. Important dimensions—that is, dimensions that you will use to quote or construct from—should always be checked with the architect, draftsperson or client.

Retractable metal tape

The retractable metal tape is available in lengths from 1 metre to 10 metres and is most suitable for measuring or marking out dimensions from 1 metre to 10 metres (Figure 5.5). Metal retractable

Figure 5.5 A typical retractable metal tape

tapes have a hook on the end, which adjusts depending on whether the measurement to be taken or marked out is internal or external.

The hook slides a distance equal to its own thickness, so that an internal measurement, for example between two walls, begins from the outside of the hook. But an external measurement, for example from the end of a piece of timber, begins from the inside of the hook. Care must be taken when returning the blade to the case not to allow the hook to slam against the case (which will stretch or distort the hook, making the tape inaccurate) nor to tear the hook from the end of the tape (which will make it useless and in need of replacement).

QUANTITIES AND COSTS

Basic method

Calculation of material quantities is a requirement for all projects undertaken in the building industry.

The principles of quantity calculation for materials are the same for a garage as they are for a multistorey development. A method must be adopted that allows for the planning and calculation of material quantities in sequential order, to prevent mistakes and/or omission of some components (Figure 5.6).

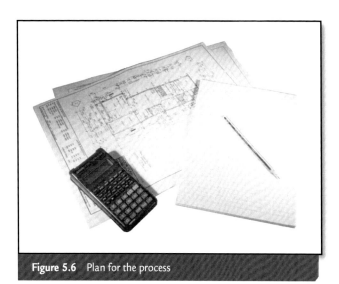

Figure 5.6 Plan for the process

Material units

It's one thing to be able to identify the various materials required, quite another to determine how they are available (Figure 5.7). For example:

- General timber is purchased by the linear metre (length) in increments of 300 mm (which leads to the interesting fact that if you buy 1.0 m of timber, you will have to pay for 1.2 m...).
- Sheet material is available individually at various lengths and widths, or in pack lots. Plaster sheet, for example, can be purchased at widths of 1200 mm or 1350 mm, and in various lengths.
- Tiles may also be purchased individually or by the box.
- Adhesives, sealants and stopping off materials may be purchased either by the tube, tub, pot, drum or bag depending upon whether they are in dry or liquid form. Hence they may be purchased either by the mass (kilograms—kg) or volume (litre—L). In some cases, such as some grouts and tile adhesives, they may need to be purchased as a combination of both (so many litres of liquid and so many kilograms of powder).
- On occasions these materials may be purchased based upon the area, or square metres of coverage, required.
- Paint is purchased similarly by the can or drum in quantities as little as 200 ml upwards to 20-litre drums. Again, while you may purchase by the litre, it is the area of coverage, and so the number of square metres to be painted, that is the determining factor.

Generally, these materials are priced or costed using the same units of measure.

The units of measurement adopted for the building industry are metric units taken from the

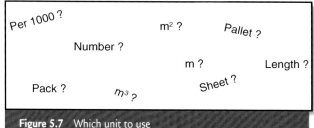

Figure 5.7 Which unit to use

Table 5.1 SI units commonly used in construction

Unit	Function	Symbol	Use
metre	unit of length	m	Length of a building block, length of timber
millimetre	1000 millimetres = 1 metre	mm	Section size of material, dimensions and lengths
square metre	unit of area (m x m)	m^2	Surface area of wall, area of floor
hectare	unit of area (10 000 m^2)	ha	Land subdivision, building site area
cubic metre	unit of volume (m x m x m)	m^3	Excavation of soil, concrete quantity
litre	unit of volume or capacity (100 mm x 100 mm x 100 mm) 1000 litres = 1 m^3 1 litre of water = 1 kg	L	Paint quantities, tile adhesives, cleaning agents, mortar additives, etc.
kilogram	unit of mass	kg	Cement is sold in 40 kg bags, building design loads
gram	1000 grams = 1 kilogram	g	Quantity of small nails
tonne	1000 kilograms = 1 tonne	t	Mass of building material, e.g. reinforcement, safe working load of lifting equipment, e.g. cranes
newton	unit of force	N	Calculation of design loads of buildings
pascal	unit of pressure = 1 N/m^2	Pa	Calculation of design loads to floors
kilopascal	1000 pascals	kPa	Calculation of pressure exerted in building design
megapascal	1 000 000 pascals	MPa	Calculation and specification of design strength of concrete

Système International d'Unités (SI). The **SI units** commonly used in construction, together with an example of typical use, are shown in Table 5.1.

MEASUREMENT, CALCULATIONS AND QUANTITIES

Construction workers regularly take measurements or make calculations such as additions or subtractions, calculate **areas** or **volumes**, measure **quantities**, and calculate the cost of materials. When using metric units, small letters are used for all symbols except where the value is over one million, as in megapascals (MPa), or where the unit name is derived from a proper name, as in newtons (N) or pascals (Pa) (see Table 5.1).

Linear measurements in the building industry are in metres (m) and millimetres (mm); 1 metre is equal to 1000 millimetres. Most plans and working drawings are dimensioned in millimetres, but where long lengths are recorded, such as on a site plan, the length may be shown as metres. Where millimetres are used it is generally accepted that the symbol (mm) need not be used, and for metres a decimal point is used to separate the metres from the millimetres, again without the need for the symbol (m). For example, if a figure of 3600 is used it is accepted as millimetres, whereas 3.600 would indicate metres.

Surface area measurements must be accompanied by the symbol (m^2) and volume measurements must be accompanied by the symbol (m^3) to ensure correct interpretation of the calculation. For example, 2.500 × 5.000 = 12.5 m^2 indicates surface area and 2.500 × 5.000 × 0.150 = 1.875 m^3 indicates volume.

Measurement will be a major part of everyday activity in each section of the building industry. For example, tilers constantly measure and cut both tiles and metal/plastic trims as well as measure for quantities; bricklayers and plasterers measure for heights, window and door positions and the set out of complex curves or other geometric shapes; and painters measure for estimating quantities or setting out detailed pattern work. Accurate measurement saves time and money, and reduces waste.

Linear measurement

Many materials, such as timber, metal trims and plaster cornice are sold by the metre. Determining the total **linear metres** required is important to any trade. In addition the easiest way to quote some work, such as painting fences or walls, can be to work out the total linear metres first.

However, suppliers hold many materials in standard lengths. This applies to timber, plastic or metal tile trims, rolls of tape, plaster cornice or galvanised lintels for openings in brickwork. In most cases these lengths will be in increments of 0.3 metres, though not all incremental lengths will be available. For example, cornice may only be readily available in 4.8-metre lengths (though lengths down to 2.4 metres might be ordered), and most metal tile trims are 3.0 metres long, while rolls of jointing tape or masking tape come in lengths of up to 160 metres.

When ordering it is important to make out the order using standard sizes and lengths. Sectional sizes (cornice or tile trims), or tape width, must be stated, along with the desired length, and type of product or material required. Where there are multiple differing lengths, the list starts with the longest and progresses down to the shortest. For example:

- 55 mm blue masking tape: 2–50 m (sometimes written as 2/50 m) ;
- 10 mm square section aluminium tile trim: 5/3.0 m;
- 90 mm cove cornice: 5/4.8 m, 6/2.4 m;
- 100 × 100 galvanised angle lintel: 4/2.4 m, 3/2.1 m.

It is from lists such as these that you will need to work out the total linear metres for quoting. For example, the total linear metres of 90 mm cornice is calculated by:

Total linear metres = (5 × 4.8 m) + (6 × 2.4 m)
= 24 + 14.4
= 38.4 m (sometimes written as 38.4 lm).

Example 1
Total linear metres of 8 mm square tile trim when 7 lengths at 3.0 m is required:

Total minear metres = 7 × 3.0 m
= 21 m.

Example 2
Total length of fencing required to be painted when the fence lengths are 35.7, 24.2, 19.6, 15.1.

Total linear metres = 35.7 + 24.2 + 19.6 + 15.1
= 94.6 m.

Example 3
Find the total linear metres of screed boarding for circular paving having a radius of 2.25 m.
Formula for the circumference of a circle:
$\pi \times$ diameter
$\pi \times 4.5 = 14.137$ m board required.

Area measurement

Square measurement can be of areas having regular or irregular shape, and is determined by the number of square metres a figure contains. The area measurement hectare (ha) is used for land subdivision because of the larger areas involved, but when measuring or describing building blocks the unit of measure is square metres (m^2). Area measurement in square metres (m^2) is used in building construction to determine:

- total floor area of buildings;
- specific floor areas for painting, polishing or tiling;
- wall areas for tiling, painting, sheeting, brick quantities;
- ceiling areas for sheeting or painting;
- external areas for paving or planting out.

Square measurement is also used for the calculation of the number of bricks required to construct a wall, or for the number of pavers for a path or driveway. This number is then multiplied by the number of bricks or pavers that will cover 1 square metre. For example, if using metric standard bricks (230 × 110 × 76 mm) for a wall that will be 110 mm in thickness, the total metres squared of wall area will be multiplied by 50, as there are approximately 50 bricks needed per square metre.

The area of plane figures and rectangles is found by multiplying the length by its width or breadth. The following are typical examples of area measurement for regular shapes.

Example 4
A large entertainment room measures 14.6 m × 9.1 m. It needs to be painted with an oil-based clear finish.

To know how much oil finish is required the area must first be calculated and an allowance of 10% for waste added.

$$\text{Floor area} = \text{length} \times \text{width}$$
$$= 14.600 \times 9.100$$
$$= 132.860 \text{ m}$$
$$\text{Allow 10\%} = 132.860 \times 1.1$$
$$= 146.146 \text{ m}^2$$

(To understand percentage calculations go to page 185, 'Calculation of percentages'.)

Example 5
The oil finish to be used in the above example has a stated coverage of 12 m²/L. Two coats will need to be applied and the oil finish is only available in 4-litre tins. Calculating the number of tins required may be achieved in several ways, one of which is shown below:

$$\text{Floor area} = 146.146 \text{ m}^2$$
$$\text{Coverage of 1 tin} = \text{Litres in tin} \times \text{coverage rate}$$
$$= 4 \times 12 \text{ m}^2$$
$$= 48 \text{ m}^2$$
$$\text{Number of 4-litre tins} = \text{Area} / \text{coverage of 1 tin}$$
$$= 146.146 / 48$$
$$= 3.045$$
$$\text{Multiply by number of coats} = 3.045 \times 2$$
$$= 6.090$$
$$= 7 \text{ tins}.$$

Example 6
A wall 5.400 m long and 2.400 m in height is to be covered with plasterboard. Calculate the net wall area if the wall has a window 2.400 m × 1.200 m and a door 2.100 m × 0.900 m.

Area of wall 5.400 × 2.400 = 12.960 m²
Deductions for:
  Area of window 2.400 × 1.200 = 2.880 m²
  Area of door 2.100 × 0.900 = 1.890 m²
  Total deductions = 4.770 m²
  Net wall area 12.960 − 4.770 =
  8.190 m² plasterboard will be required.

Example 7
The bathroom of a timber floored house is to be tiled. The tiles will be laid on 6 mm fibre cement sheets that are 2.4 m × 1.2 m in size. The room is 4.2 m × 3.0 m. To work out how many sheets will be needed requires the calculation of two areas: the area of the room and the area of a sheet.

$$\text{Floor area} = \text{length} \times \text{width}$$
$$= 4.2 \times 3.0$$
$$= 12.6 \text{ m}^2$$
$$\text{Sheet area} = \text{length} \times \text{width}$$
$$= 2.4 \times 1.2$$
$$= 2.88 \text{ m}^2$$
$$\text{Number of sheets} = \text{Floor area} / \text{Sheet area}$$
$$= 12.6 / 2.88$$
$$= 4.375$$
$$= 5 \text{ sheets required}.$$

Example 8
Calculate the area of paving for a triangle-shaped courtyard with a base length of 8.950 m and a perpendicular length of 10.250 m.

Area of a triangle = ½ base × perpendicular length.

$$\left(\frac{8.950}{2}\right) \times 10.250 = 45.869 \text{ m}^2.$$

Area measurement for irregular shapes in the following examples are for shapes having straight sides, such as irregular **quadrilaterals** (four sides) or irregular **polygons** (five or more sides). Irregular figures may be subdivided into triangles, **parallelograms** or rectangles. To find the area of the whole figure, add the areas of the parts into which it has been divided.

In some cases, areas may be determined by considering the figure as a rectangle from which parts have been removed. For example, if a quadrilateral has only two parallel sides, its area may be found provided that the length of the two parallel sides and the perpendicular distance between them is known. The formula to calculate the area of this figure would be:

½ the sum of the parallel sides × perpendicular distance between parallel sides.

Typical examples of calculations involving irregular shapes are as follows.

Example 9
Find the area of Figure 5.8.

$$\text{Area} = \left(\frac{20.000 + 36.000}{2}\right) \times 10.000$$

$$= 28.000 \times 10.000$$

Total area = 280 m²

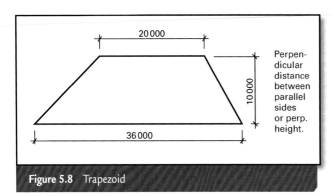

Figure 5.8 Trapezoid

Example 10
Find the area of the block of land shown in Figure 5.9.

$$\text{Area} = \left(\frac{90.000 + 96.000}{2}\right) \times 46.000$$

$$= 93.000 \times 46.000$$

Total area = 4278 m² of land.

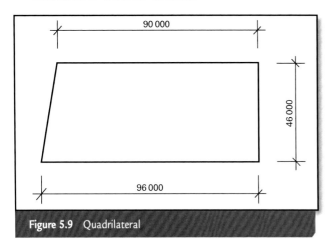

Figure 5.9 Quadrilateral

Example 11
Calculate the area of the irregularly shaped concrete path in Figure 5.10. Divide the shape into separate sections and find the area for each section, then add the two areas together to find the total area.

Section one (parallelogram)
3.600 × 0.762 = 2.743 m²

Section two (rectangle)
2.134 × 1.000 = 2.134 m²

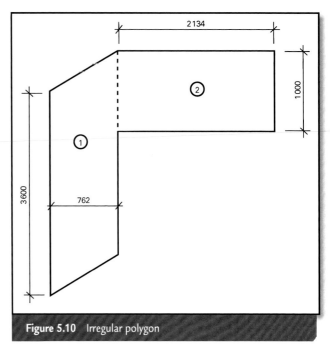

Figure 5.10 Irregular polygon

Total area
2.743 + 2.134 = 4.877 m².

The areas of quadrilaterals with sides of different lengths, and irregular polygons, can be found by drawing in diagonals on the figure, finding the areas separately, and then adding them together to find the total area.

Typical examples of this type of calculation are as follows.

Example 12
Find the area of the building block shown in Figure 5.11.

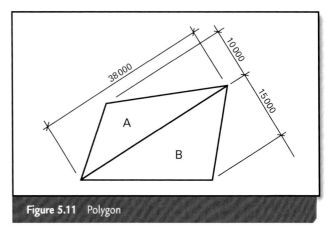

Figure 5.11 Polygon

Area of a triangle = ½ base × perpendicular height

Area of triangle A $= \left(\dfrac{38.000}{2}\right) \times 10.000$

$= 190 \text{ m}^2$

Area of triangle B $= \left(\dfrac{38.000}{2}\right) \times 15.000$

$= 285 \text{ m}^2$

Total area of the block is 190 + 285 = 475 m².

Example 13

Calculate the area of floor tiles required for a shop having the dimensions shown in Figure 5.12. Add to the total area calculated 15% for waste in cutting the tiles.

Area of triangle A $= \left(\dfrac{14.400}{2}\right) \times 4.600$

$= 33.120 \text{ m}^2$

Area of triangle B $= \left(\dfrac{14.400}{2}\right) \times 1.600$

$= 11.520 \text{ m}^2.$

Total area of floor tiles required
33.120 + 11.520 = 44.640 m²
Add 15% for waste = 6.696 m²
Total tiles required = 51.336 m².

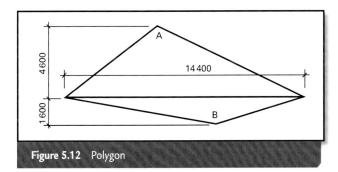

Figure 5.12 Polygon

Volume measurement

Volume is measured in cubic metres (m³), and is found by multiplying the length by the width by the depth (or thickness).

Calculations of volume in the industry are used to determine:
- the volume of fluids held in containers;
- concrete quantities;
- materials for mixing mortar;
- orders for paints and other coatings;
- quantities of tile adhesive and other compounds such as plaster stopping off materials;
- bulk timber orders (for formwork etc.);
- soil (known as spoil) to be excavated for footings.

As Example 5 showed ('Area measurement', page 177), liquid finishes are ordered by volume or litres. This is based upon the coverage or spreading rate for a particular product (as is the case with many tile adhesives; mortar for bricks works on a similar basis). That is, an area is calculated, and then the volume of coating or adhesive is calculated simply by multiplying this area by the number of litres required to cover each square metre.

However, should you need to know how much a particular container is capable of holding (for mixing or storing perhaps; even how much sand may be held in a trailer), or how much screeding material is needed to cover a bathroom floor before tiling, then a 'true' volume calculation is required
(i.e. length × width × depth).
Note: It is important to know when dealing with fluid volumes (i.e. litres) that one litre of fluid (at room temperature) fills a box 0.1 m × 0.1 m × 0.1 m as shown in Figure 5.13.

So the volume occupied by 1 litre of fluid is:
= 0.100 m × 0.100 m × 0.100 m = 0.001 m³.

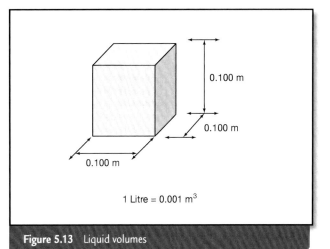

Figure 5.13 Liquid volumes

Example 14
You need a container big enough to mix up 25 litres of tile adhesive. You have a found a container that looks as if it might do the job but are not sure. The container is box-like and measures 350 mm long, 300 mm wide and 400 mm high. The volume is found by:

$$\begin{aligned}\text{Volume of container} &= \text{length} \times \text{width} \times \text{depth} \\ &= 0.350 \times 0.300 \times 0.400 \\ &= 0.042 \text{ m}^3\end{aligned}$$

$$\begin{aligned}\text{Number of litres it} &= \text{Volume of container} / \\ \text{can contain} &\quad \text{Volume of 1 litre} \\ &= 0.042 / 0.001 \\ &= 42 \text{ L}.\end{aligned}$$

Note: Always change millimetres (mm) to metres (m) when doing calculations. This is a 'habit' that all tradespeople must adopt. If you don't you will end up with a lot of very large numbers that are quite difficult to interpret.

Example 15
Calculate the volume of concrete for a strip footing for a trench if the depth of the footing is 450 mm.

$$\begin{aligned}\text{Volume} &= \text{length} \times \text{width} \times \text{depth} \\ \text{Amount of concrete required} & \\ &= 18.500 \times 0.450 \times 0.450 \\ &= 3.746 \text{ m}^3.\end{aligned}$$

Example 16
You have to screed the base of a shower for tiling. Being a very modern and stylish home, the base is round and set in the middle of the room with the shower rose coming from the ceiling. The average depth of screeding will be 20 mm thick. The radius of the base is to be 750 mm.

$$\begin{aligned}\text{Area of a circle} &= \pi \times \text{radius of circle}^2 \\ \text{Volume of a cylinder} &= \pi \times \text{radius}^2 \text{ of circle} \\ &\quad \times \text{depth}\end{aligned}$$

where π is taken as being 3.142 (see 'Calculation of various solid shapes' on page 183 for more information on using π).

$$\begin{aligned}\text{Volume of screeding required} & \\ &= 3.142 \times 0.750^2 \times 0.020 \\ &= 3.142 \times 0.563 \times 0.020 \\ &= 0.035 \text{ m}^3.\end{aligned}$$

Allowing for waste we might round this off to 0.04 m³.

Note: When calculating the volume of fluid materials such as mortar (or screeding), concrete, and even paint and adhesives, it is usual to make an allowance for waste. Generally this will be in the order of 10%. This may be reduced to 5% if the base material to which it is applied, or the formwork into which it is to be placed, is sound. Likewise, more material will be wasted when having to do lots of small areas (a bit is lost at each mixing) compared to doing one large area.

Quantities and costs

Quantities is a term used in the building industry for the calculation of materials required for a particular task, and may also include the cost of supply of the materials or a total figure that would be the total cost of supply and fixing of the materials.

Typical examples of calculations of quantities and costs are as follows.

Example 17
Figure 5.14 shows the wall of a small room with a door in it. This wall needs to be plastered with 10 mm sheet, then painted. The door is a standard 2040 × 820 flush panel.

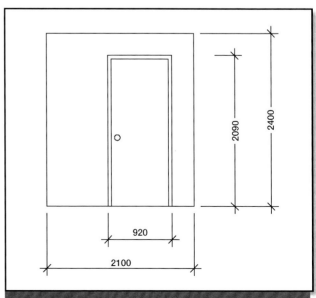

Figure 5.14 Wall quantities

Costs of materials are as follows:
10 mm plaster sheet $4.95/m²
Joint tape (paper) $0.07/m
Base coat $2.50/kg
Top coat $1.75/kg
Stud adhesive $5.50/kg
Fixings $5.20/kg

1 Calculate quantities and costs.
$$\text{Total wall area} = 2.4 \text{ m} \times 2.1 \text{ m}$$
$$= 5.04 \text{ m}^2$$
$$\text{Area of opening} = 2.040 \times 0.820$$
$$= 1.673 \text{ m}^2$$
$$\text{Plaster sheets} = 5.04 - 1.6728$$
$$= 3.367 \text{ m}^2$$
$$\text{Cost of sheeting} = 3.3672 \times 4.95$$
$$= \$16.67$$
$$\text{Jointing tape} = 2.100 - 0.820$$
$$= 1.28 \text{ m}$$
$$\text{Cost of tape} = 1.28 \times 0.07$$
$$= \$0.09$$

2 Calculate the stud adhesive and fixings required.
The manufacturer's Application Guide suggests that 4.2 kg of adhesive and 1.0 kg of nails will be needed for every 100 m² of wall. That is:
0.042 kg/m² for adhesive
0.01 kg/m² for nails
In this case:
$$\text{Adhesive required} = 3.367 \times 0.042$$
$$= 0.141 \text{ kg}$$
$$\text{Cost of adhesive} = 0.141 \times 5.50$$
$$= \$0.78$$
$$\text{Fixings required} = 3.367 \times 0.01$$
$$= 0.034 \text{ kg}$$
$$\text{Cost of fixings} = 0.034 \times 5.20$$
$$= \$0.18$$

3 Calculate the jointing compound required.
The manufacturer's Application Guide suggests that 16 kg of base coat will be needed for every 100 m² of wall (allows for tape and second coats), and 7 kg of top coat. That is:
0.16 kg/m² for base coat (16 / 100)
0.07 kg/m² for top coat (7 / 100)

In this case:
$$\text{Base coat required} = 3.367 \times 0.16$$
$$= 0.539 \text{ kg}$$
$$\text{Cost of base coat} = 0.539 \times 2.50$$
$$= \$1.35$$
$$\text{Top coat required} = 3.367 \times 0.07$$
$$= 0.237 \text{ kg}$$
$$\text{Cost of top coat} = 0.237 \times 1.75$$
$$= \$0.41$$

4 Calculate the total cost of materials.
$$\text{Total cost (materials)} = 16.67 + 0.09 + 0.78 + 0.18 + 1.35 + 0.41$$
$$= \$19.48$$

Example 18
The wall that was plastered in Example 17 now needs to be painted. The architrave will be painted a different colour and is 50 mm wide and 20 mm thick. The 2040 × 820 door will be painted a different colour again. All finish coats will be applied twice.

Costs and coverage of materials are as follows:
Primer/sealer/undercoat: $15.45/L @ 14 m²/L
Finish coats—walls: $23.75/L @ 10 m²/L
Finish coats—architrave: $21.50/L @ 18 m²/L
Finish coats—door: $20.00/L @ 16 m²/L

1 Calculate the area to be painted.
From Example 17 (on p. 180):
$$\text{Total area of wall} = 5.04 \text{ m}^2$$
$$\text{Area of door} = 1.673 \text{ m}^2$$
$$\text{Area of wall to be painted} = 3.367 \text{ m}^2$$
$$\text{Surface area of architrave} = \text{total length of architrave} \times \text{total width}$$
$$= (2.090 + 2.090 + 0.920) \times (0.020 + 0.050 + 0.020)$$
$$= 5.100 \times 0.090$$
$$= 0.459 \text{ m}^2$$

2 Calculate the quantity and cost of the primer/sealer/undercoat.
The primer/sealer/undercoat will be applied to all surfaces, so:
$$\text{P/S/U required} = 5.04 / 14$$
$$= 0.36 \text{ L}$$
$$\text{Cost of P/S/U} = 0.36 \times 15.45$$
$$= \$5.56$$

3 Calculate the quantity and cost of the finish coat.
 Finish coat to walls:
  Wall finish required = 3.367 / 10
   = 0.34 L
  Cost of wall finish = 0.34 × 23.75
   = **$8.08**
 Finish coat to door:
  Door finish required = 1.673 / 16
   = 0.10 L
  Cost of door finish = 0.1 × 20.00
   = **$2.00**
 Finish coat to architrave:
  Architrave finish required = 0.459 / 18
   = 0.03 L
  Cost of architrave finish = 0.03 × 21.50
   = **$0.65**
4 Calculate the total cost of materials.
  Total cost (materials) = 5.56 + 8.08 + 2.00 + 0.65
   = **$16.29**

Example 19
The clients have changed their minds! The wall in Example 17 now needs to be tiled to a height of 1900 mm. A 100 mm² patterned tile is at mid height and a 10 mm² metal edge trim is on the top edge. The wall is to be waterproofed to the full height.
 Costs and coverage of materials are as follows:
 Main tile 200 × 200 porcelain $45.45/m²
 Patterned tile $8.75 each
 Tile adhesive $2.50/kg @ 3.5 kg/m²
 Liquid flashing $13.50/L @ 1.4 L/m²
 Metal trim $7.20/m
 Sanded grout $5.25/kg @ 2.0 kg/m²
 From Example 17 (on p. 180)
 Total area of wall = 5.04 m²
 Area of door = 1.673 m²
 Area of wall to be tiled = 3.367 m²
1 Calculate the quantity and cost of the liquid flashing.
  Flashing required = 3.367 × 1.4
   = 4.71 L
  Cost of flashing = 4.71 × 13.50
   = **$63.59**

2 Calculate the quantity and cost of the tile adhesive.
  Adhesive required = 3.367 × 3.5
   = 11.78 kg
  Cost of flashing = 11.78 × 2.50
   = **$29.45**
3 Calculate the quantity and cost of the main tiles (200 mm × 200 mm).
  Number of tiles/m² = 1.000 / (0.2 × 0.2)
   = 1.000 / 0.04
   = 25
  Main tiles required = No. of tiles/m² × Area to be tiled
   = 25 × 3.367
   = 84.18
  Allow 10% for cutting = 84.18 × 1.1
   = 92.60
   = 93 tiles
  Cost of main tiles = Area × price/m²
   = 3.367 × 45.45
   = **$153.04**
4 Calculate the quantity and cost of the pattern tiles.
  Length of wall to be tiled = 2.1 − 0.920
   = 1.18 m
  Pattern tiles required = Length to tile / tile width
   = 1.18 / 0.100
   = 11.8
  Allow 10% for cutting = 11.8 × 1.1
   = 12.98
   = 13 tiles
  Cost of patterned tiles = 13 × 8.75
   = **$113.75**
5 Calculate the quantity and cost of the 10 mm metal trim
  Length of trim = 2.1 − 0.920
   = 1.18 m
  Cost of trim = metres of trim × cost/m
   = 1.18 × 7.20
   = **$8.50**
6 Calculate the quantity and cost of the sanded grout.
  Quantity of grout = Area to grout × coverage rate/m²
   = 3.3672 × 2.0
   = 6.73 kg

Cost of grout = 6.73 × 5.25
= **$35.36**

7 Calculate the total cost of materials.
Total cost (materials) = 63.59 + 29.45 + 153.04
+ 113.75+ 8.50 + 35.36
= **$403.64**

Example 20
Calculate the cost to lay the bricks for a wall 4.500 long by 2.100 high (Figure 5.15). The wall is one thickness (110 mm) and there are 50 bricks per square metre. The cost to supply and lay the bricks is $1345.00 per 1000 bricks.
1 Calculate the area in square metres of wall surface.
Area = 4.500 × 2.100
= 9.450 m².
2 Calculate the number of bricks.
Number = 9.450 × 50 bricks/m²
= 472.5 (473 bricks are required with no allowance for waste).
3 Calculate the cost. Divide the number of bricks by 1000 as the total number of bricks is less than 1000.
Cost = $1345.00 × 0.473.
= $636.19 to supply material and labour to lay 473 bricks.

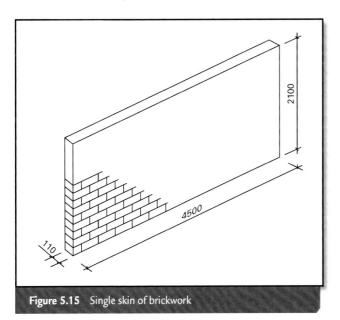
Figure 5.15 Single skin of brickwork

CALCULATION OF VARIOUS SOLID SHAPES

Types of solid shape

Prisms

Prisms are solid objects with two ends formed by straight-sided figures that are identical and parallel to one another. The sides of the prisms become parallelograms. The ends may be formed by common plane geometric shapes (e.g. the square, rectangle, triangle, pentagon, hexagon, octagon) (Figure 5.16).

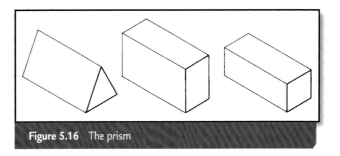
Figure 5.16 The prism

Cylinders

Cylinders are solid shapes that have their ends formed by circles of equal diameter. The ends are parallel and joined by a uniformly curved surface (Figure 5.17).

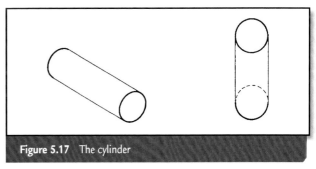
Figure 5.17 The cylinder

Cones

Cones are solid shapes that have a circular base and a uniformly curved surface, which tapers to a point called the apex (Figure 5.18).

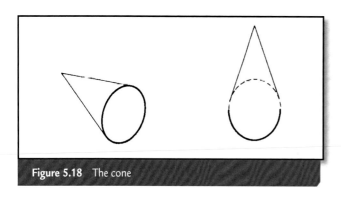

Figure 5.18 The cone

Pyramids

Pyramids are solid shapes with a base consisting of a straight-sided figure (square) and triangular sides, which terminate at the apex (Figure 5.19).

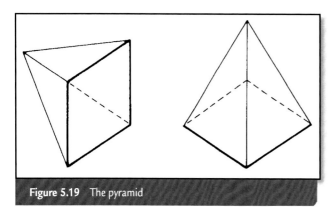

Figure 5.19 The pyramid

Surface development and area of solid shapes

When the surface of a solid object requires measuring, or a true shape is required to create a template, the simplest way to provide an accurate detail is to develop the surface. This requires the true shape of all sides to be laid out flat as a continuous surface.

Drawing the detail in a two-dimensional view is not the only way to determine the amount of material required to form the surface shape. For example, calculation of the surface area would be a more effective method. The idea of laying the shape of each surface out flat still applies, but it's the calculation of each area added together that provides the information required (Figure 5.20). The calculation of these surfaces is required when a roof tiler needs to order roof tiles for a conical or pyramidal style roof surface, or a formworker needs to work out how much formply is required for a prism-like member, or a tank maker needs to work out how much Colorbond® custom orb is required to create a water storage tank.

Example 1

Calculate the surface area of a rectangular prism having a length of 5.540 m, a width of 1.250 m and a height of 850 mm:

Area = (area of base × 2) + (area of one side × 2) + (area of one end × 2)
= [(5.540 × 1.250) × 2] + [(5.540 × 0.850) × 2] + [(1.250 × 0.850) × 2]
= 13.850 + 9.418 + 2.125
= 25.393 m²

Example 2

Calculate the surface area of a cone having an inclined length of 1.800 m and a base radius of 600 mm:

Area = (πr × length of incline) + (πr^2)
= (3.142 × 0.600 × 1.800) + (3.142 × 0.360)
= 3.393 + 1.131
= 4.524 m²

Example 3

Calculate the surface area of a cylinder having a height of 2.100 m and a base radius of 850 mm:

Area = (πr^2 × 2) + (2πr × height)
= [(3.142 × 0.723) × 2] + [2 × (3.142 × 0.850)] × 2.100
= 4.540 + 11.217
= 15.757 m²

Example 4

Calculate the surface area of a pyramid having a square base side of 1.200 m and an inclined perpendicular height of 1.800 m:

Area
= (area of base) + $\left(\dfrac{\text{length of base side}}{2} \times \text{inclined height} \times 4\right)$

= (1.2 × 1.2) + $\left(\dfrac{1.2}{2} \times 1.8 \times 4\right)$

= 1.440 + 4.320
= 5.760 m²

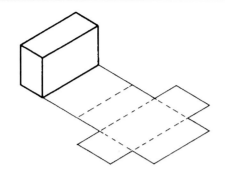

Prism
Area = (Area of base x 2) + (Area of 1 side x 2)
         + (Area of 1 end x 2)
     = (L x W x 2) + (L x W x 2) + (L x W x 2)

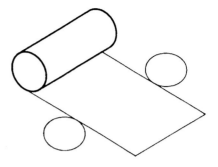

Cylinder
Area = (Area of 1 end x 2) + (Area of surface)
     = (πr^2 x 2) + ($2\pi r$ x height)

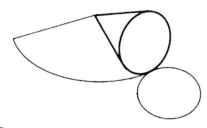

Cone
Area = (πr Length of incline) + (πr^2 base)
     = (πrL) + (πr^2)

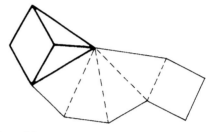

Pyramid
Area = (Area of base) + [(Area of 1 side) x 4]
     = (L x W) + [(½ base x height) x 4]

Figure 5.20 Surface area formulae

CALCULATION OF PERCENTAGES

Percentage literally means 'parts of' or 'parts in' one hundred (100). When using percentages, we divide whatever the 'whole' is into 100 parts, and then state how many of those parts we want.

Mathematically the easiest way to do this is multiplying using decimals. In the construction industry, if we want to find 10% of something we simply multiply it by 0.10. If we want to add 10% to something (such as a waste allowance) we simply multiply it by 1.10.

Let's see why this works. If you want something divided into 100 parts and find out what size that part is you could do it two ways:

100 / 100 = 1
or 100 × 0.01 = 1 (i.e. 1 is 1%—or one hundredth part—of 100).

400 / 100 = 4
or 400 × 0.01 = 4 (i.e. 4 is 1%—or one hundredth part—of 400).

The advantage of the decimal approach becomes clearer when we want to add a percentage on to something (such as for the waste described above). First, however, we should look at how we find more than just 1% of something.

Say we want to add 10% for waste onto the total price of a job. Let the job price be $400.00. We could begin by finding 10% of 400 and then adding it, like this:

400 / 100 = 4

4 equals 1%, so if we multiply 4 by 10 we will get 10%:

10 × 4 = 40

So $40.00 is 10% of $400.00. If we add these together we will have our answer:

400 + 40 = $440.00

But this is rather slow. What if we use the decimal approach?

400 × 0.01 = 4

—and oops, we are into the same cycle.

So try this instead:

0.01 = 1%
0.02 = 2% (or 2 × 1%)
0.05 = 5% (or 5 × 1%)
0.10 = 10% (or 10 × 1%)
0.15 = 15% (or 15 × 1%)

and so on until till you get to:

1.0 = 100% (100 × 1% or the whole thing)
1.01 = 101% (the whole thing *plus* 1% more)
1.10 = 110% (the whole thing *plus* 10% more)
1.20 = 120% (the whole thing *plus* 20% more) etc.

Therefore, if we want to add 10% to a job we can do so by simply multiplying our price or quantity by 1.10. This gives us the existing amount, and automatically adds on the bit we wanted, in this case 10%.

So now our equation is rather simple:

$400.00 × 1.10 = $440.00.

Other applications of percentages

Perhaps you are told that from a job worth $3000.00 you will be get $850.00. What percentage of the job are you getting?

To find this out you must first find out what 1% of $3000.00 is; that is:

3000.00 / 100 = 30.00

We now need to find out how many of these 1% amounts are in $850.00

850 / 30 = 28.33

This means you are getting 28.33% of the total job amount.

A quick alternative to this is to add two zeros to the amount you are getting (i.e. multiply by 100) and then divide the result by the total job value. For example:

85000 / 3000 = 28.33%

Of course if you are told you are getting 28.33% of $3000.00 you would simply do as was outlined in the first section:

3000 × 0.2833 = $850.00

Remember, 0.01 is 1%, 0.10 is 10%, 0.20 is 20%, so 0.2833 is 28.33%.

RATIOS

A sound understanding of ratios is critical to trades such as ours, as they are used in the mixing of adhesives, mortar, grout, sealers and additives to either.

A **ratio** is the amount of one (or more) elements in relation to another. A simple example is when you are required to mix up tile grout with water. A common mix is 5 parts powder to 1 part water. This is expressed as a ratio of 5:1.

The problem is that often you are given a mix based upon a particular weight matched to a given volume. Since this weight might be quite large compared to the amount you may desire to mix, you need to be able to find the ratio and 'scale' it down.

Example 1

The bag states that 20 kg of grout powder requires 4 litres of water. By dividing 20 by 4 we get a ratio of powder to liquid:

20 / 4 = 5

This means that for every 5 parts of powder by weight (kg), you need 1 part of water by volume (L); that is, a ratio of 5:1.

We can now use this to mix smaller amounts based upon another ratio: the coverage.

Example 2

You have a small area to grout. The area is only 0.6 m². The coverage rate for the grout you are using is 2.0 kg/m² (a ratio of 2:1).

Grout required = kg/m² × area to grout
= 2.0 × 0.6
= 1.2 kg.

The amount of water is based on a 5:1 ratio, so:

Water required = 1.2 / 5
= 0.24 L.

Worksheet 1

Student name: _____

Enrolment year: _____

Class code: _____

Competency name/Number: _____

To be completed by teachers:	
Student competent	☐
Student not yet competent	☐

Task: Read through the sections beginning at *Correct units of measure* and up to and including *Material units*, then complete the following.

Q. 1 How are linear measurements usually expressed in the building industry?

_____
_____

Q. 2 State the units of measure that tradespeople usually use on-site in relation to dimensions and measurements:

_____
_____

Q. 3 Name the two measuring tools most tradespeople carry on-site?:

1. _____ 2. _____

Q. 4 Describe the uses of each of the measuring tools above:

1. _____
_____

2. _____
_____

Worksheet 2

Student name: _____

Enrolment year: _____

Class code: _____

Competency name/Number: _____

To be completed by teachers:

Student competent ☐

Student not yet competent ☐

Task: Read through the sections beginning at *Measurement, calculations and quantities* and up to and including *Volume measurement*, then complete the following.

Q. 1 Using the conversion examples shown, convert the measurements in the exercise from millimetres to metres and metres to millimetres respectively:

Conversion examples

100s of mm in m	10s of mm in m	Individual mm in m
1000 mm = 1.0 m	90 mm = 0.09 m	9 mm = 0.009 m
900 mm = 0.9 m	80 mm = 0.08 m	8 mm = 0.008 m
800 mm = 0.8 m	70 mm = 0.07 m	7 mm = 0.007 m
700 mm = 0.7 m	60 mm = 0.06 m	6 mm = 0.006 m
600 mm = 0.6 m	50 mm = 0.05 m	5 mm = 0.005 m
500 mm = 0.5 m	45 mm = 0.045 m	4 mm = 0.004 m
400 mm = 0.4 m	40 mm = 0.04 m	3 mm = 0.003 m
300 mm = 0.3 m	35 mm = 0.035 m	2 mm = 0.002 m
200 mm = 0.2 m	30 mm = 0.03 m	1 mm = 0.001 m
100 mm = 0.1 m	25 mm = 0.025 m	0.5 mm = 0.0005 m

Exercise

Convert from millimetres to metres			Convert from metres to millimetres		
745 mm	=		6.0 m	=	
107 250 mm	=		536.45 m	=	
50 248 mm	=		27.01 m	=	
3 mm	=		0.052 m	=	
67 mm	=		54.209 m	=	
128 mm	=		0.002 m	=	
7 002 mm	=		9.6 m	=	
22 045 mm	=		11.08 m	=	
556 mm	=		457.02 m	=	
33 333 mm	=		3.44 m	=	

CHAPTER 5 BASIC MEASURING AND CALCULATIONS

Q. 2 Add the following measurements to find total metres:

1. 5.35 + 0.345 + 11.5 =

2. 27.467 + 0.004 + 3.32 =

Q. 3 Subtract the following measurements to find resulting metres:

1. 7.005 − 0.456 =

2. 345.450 − 15.01 =

Q. 4 Multiply the following measurements to find total metres squared:

1. 16.7 × 5.433 =

2. 45.00 × 0.055 =

Q. 5 Divide the following measurements to find the number of times:

1. 15.00 / 0.5 =

2. 72.40 / 6.2 =

Q. 6 Calculate the area of a square with sides 4.5 m long:

_____

Q. 7 Calculate the metres of perimeter of a square with sides 2.750 m long:

_____

Q. 8 Calculate the square metres of area for a rectangle 4.250 m long by 3.3 m wide:

_____

Q. 9 Calculate the metres of perimeter for a rectangle with sides 17.5 m long by 6.25 m wide:

_____

Q. 10 Calculate the square metres of area for a parallelogram 4.65 m long and having a perpendicular height of 1.5 m:

_____

Q. 11 Calculate the metres of perimeter for a parallelogram 5.520 m long with inclined sides 2.750 m long:

_____

Q. 12 Calculate the square metres of area for a triangle with a base 16.6 m long and a perpendicular height of 3.2 m:

_____

continued ▶

Q. 13 Calculate the metres of perimeter for a triangle with a base 4.5 m long and equal-length sides of 3.650 m long:

_____

Q. 14 Calculate the metres of circumference for a circle with a radius of 1.5 m:

_____

Q. 15 Calculate the square metres of area for a circle with a radius of 2.55 m:

_____

Q. 16 Calculate the cubic metres of volume of a cube with sides 2.5 m long:

_____

Q. 17 Calculate the surface area of a cube with sides 3.2 m long:

_____

Q. 18 Calculate the cubic metres of volume for a rectangular prism with a length of 2.4 m, a width of 1.2 m and a height of 0.9 m:

_____

Q. 19 Calculate the surface area in square metres for a rectangular prism with a length of 2.4 m, a width of 1.2 m and a height of 0.9 m:

_____

Q. 20 Calculate the square metres of area for an isosceles triangle having a base of 3.2 m and a perpendicular height of 5.4 m:

_____

Q. 21 Calculate the cubic metres of volume for a cylinder having a base diameter of 1800 mm and a perpendicular height of 3.6 m:

_____

SUGGESTED ACTIVITY one

Task: Using a four-fold rule and a retractable tape, measure and record accurate lengths of various materials and surfaces as directed by your teacher.

Suggested materials for practice measuring may include the following:
- various lengths of timber;
- section sizes of various timber;
- spacings between exposed framing;
- surface of walls;
- surface of a concrete path or driveway;
- height and width of nominated windows;
- height and width of nominated doors;
- length, width and thickness of standard MDF or plywood sheets, etc.

Note: Practice measuring should also take place during practical exercises.

CHAPTER 5 BASIC MEASURING AND CALCULATIONS   191

Worksheet 3

Student name: _____

Enrolment year: _____

Class code: _____

Competency name/Number: _____

To be completed by teachers:
Student competent ☐
Student not yet competent ☐

Task: Read through the sections beginning at *Quantitites and costs*, then complete the following.

Q. 1 **Brickwork**

Calculate the number of bricks in a single-skin wall 8.5 m long by 1.8 m high, when there is an average of 50 bricks per m²:

No. of bricks =

_____

Q. 2 How much will it cost to purchase the bricks if they are $865.00 per 1000?

Cost of bricks =

_____

Q. 3 Calculate the quantity of bricks in a solid 230 mm thick front fence 15.0 m long by 900 mm high, when there is an average of 50 bricks per m² per skin of brickwork:

No. of bricks =

_____

Q. 4 **Concrete**

The house has a portico entrance with two grand round columns supporting the roof. The columns are 600 mm in diameter and 4.95 high. How much concrete would be required to build them?

Volume of concrete =

_____

Q. 5 How much will it cost for the concrete when it is $147.00 (including GST) delivered to the site?

Cost of concrete =

_____

Q. 6 If the bricklayer building a 3.6 m diameter fishpond pours a concrete slab base 125 mm thick, how much concrete will be required?

Volume of concrete =

_____

continued ▶

Q. 7 The home owner has a number of ornamental garden statues to be placed around the garden on six triangular prism-shaped plinths. The height of the plinths is 1.2 m and the triangular-shaped ends have a base of 500 mm and a perpendicular height of 433 mm. How much concrete will be required to make the bases?

Volume of concrete =

Q. 8 To build the piers to support the floor of the house, the builder must excavate 24 holes 400 mm × 400 mm × 300 mm deep. How much soil must be removed?

Volume of soil =

Q. 9 If the builder places concrete 200 mm deep in the holes, how much concrete will be required?

Volume of concrete =

Q. 10 What will be the total cost to excavate and pour the concrete to the pier holes if the cost to excavate each hole is $12.00, the cost of the concrete is $145.00 per m³ (including GST), and a labourer is required for 2¼ hours to pour the concrete at $26.00 per hour?

Cost to excavate and pour concrete =

Worksheet 4

Student name: _____

Enrolment year: _____

Class code: _____

Competency name/Number: _____

To be completed by teachers:	
Student competent	☐
Student not yet competent	☐

Task: Read through the sections beginning at *Quantities and costs*, then use Examples 17 to 20 as a guide to complete the following.

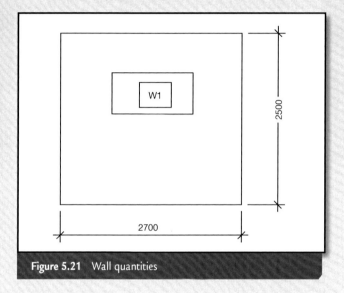

Figure 5.21 Wall quantities

Figure 5.21 shows the inside of a wall with a window opening. The client is unsure if they want to paint or tile the wall. It will need to be priced for both options.

In either case it is to be sheeted with FC sheet, tapped and stopped off as for painting but screwed for tiling. The specifications state that:
- W1 is a window 600 mm high x 1200 wide.
- Head height of window is 2100 mm.
- The architrave is 65 mm wide x 20 mm deep.

Plastering and sheeting materials are as follows:

6 mm FC sheet	$6.95/m²
Joint tape (paper)	$0.07/m
Base coat	$2.75/kg @ 0.16 kg/m²
Top coat	$1.50/kg @ 0.07 kg/m²
Stud adhesive	$4.95/kg @ 0.03 kg/m²
Fixings	$6.25/kg @ 0.05 kg/m²

If tiled:
- Tiles are to be to 1800 mm high.
- A 100 mm wide strip of stone trim tile is to run vertically up the middle of the wall to the underside of the window.

continued ▶

194 PAINTING AND DECORATING, AND MORTAR TRADES

- No waterproofing is required.
  Cost and coverage of materials are as follows:
  Main tiles 300 x 300 $54.00/m²
  Stone trim tiles 300 x 100 $17.00 each
  Tile adhesive: $2.50/kg @ 3.5 kg/m²
  Sanded grout $5.25/kg @ 2.0 kg/m²

If painted:
- A Tuscany style of paint is to be applied up to the underside of the window (sill height).
- A washable flat is to be applied to the rest of the wall.
- A high gloss enamel is to be applied to the architrave around the window.

  Excluding the primer/sealer/undercoat, all other finishes require two coats.
  Cost and coverage of materials are as follows:
  Primer/sealer/undercoat $18.25/L @ 14 m²/L
  Tuscany finish wall $28.35/L @ 9 m²/L
  Washable flat to wall $22.35/L @ 16 m²/L
  Gloss enamel to architrave $19.50/L @ 17 m²/L

Q. 1 Calculate the total quantity and cost of the sheeting and plastering materials required to sheet, fix and stop off the wall. Allow 5% waste for sheeting, and 10% waste for all other materials.

Q. 2 Calculate the total quantity and cost of tiling materials, including adhesive and grout. Allow 10% waste for all materials except grout. For grout allow 15% waste.

Q. 3 Calculate the total quantity and cost of the various paints required to paint the wall. Allow 5% waste for all finishes except the Tuscany finish. For the Tuscany finish allow 10% waste.

Worksheet 5

Student name: _____

Enrolment year: _____

Class code: _____

Competency name/Number: _____

To be completed by teachers:

Student competent ☐

Student not yet competent ☐

Task: Read through the section beginning at *Calculations of various solid shapes* and up to and including *Calculation of percentages*, then complete the following.

Q. 1 Calculate the surface area of a rectangular prism having a length of 6.520 m, a width of 1.125 m and a height of 750 mm:

Area = [(area of wide side) × 2] + [(area of narrow side) × 2] + [(area of one end) × 2]

_____

Q. 2 Calculate the surface area of a cone with an inclined length of 2.580 m and a base radius of 1.200 mm: Area = (πr × length of incline) + (πr^2)

_____

Q. 3 Calculate the surface area of a pyramid having a square base side of 1.550 m and an inclined perpendicular height of 2.7 m: Area = (area of base) + $\left(\dfrac{\text{length of base side}}{2} \times \text{inclined height} \times 4\right)$

_____

Q. 4 Convert the following percentages to fractions:

75% =   7% =

35% =   140% =

Q. 5 Convert the following fractions to percentages:

$\dfrac{3}{5}$ = $\dfrac{7}{20}$ =

Q. 6 Convert the following percentages to decimals:

60% =   22.5% =

Q. 7 A four litre container of paint can cover 60m². You only need to paint 38m².

a. What is the coverage ratio per m²? Coverage ratio = Area ÷ litres

_____

b. How many litres do you need to cover 38m²? Litres required = Required area ÷ coverage ratio

_____

Q. 8 A 10 kg bag of grout requires 3 litres of water for mixing. It has a stated coverage ratio of 1.5 kg/m².

a. What is the weight to volume ratio for this grout? Ratio = kg ÷ litres

_____

b. How many kg of dry grout do you need to cover 4.5 m²? Grout required = kg/m² × area to grout

_____

c. How much water will you need to make up this amount of grout? Water require = kg of dry grout ÷ weight to volume ratio

_____

REFERENCES AND FURTHER READING

Acknowledgment

Reproduction of the following *Resource List* references from *DET, TAFE NSW C&T Division (Karl Dunkel, Program Manager, Housing and Furniture)* and the *Product Advisory Committee* is acknowledged and appreciated.

Texts

Australian Institute of Building (1985), *Code of estimating practice for building work*, AIB, Canberra

Australian Institute of Quantity Surveyors and the Master Builders' Association (1987), *Australian standard method of measurement of building works*, Australian Institute of Quantity Surveyors and the Master Builders' Association, Canberra

Marsden, P.K. (1998), *Basic building measurement*, New South Wales University Press, Sydney

Milton, H.J. (1992), *Australian building & construction definitions*, Standards Australia, Sydney

Sierra, J.E.E. (1998), *The A-Z guide to builders' estimating*, Australian Institute of Quantity Surveyors, Canberra

Smith, J. (2000), *Building cost planning in action*, Deakin University Press, Melbourne

Web-based resources

Resource tools and VET links

<www.resourcegenerator.gov.au> ANTA resource generator

<http://toolbox.flexiblelearning.net.au/search.asp> ANTA

<www.ntis.gov.au> National Training Information Service

<www.hsc.csu.edu.au/construction> NSW HSC Online—Construction

CHAPTER 6
PLAN INTERPRETATION AND SPECIFICATIONS

This chapter covers building working drawings and the specifications related to them. In the building industry the majority of descriptions and instructions passed on to the people performing the work originate from a plan or drawing detail. To be able to successfully carry out construction work it is critical to be able to interpret plans, drawings, details and specifications correctly. The skill of plan reading is best acquired by practice and exposure to a variety of details and situations.

Drawings are presented in a scaled form showing views from above, the front, the sides and even inside the building or structure. The work shown on plans is also described in words within a document called the specification. Both the plans and specification must be submitted to the local council for approval prior to contracts being signed or work commencing.

The details and measurements shown or described in these documents allow for quoting procedures to take place and for material orders to be placed. Therefore, plan and specification reading and interpretation form the basis for the successful completion of building contracts.

Areas addressed from the unit of competency include:

- identification of types of drawings and their functions;
- recognition of amendments;
- recognition of commonly used symbols and abbreviations;
- key features of site plans;

- specific project requirements; and
- reading and interpreting job specifications.

INTRODUCTION TO PLAN AND SPECIFICATION READING

In the building industry, drawings are used to convey great amounts of technical information between the designer of a building or structure and the builder. This technical information must be able to be conveyed without the risk of any misunderstanding. This can be done successfully only if the technical language of drawings is universally understood by everyone who is required to use them (Figure 6.1). For this reason the technical language of drawings is a standard language, just as written and spoken language is standard. And just as written and spoken language must be learned, so must the technical language of drawings be learned.

The technical language of drawings is expressed through the use of a standardised drawing layout, standardised symbols and standardised abbreviations of terms used. With study, practice and experience the technical language of drawings may be as quickly and easily understood as the written word.

KEY USERS OF DRAWINGS

When a drawing of a proposed building or structure has been prepared, many copies are made for the people who will use them.

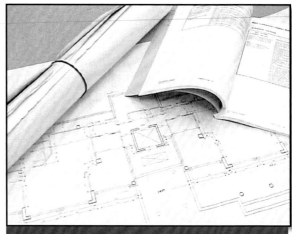

Figure 6.1 Plans and specifications

- The owners of the proposed building need a drawing to see that the design is as they imagined it.
- Structural, electrical and mechanical engineers need copies so they can design their part of the structure.
- Council health and building surveyors will require copies of drawings and specifications to make sure that the building conforms to building codes and council regulations.
- Council town planning officers will require copies of drawings and specifications to make sure that the building conforms to council planning regulations.
- Bank or building society officers will require copies of the drawing before giving approval for finance for construction.
- A builder will require copies of the drawing in order to cost the building and to prepare a quotation and then to construct the building (Figure 6.2).
- A builder will pass copies of the drawing to subcontractors such as concreters, bricklayers, electricians, plasterboard fixers, tilers and

CHAPTER 6 PLAN INTERPRETATION AND SPECIFICATIONS

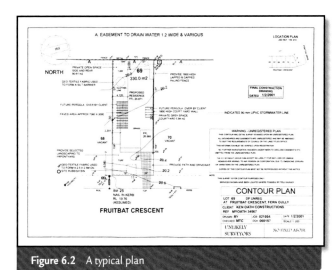

Figure 6.2 A typical plan

Pictorial representation

Pictorial drawings are used by architects and designers to determine, with the client, the final design or appearance of the project. A **perspective view** (Figure 6.3), or pictorial representation (Figure 6.4), is often used to assist in visualising the object (the faces or sides of the object appear to taper away or recede to a vanishing point). Parallel lines seem to converge, as is the effect when viewing railway tracks in the distance.

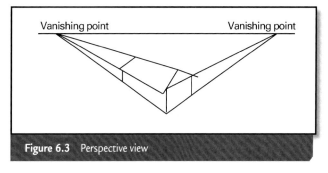

Figure 6.3 Perspective view

painters to obtain their prices to carry out work during construction.
- Suppliers of prefabricated building components such as wall frames, roof trusses, windows, doors, air-conditioning and heating will need copies of the drawing in order to calculate their prices for their part of the job.

Drawings are important documents, and should be carefully looked after.

DRAWINGS AND THEIR FUNCTIONS

So that construction methods and ideas for building can be carried out, they are first set down as drawings, and, together with a specification, become the instructions for the different trades to follow without the need for further referral.

Completed drawings and specifications must be approved by the local council before work is begun. Drawings must be in a standard format for accurate interpretation by builders and tradespeople from any area or background.

Drawings can be divided into two groups:
1. those giving a pictorial representation of the project as design drawings;
2. those best described as working drawings.

Figure 6.4 Pictorial representation

Perspective drawings are the closest to what the eye would see, and are often used by builders to present a project for sale.

Isometric projection (Figure 6.5) is also a pictorial view, with lines drawn parallel to the axis at 30° because of the ease in using 60°–30° set squares. Perspective view and isometric projection are used as working or freehand sketches for ease of interpretation of house design or building construction technique.

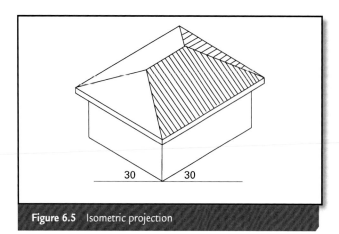

Figure 6.5 Isometric projection

Working drawings

Working drawings are produced so that users may:
- gain an overall picture of the layout and shape of the building;
- determine setting-out dimensions for the building as a whole;
- locate and identify the spaces and parts of the building, e.g. rooms, doors, cladding panels, drainage;
- pick up references leading to more specific information, particularly about junctions between parts of the building.

A basic form of working drawing is **orthographic projection** (single angle), which consists of three related views—plan, elevation, and section—to give a complete understanding of the building.

Floor plan

The **floor plan** is a horizontal section of the building as viewed from above and is the more important of the related views, as it contains most of the information for construction (Figure 6.6).

Items shown on a floor plan would include:
- overall dimensions to the outside of the walls;
- door and window positions and opening sizes;
- thickness of external and internal walls;
- internal room dimensions;
- position of cupboards, stoves, laundry tubs etc.;
- function of each room, such as kitchen or bedroom;
- floor surface and type of floor covering;
- position and direction from which section lines are taken for sectional elevations.

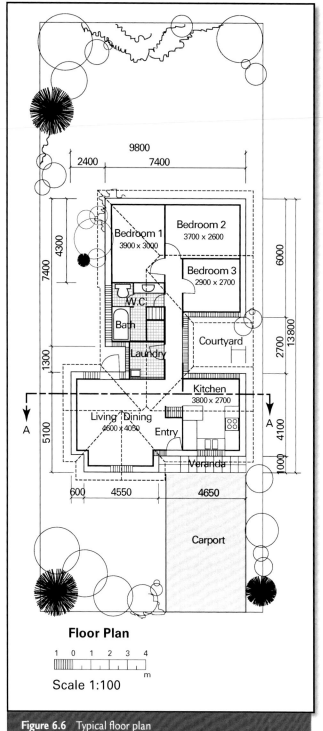

Figure 6.6 Typical floor plan

Elevation

Elevations provide information relating to vertical measurements and external finishes. Each view is identified according to the direction it faces, in relation to the points of the compass, based on the north point shown on the site plan.

Elevations give a projection of the building at right angles, and show:
- height of finished floor level (FFL) to finished ceiling level (FCL);
- design of the building;
- roof shape and width of roof overhang;
- position of doors and windows;
- window sill height above floor level;
- type or function of windows;
- roof covering and slope;
- floor height above ground;
- finish to external walls.

Standard working drawings in orthographic projection generally require a minimum of two elevations—the front and side of the building—thus enabling correct interpretation of design. Typical information may be indicated on one or both elevations (Figure 6.7).

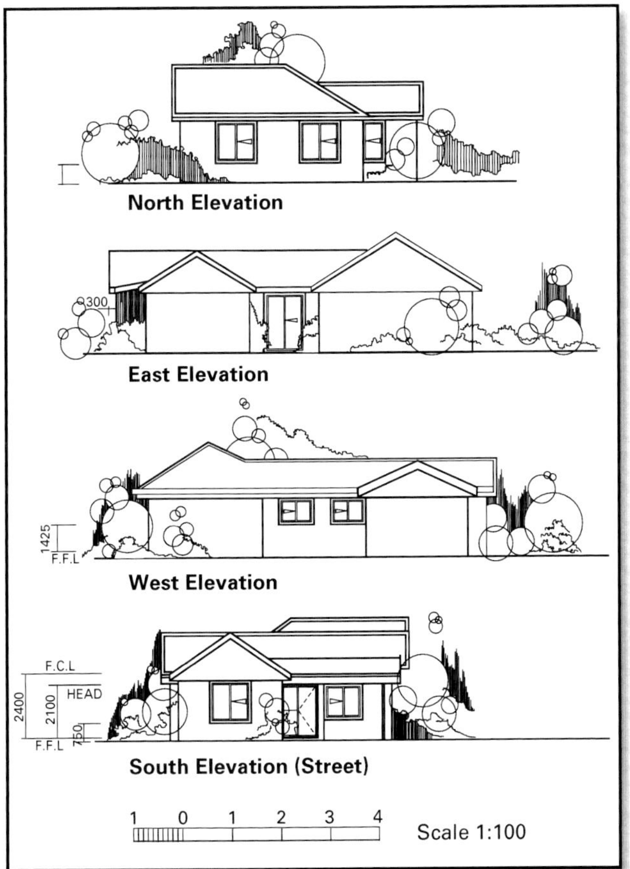

Sections

Section drawings are elevations cut through the building in the position and direction indicated on the floor plan. The section is a cross-section from the bottom of the footings, through the walls, ceilings and roof structure (Figure 6.8). Sections give information such as:
- footing sizes;
- wall thickness and construction;
- design of sub-floor;
- floor construction;
- roof construction (e.g. trussed);
- roof pitch;
- section sizes and spacing of structural members.

Other information to be shown on working drawings includes:
- details;
- site or block plan;
- scales used for drawings.

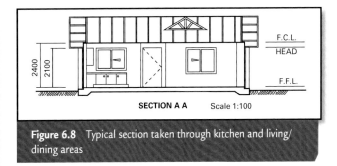

Figure 6.8 Typical section taken through kitchen and living/dining areas

Details

Details are sectional views drawn to a larger scale than sectional elevations, showing detail specific requirements that cannot be drawn accurately to scale on sectional elevations (Figure 6.9).

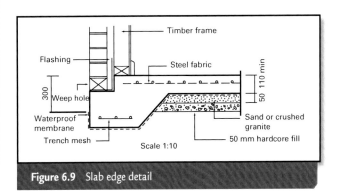

Figure 6.9 Slab edge detail

Figure 6.7 Typical details of elevations

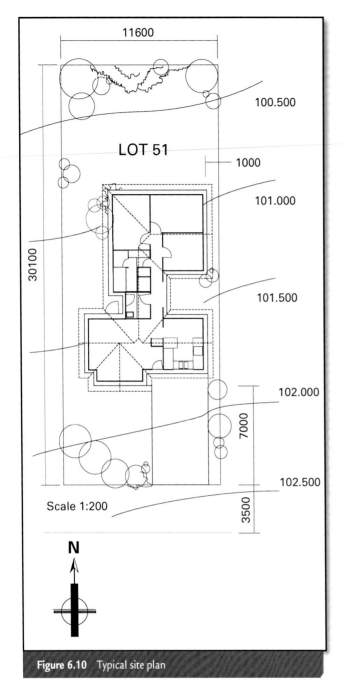

Figure 6.10 Typical site plan

Site or block plan
Site or **block plans** are essential for determining the location of the building on the building block. Information contained on a site (Figure 6.10) or block plan would be:
- boundary dimensions of the block;
- distance from street to boundary;
- set-back distance from front boundary to building line;
- distance from side boundary to building;
- contour lines and their height;
- position of paths and driveways;
- trees;
- direction of north;
- lot number.

Special details
These details may include:
- a site plan to indicate landscaping details;
- details of retaining walls;
- a site layout plan for storage of materials and environmental compliance/water drainage etc.;
- large-scale details of special construction requirements; and so on.

Scale drawings
A **scale drawing** is the reduction of a full-sized object to a suitable scale to enable its reproduction on drawing sheets. Working drawings state the scale or scales that have been used on the drawing. However, in some cases there is also a warning to the effect that all measurements should be taken as read from the drawing, rather than determining lengths using scale rules.

A scale of 1:50 (Figure 6.11) indicates that the drawing is reduced to 50 times smaller than full size.

Table 6.1 shows standard scales adopted when producing working drawings.

Table 6.1 Standard scales for working drawings			
Drawing element	Common scale	Alternatives	
Site plans	1:500	1:200	
Plan views	1:100		
Elevations	1:100		
Sections	1:100		
Construction details	1:20	1:10	1:5

Calculating scales
When a scale rule with the appropriate scale cannot be found, or an unusual scale needs to be used, it may be necessary to make your own. This may be done simply by using a calculator and a rule with standard size millimetres or a scale of 1:1, which is full size.

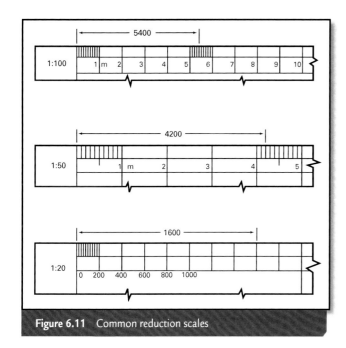

Figure 6.11 Common reduction scales

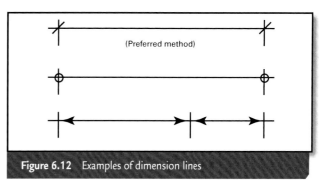

Figure 6.12 Examples of dimension lines

Drawing sheet sizes

Plans such as working drawings are produced to scale on standardised drawing sheets, which range in size from A0 (1189 × 841 mm) to A4 (210 × 297 mm), with margin or border lines and title blocks (Figure 6.13). The margins assist in the folding or filing of the plans, while the title block contains information essential to the project.

Title block

The title block on a set of plans for a house may contain a minimum of information, such as the name of the owners, the lot number and street number, street name, suburb, and scales used in the drawing (Figure 6.14).

The title block for a commercial project may contain:
- the name of the client or company for whom the project is to be constructed;
- the lot number and address of the project;
- scale or scales used on the plans (e.g. 1:100);
- numbers on the drawing sheets, if more than one is used (e.g. sheet 1 of 3);
- the name of the person or drafting service that has prepared the drawings;
- a filing system in the form of numbers or letters for use by the person who prepared the drawings;
- warnings (in some cases) against scaling from drawings to prevent incorrect measurement when using scales rather than using figured dimensions.

Specifications

A **specification** is a precise description of all construction and finishing, including workmanship, which is not shown on the drawings (Figure 6.15

Example 1
Scale required = 1:25
Measurement to be scaled = 6.200 m.
1 Change the measurement from metres to millimetres (i.e. change 6.200 m to 6200 mm).
2 Divide the millimetre measurement by the desired scale (i.e. 6200 / 25 = 248).
Therefore the 1:25 scaled measurement = 248 mm (full-size mm).

Example 2
Scale required = 1:75
Measurement to be scaled = 1.500 m.
1 Change the measurement from metres to millimetres (i.e. change 1.500 m to 1500 mm).
2 Divide the millimetre measurement by the desired scale (i.e. 1500 / 75 = 20).
Therefore the 1:75 scaled measurement = 20 mm (full-size mm).

Dimensions

Dimension lines on drawings enable scales to be used to determine lengths that are not shown. The forms of dimension lines vary (Figure 6.12), but all are shown as a line parallel to the drawing. Lines at right angles to the main line indicate the position at which the dimension is taken.

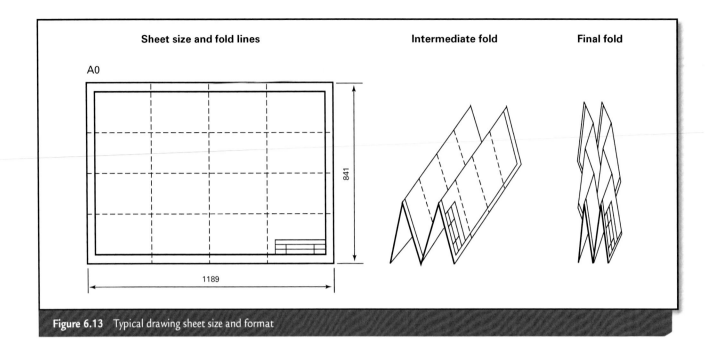

Figure 6.13 Typical drawing sheet size and format

Figure 6.14 Typical title block showing required information

on p. 205). This includes composition of concrete in footings, species and grades of timber, brick type, mortar quality, paint colours, size and type of hot water system, and number and position of power points. The specifications are arranged according to trades, and cover all information relevant to the particular trade in the sequence of construction. Specifications must be kept with the drawings and are read in conjunction with them.

Drawing symbols and abbreviations

Symbols and abbreviations appear on plans, elevations and in sections. Common symbols in use are shown in Figures 6.16–6.25 which follow on pp. 206–209.

They represent common construction practice for standardised details that are common knowledge in the building industry. They are of great importance in enabling the transfer of technical information between designer and builder, as they reduce the risk of any misunderstanding of the intention of the designer.

Symbols used in drawings for restoration of or extension to existing structures are coloured to distinguish the new work from the existing structure.

Some examples of the use of symbols would be to show:
- direction of opening and operation of doors and windows;

Items	
clothes lines	see drawings ...
letterboxes	1 per dwelling
shower screens	see drawings
bathroom cabinets	minimum 1 per bathroom
garage doors	see drawings

05 CARPENTRY & JOINERY

5.1 GENERALLY
Timber inspections and bandings: Refer PRELIMINARIES—Materials and workmanship

To AS 1684/NSW TFM

NSW Timber Framing Manual: The current edition of the NSW Timber Framing Manual may be used in lieu of AS 1684.

To AS 1720. AS/NZS 1748. Refer table in ASDC.

5.2 MATERIALS
Timber stress grades: The timbers used must comply with the Timber Marketing Act, and be graded to the appropriate SAA grading specification.

Timber Species: Do not use tropical rainforest timbers. Refer to SCHEDULES— Schedule of Timber Species and Durability Rating.

To AS 1604. Obtain Superintendent's approval for treatment details.

Preservative Treatment:
—All Lyctus susceptible sapwood in local rainforest timbers.
—All Lyctus susceptible sapwood in hardwoods other than milled exceeding 20% of the perimeter of the piece.
—All Lyctus susceptible sapwood in milled hardwood products.
—Radiata pine used externally and for bearers and joists.

To AS/NZS 1859. Refer SAPPC.

Panel and sheet products:
Plywood and Blackboard: Interior use, type D: exterior use Type A. Use particle-board grades designated by the manufacturer to have moisture resistance appropriate to the conditions of use. Melamine surfaced particle-board shall be finished with melamine, surface bonded to all faces.

Refer SAPPC.

External cladding: hardboard planks, fibre cement flat sheets, fibre cement planks.

To AS/NZS 2924, AS 2131.

Laminated plastic sheet: Fix to background with contact adhesive.

5.3 WORKMANSHIP
General: Perform the operations and provide the accessories necessary for the completion of woodwork items. Ease and adjust moving parts, lubricate hardware, and leave the completed work in a sound, clean, working condition.

Joinery:
—mortice and tenon joints in doors, frames, sashes and other parts.
—mitre joints in mouldings, skirting, etc., but scribe internal angles
—dress joinery stock and mouldings, hand finish exposed surfaces and remove arrises to provide smooth surface for painting.
—all moulded runs of 800 mm and less must be in single lengths.

To AS 1684.

Sizes and tolerances:
—Maximum possible tolerance for dressing to be 3 mm per face.

Reference Specification for Detached Dwellings—Public Housing
March 1994
Page 8

Figure 6.15 Extract from a standard Department of Housing specification

- wall or floor construction, such as timber, brick, reinforced concrete;
- foundation soil composition, earth or filled material.

Abbreviations are used on drawings to reduce the written content, thereby minimising the congestion of information necessary to convey the correct interpretation of the drawings. The abbreviated form should be used only where confusion or misinterpretation are not likely to occur.

Table 6.2 is a list of common abbreviations in use.

Table 6.2 Common abbreviations

Word	Abbreviation	Word	Abbreviation
aggregate	aggr	joist	J
angle	L		
approximate	APPROX	kitchen sink	KS
at	@		
average	AV, AVG	level	LEV
		longitudinal	LONG
bench mark	BM		
bottom	BOT	main	
brick	BK	switchboard	MSB
brick veneer	BV	minimum	MIN
brickwork	BWK		
building	BLDG	north	N (NTH)
		not to scale	NTS
ceiling joist	CJ		
ceiling level	CL	out of	O/O, ex
centre line	LC or C	over-all	OA
concrete	CONC		
concrete		pad footing	F
reinforced	RC	prefabricate	PREFAB
countersink	CSK		
		quadrant	
damp-proof		moulding	quad
course	DPC	quantity	QTY
detail	DET		
diagonal	DIAG	rainwater pipe	RWP
diameter	DIA	reduced level	RL
dimension	DIM	reinforced	
distance	DISt	concrete	Rc
		retaining wall	RW
equal to	EQ		
existing	EXST	sewer	SEW
expansion	EXP	shower	SHR
external	EXT	sliding door	SD
finished floor		temporary	
level	FFL	bench mark	TBM
fixed glazing	FG	tongue and	
floor level	FL	groove	T&G
floor waste	FW	typical	Typ
ground floor	GF	underside	U/S
ground level	GL	universal beam	UB
ground floor		universal	
level	GFL	column	UC
hand left	LH	vent pipe	VP
hand right	RH	vertical	VERT
hardwood	HWD	volume	VOL
height	HT or HGT		
horizontal	HORIZ	waste pipe	WP
hot water		water closet	WC, W
supply	HWS	window	
		(number)	W
include	INCL		

A more detailed list can be obtained from Australian Standards:

- AS 1100.101-1992, 'Technical drawing—General principles';
- AS 1100.301-2008, 'Technical drawing—Architectural drawing'.

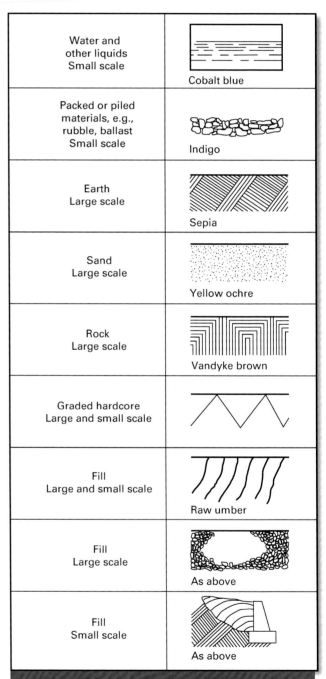

Figure 6.16 Symbols for sections in-ground

CHAPTER 6 PLAN INTERPRETATION AND SPECIFICATIONS 207

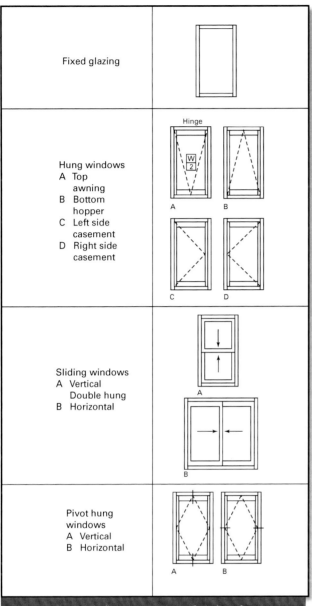

Figure 6.17 Symbolic representations of windows for elevations

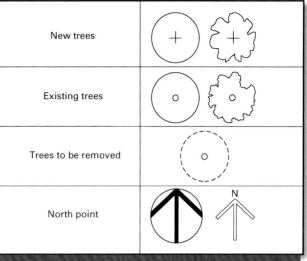

Figure 6.18 Graphics for use on site plans

Timber stud	Chrome yellow (shaded grey)
Brickwork single skin	Vermilion
Brickwork cavity wall	Vermilion
Brick veneer	Chrome yellow and vermilion
Concrete block single skin	Prussian green
Concrete block cavity wall	Prussian green
Concrete	Hooker's green deep
Stone	Vandyke brown
Existing wall (alterations & additions) Alternatives a, b Note: Colour new work	Heavy black for prominence Light outline ▼ New work Ghosted

Figure 6.19 Symbolic representations for floor plans and details

208 PAINTING AND DECORATING, AND MORTAR TRADES

Fitting	Symbol	Fitting	Symbol
KITCHEN FIXTURES & FITTINGS All coloured French ultramarine		Water closet	
Stove	S	Bidet	B B
Wall oven / Hot plates	O ⚬⚬⚬	**LAUNDRY FITTINGS** All coloured French ultramarine	
Refrigerator	R	Sink/Tub	
Dishwasher	DW	Washing machine	WM
Sinks (with drainers)		Clothes dryer	CD
		MISCELLANEOUS FITTINGS All coloured French ultramarine	
SANITARY FITTINGS All coloured French ultramarine		Cleaner's sink	CS
Bath		Slop sink	SS
Basin		Hot water unit Alternatives	HW HW
Shower recess	⊠	Rainwater tank	RWT 2000 litre
Urinal		Fire hose rack/reel Recessed	FHR
Urinal stall		Fire hose rack/reel Free standing	FHR

Figure 6.20 Symbolic representations of fixtures and fittings

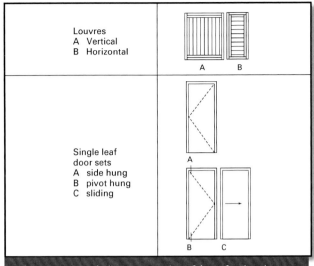

Figure 6.21 Symbolic representation of doors for elevation

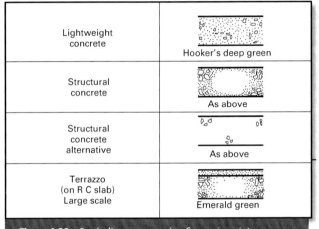

Figure 6.22 Symbolic representation for sections (a)

Figure 6.23 Symbolic representation for sections (b)

Figure 6.24 Symbolic representation for floor plans and horizontal sections

Figure 6.25 Symbolic representation for floor plans and horizontal sections

PLAN AND DOCUMENT READING

Plan and document reading is a skill that requires practice. Accurate and quick interpretation of plans and specifications requires a sound understanding of basic drawing symbols and abbreviations. Knowing where to look for the information required is a definite advantage.

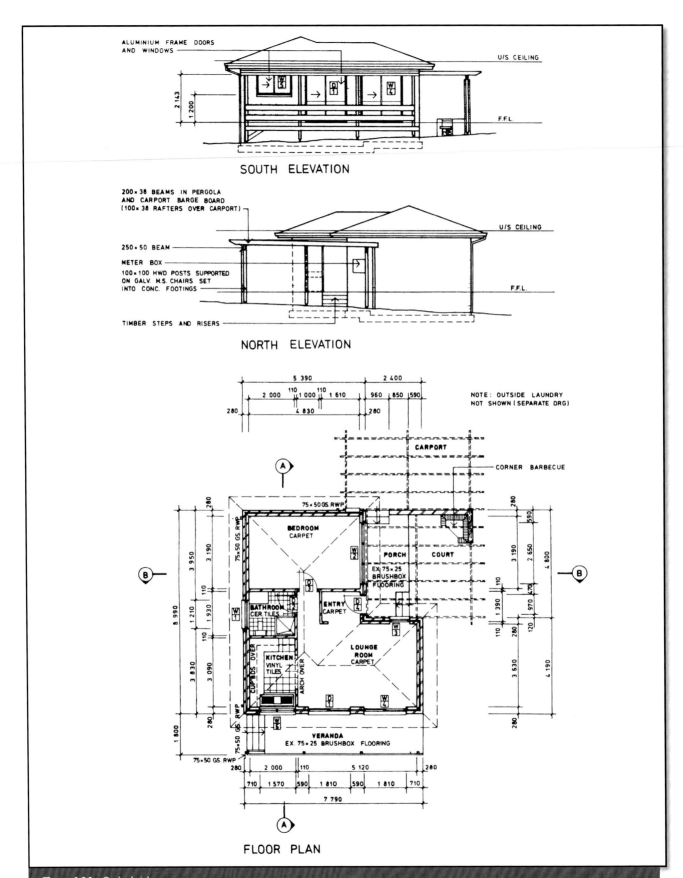

Figure 6.26 Cavity brick cottage

CHAPTER 6 PLAN INTERPRETATION AND SPECIFICATIONS

Table 6.3 Sample question sheet

	Statement	Answer	Where found
1	The meter box is found on which detail?	South Elevation	North elevation
2	The section size of the brushbox flooring to the porch is	Ex 75 × 25	Floor plan
3	The window frames are	Aluminium	South elevation
4	The internal dimensions of the bedroom are	4830 × 3190	Floor plan
5	The thickness of the internal walls is	110	Floor plan
6	The kitchen floor is covered with	Vinyl tiles	Floor plan
7	The sill height of window 5 off the floor is	1200	South elevation
8	Window 1 is on the east elevation of the	Bathroom	Floor plan
9	The overall size of the external walls is	280	Floor plan
10	The section size of the galvanised steel rainwater pipes is	75 × 50	Floor plan
11	The head height of window 4 off the floor is	2143	South elevation
12	The width of door 4 is	970	Floor plan
13	The width of window 2 is	3190	Floor plan
14	The width of window 1 is	1210	Floor plan
15	The section size of the rafters over the carport is	100 × 38	North elevation

Look on the floor plan for the external size of the building, and on the site plan for the position of the building on the building block. The heights of walls and windows are found on elevations, while specific construction practice can be found in sections or details. Technical information such as material size and spacing not found on the plans will be described in the specifications.

The answers to the incomplete statements in Table 6.3 are typical of those required to be found on plans before and during construction. Refer to Figure 6.26 to confirm that the statements and views are correct.

ISSUES FOR PAINTERS, TILERS AND PLASTERERS

For bricklayers and other trades aligned with the construction of the main 'structure' of a building, plans offer most of the detail they require, and the associated specifications and schedules provide additional information on materials and acceptable standards of work.

For painters, tilers and plasterers, however, the plans offer little other than the size of a room, and even that will need to be confirmed after the frames are up. Tilers are at a particular disadvantage, for while the plasterer can at least obtain wall heights from an elevation, the area to be tiled on walls is generally not stated.

It is to the specifications and schedules that painters, tilers and plasterers must look very carefully for this information. They may also have to talk directly with the client or designer. Painters and tilers, particularly, may find that they have to sketch their own 'plan' of a room. This is called a **fold out**.

Figure 6.27 shows the floor plan of a modern ensuite, while Figures 6.28 and 6.29 show how it is ultimately to look. 'Seeing' this from plans and specifications is not always easy, hence the 'fold out' approach. (Figure 6.29 is an elaborate fold out; a 'working' version is never so detailed.)

Creating a fold out

1 From the plan (or by direct measuring), draw the floor of the room as shown in Figure 6.30. Include the dimensions, and label the walls A, B, C and so on.

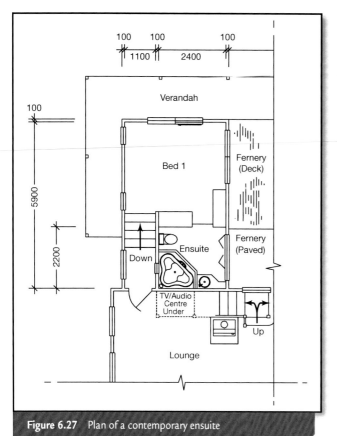

Figure 6.27 Plan of a contemporary ensuite

Figure 6.28 A contemporary ensuite

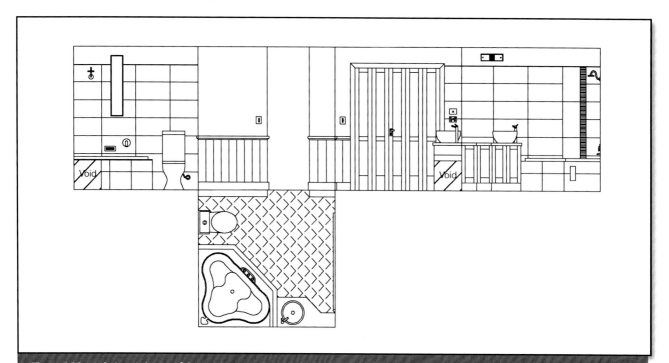

Figure 6.29 A designer's image of the ensuite

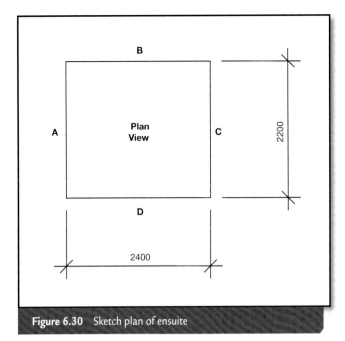

Figure 6.30 Sketch plan of ensuite

2 Now create the fold-out view as shown in Figure 6.31. It looks much like a paper box cut-out that you might have made as a child.

Now we have a view that allows:
- easier calculations of areas;
- the tiled areas to be clearly shown;
- the client to see what work you will need to do.

3 On this fold out you now draw as much information as is relevant. This may include:
- wall heights;
- windows, doors and other openings;
- baths, spas, hobs etc.;
- vanity units;
- any panelled or painted areas that will not be tiled;
- the direction of the pattern in which tiles are to be laid (diagonal, square, stretcher etc.);
- the location of trims or pattern tiles.

Figure 6.32 shows how a fold out might look in this case (due to size, some information has been excluded). Note that non-tiled areas have been simply lined out, and tiled areas are shaded differently for each tile type. The painter would produce a similar fold out, only their focus would be on the panelling and painted areas—that is, the tiled areas would be lined out.

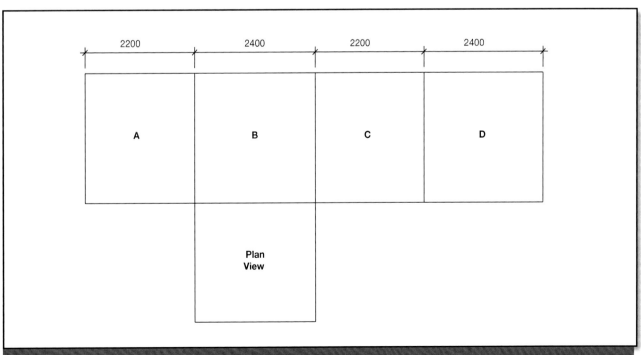

Figure 6.31 Simple fold out of ensuite

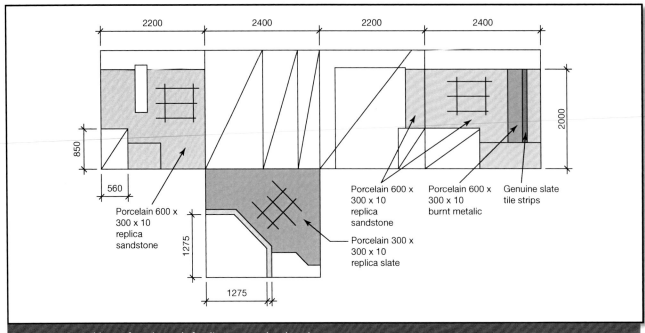

Figure 6.32 Fold out of ensuite ready for client approval and costing

RESIDENTIAL BUILDING STRUCTURE TYPES

Domestic construction

To help you understand domestic construction practice and the graphical presentation of details, the following components of a building have been identified in segments or structures which, when grouped together, become the complete project.

This section provides vital knowledge for bricklayers. Painters, plasterers and tilers, however, are not ordinarily engaged in digging footings, pouring concrete, or other matters concerned with the early phases of construction. Yet it is important to understand how a building comes together, and how these early stages can influence your work. It is also important because you need to ensure that the work you do does not have a negative influence upon the final structure as a whole.

This understanding can be critical when it comes to reading specifications. New bushfire codes and energy efficiency audits can lead to very specific installation requirements and material choices. Brick, plaster, paint, tile—even colours and textures—can have an influence on whether a design is ultimately able to work efficiently.

Footings

Footings are the lowest part of a building, designed to distribute the load of the building evenly over the foundation (Figure 6.33). There are several types, which are designed according to the load of the building to be supported and the ability of the foundation material to support that load. Common types of footing in building are as follows.

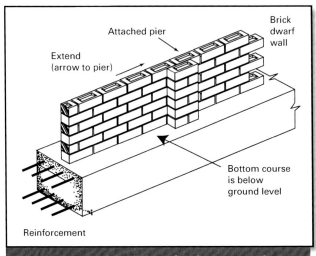

Figure 6.33 Typical detail of a reinforced concrete strip footing

Strip footings

Strip footings are a continuous reinforced strip of concrete around the outside of a building to support the external walls. The width and depth of concrete and the amount of reinforcement needed are calculated by a structural engineer, based on the design of the building and the known soil classification, as stated in AS 2870-1996, 'Residential slabs and footings'. The bottom of a strip footing is approximately 600 mm below ground level to prevent erosion of the foundation material. Timber frame, brick veneer and cavity brick are constructed on strip footings.

Slab-on-ground

Slab-on-ground combines the floor and the footing into one reinforced, monolithic concrete unit. This method of construction is used on level ground or on sloping sites that have been cut to a level surface. The edge beam can be increased to support greater loads or to pass through top reactive soils to more stable foundation material (Figure 6.34). An alternative is the 'waffle-pod' system, consisting of a series of beams

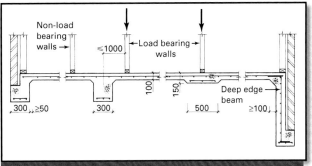

Figure 6.36 Slab-on-ground for masonry, veneer and clad frames

running at right angles to one another with a thin working slab cast over them to form a monolithic slab with a grid support system (Figure 6.35). Services such as plumbing and electrical pipes are placed in the slab before the concrete is poured, or are built into the walls during construction. Slab penetrations must be adequately protected to prevent the entry of termites, for example with the use of termite prevention products such as Granitgard, Termimesh or Kordon. Slab-on-ground has become a popular alternative in the floor construction of timber, brick veneer and cavity brick buildings (Figure 6.36).

Blob/pad footings

Blob footings are square, rectangular or round footings placed under piers or posts and may contain reinforcement, depending on the load to be carried (Figure 6.37). Their size should be calculated so that the same pressure is applied to the foundation

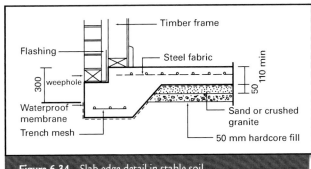

Figure 6.34 Slab edge detail in stable soil

Figure 6.35 Typical waffle-pod slab system

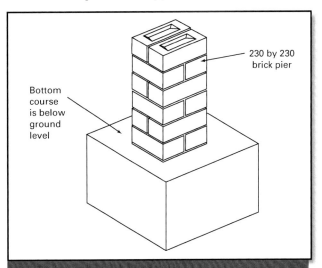

Figure 6.37 Typical blob/pad footing with a mininmum depth of 200 mm

material as to strip footings when both are used on the same building. Blob footings are commonly used for supporting brick piers, bearers and joists in timber floor construction. Alternatives to this type of support include steel adjustable piers, precast concrete piers and treated timber stumps (generally used in states other than NSW).

Flooring systems

A **flooring system** is the floor surface and the method of floor framing used as support to the ground floor. Common systems in use are as follows.

Suspended timber floor

One system of **suspended timber floor** consists of flooring boards in narrow strips laid on a timber or steel framing of bearers and joists and supported by brick walls or piers. The flooring is placed between the walls and is cramped and nailed in position when the building is advanced enough for the flooring not to be affected by the weather; that is, generally when the roof and wall cladding has been installed. (See Figure 6.38.)

Another system consists of sheet flooring of plywood or particle board laid on a timber or steel framing of bearers and joists before the walls are erected, and therefore gives the advantage of providing a platform to work on. Both of these methods require a minimum 400 mm clearance above the ground to the underside of bearers, which provides ventilation of the area beneath the floor framing to prevent decay in the timber framing. The timber bearers and joists used to support the flooring are placed at specified centres apart and are of a sectional size according to AS 1684.2-2006, 'Residential timber-framed construction—Non-cyclonic areas'.

Slab-on-ground

Slab-on-ground is a reinforced concrete floor placed directly onto the ground. One advantage of this method is the reduced building height. See Figures 6.34–6.36, and previous information on footings for details.

Suspended slab

A **suspended slab** is a reinforced concrete floor suspended above the ground and supported on

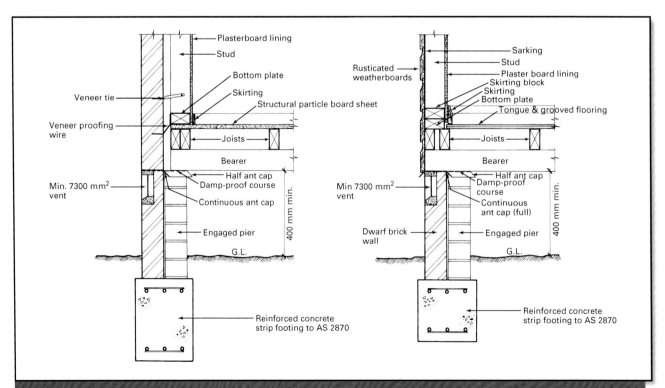

Figure 6.38 Vertical section through external walls of brick veneer and timber-frame construction

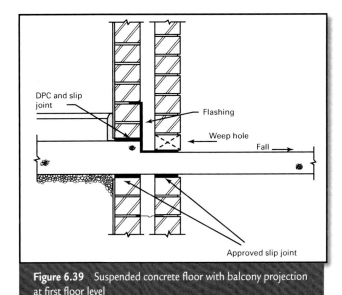

Figure 6.39 Suspended concrete floor with balcony projection at first floor level

brick walls (Figure 6.39). The amount and type of reinforcement in a suspended concrete floor will be greater than that used in a slab-on-ground as a suspended slab must carry the floor loads between supports.

Wall structures

This is the composition of the external and internal walls. The walls may be constructed from one or several materials. Common wall structures are:

Timber frame

External walls are constructed of timber framing, with a cladding on the outside of timber or fibre-cement weatherboards or sheets, and a lining on the inside of plasterboard. Internal walls are constructed of timber framing and are lined on both sides with plasterboard (Figure 6.40).

Brick veneer

Brick veneer is for external walls only, and consists of a timber frame lined internally with plasterboard and with an external skin or veneer of brick (Figure 6.41). The brick skin and the timber frame are separated by a gap (cavity) of around 40 mm (25 mm min. – 60 mm max.) to allow for ventilation and to prevent contact between the timber and the bricks. The timber wall is load-bearing, carrying the roof and ceiling, while the outer skin of brick is used for weathering, security and visual effect. The brick skin is tied to the timber frame

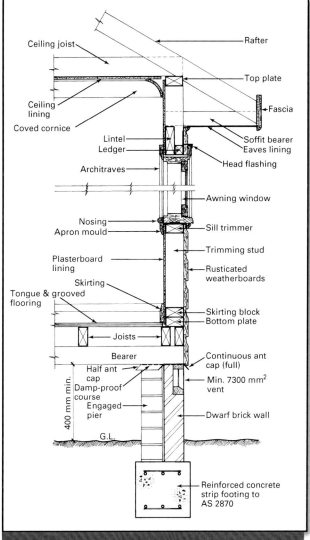

Figure 6.40 Vertical section through an external timber frame wall and timber awning window

by veneer ties built into the brickwork and nailed to the frame. A barrier of wire mesh (vermin wire) is placed at the base of the timber frame and built into the brick wall to prevent rodents from entering the frame through the cavity. Internal walls are constructed in the same way as for a timber-framed construction.

Cavity brick

Cavity brick external walls are constructed of two skins of brick, separated by a gap (cavity) of 50 mm and held by wire galvanised cavity ties, with a drip groove to prevent moisture travelling from the outer skin of brick to the inner skin when the outer skin becomes wet (Figure 6.42). The inner skin of brick

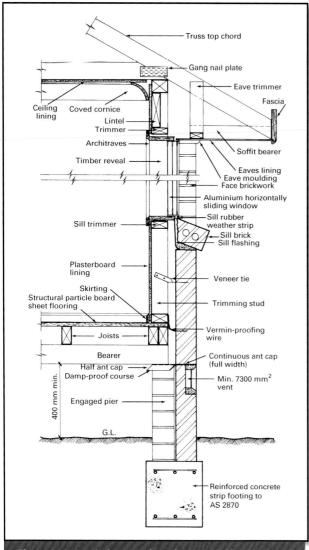

Figure 6.41 Vertical section through an external brick veneer wall and horizontally sliding aluminium window

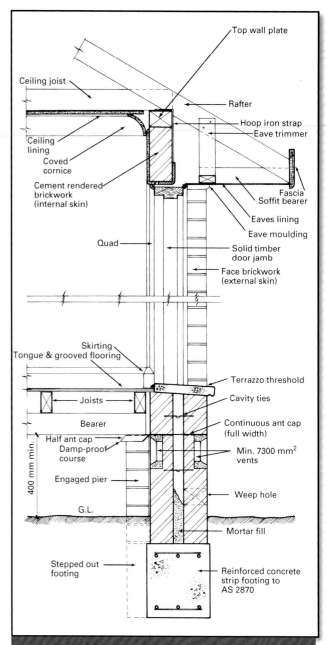

Figure 6.42 Vertical section through an external cavity brick wall and timber door jamb

is the load-bearing wall, and the outer skin is for weathering, security and visual effect. Cavity brick construction has the advantage of good thermal and sound insulation. Internal walls are constructed of one brick thickness, which can either be left as it is or have a 13 mm thick cement render applied over it and be painted or decorated.

Alternative construction methods

Current building trends allow for the conservation of timber resources, resulting in the use of metal as an alternative to timber in the manufacture of ground floor framing, wall framing and truss fabrication. Each system is prefabricated in panels or sections off-site, with final assembly carried out on-site. There is also a wide range of alternative lightweight cladding options, from fibre-cement sheeting systems to panels of styrene and autoclaved concrete systems.

Roof structures

Roof structure is the term given to the roof framing, eaves and roof covering.

Common types of roof structure are conventional roofing and trussed roofing.

CHAPTER 6 PLAN INTERPRETATION AND SPECIFICATIONS

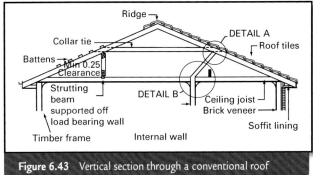

Figure 6.43 Vertical section through a conventional roof

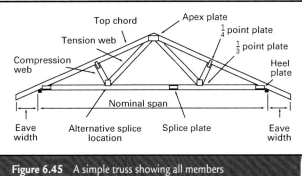

Figure 6.45 A simple truss showing all members

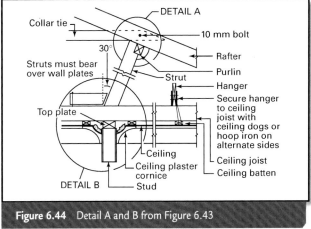

Figure 6.44 Detail A and B from Figure 6.43

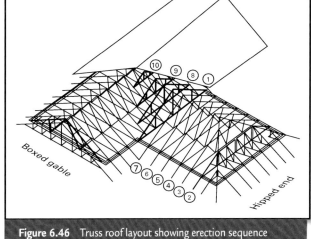

Figure 6.46 Truss roof layout showing erection sequence

Conventional roofing

In **conventional roofing** the timber roof framing is cut out and assembled (pitched) on-site. Structural framing members used in conventional roofing are rafters, ridge boards, hips, purlins, struts and collar ties (Figures 6.43 and 6.44). Ceiling framing, consisting of ceiling joists, hanging beams, strutting beams and trimmers, combines with and ties the roof framing to the walls. Sectional size and spacing of structural members are determined by the Timber Framing Code. The external finish of conventional roofing is provided by fascia boards, barge boards, gable cladding for gable roofs, and by eave or gable linings. Roof coverings, such as cement or clay tiles and corrugated metal sheeting, will influence the design and construction technique of conventional roofing.

Trussed roofing

In **trussed roofing** roof trusses are fabricated off-site, then transported to the site and lifted into position. This enables fast construction, the use of less material and the added advantage of internal design flexibility, as trussed roofs are supported on the external walls only. The structural design of a truss enables the top chord, bottom chord, struts and ties to support and distribute the roof and ceiling loads to the external walls. Trusses are designed according to the span between the external walls, the ceiling and roof loads, and the roof pitch. The roof coverings and external finish are the same as in conventional roofing. Trusses are fabricated from either timber or steel. (See Figures 6.45 and 6.46.)

Both conventional and trussed roof construction allow for freedom of design, with the floor plan of the building determining the final shape of the roof. Common shapes in either conventional or trussed roof buildings are:
- gable roof;
- hip roof;
- hip and valley roof;
- gambrel or Dutch gable;
- jerkin head.

PAINT, TILE AND PLASTER SUBSTRATES

Substrates are the materials or surfaces that another material is laid upon or adhered to. In the past, the main substrates for painters and tilers were plaster, paper-faced plasterboards, cement sheet products, timber, brick and steel. While these remain in use, there are shifts towards polymers (plastics), more metals (stainless steel and aluminium), and compressed fibre sheets such as MDF (medium density fibre board). In addition, timber products are coming pre-treated with light organic solvents, pre-primed or with some other form of preservative.

All of these materials have their own characteristics and require different techniques for ensuring a durable finish or bond. They also move differently and so have implications when choosing adhesives and/or coatings.

Being able to recognise a substrate is an important skill for all tradespeople. Knowing what that substrate is likely to do in heat or moisture is another. MDF skirting, for example, will shrink over its length, whereas timber moves very little in length, but can do so up to 5% in its width (10% if it's green hardwood).

Some doors now have plastic finishes on them, and they may need a special primer to hold a paint coat. Waterproofing coatings can add their own complexities in not always adhering well to plaster topcoat finishes (which should be avoided in areas to be tiled over); similar problems occur with silicones and other flexible sealants. Some sealants will require you to paint within 24 hours or less.

Substrates also need to be considered from the perspective of their fixing. Tiles, for example, should not be fixed to fibre cement sheets that have only been glued into place. These sheets need to be screwed at specific intervals.

Plaster sheets, while being a substrate to other finishes, are themselves fixed to other substrates such as timber and steel framing, or full masonry walls. Steel framing can be oily and therefore difficult to glue to unless cleaned with an organic solvent. Timber can shrink, causing nails and screws to 'pop' out of the face of the board after painting; and masonry may require battening to provide a suitable surface. Movement is a constant consideration that at times may be overcome only by carefully located control joints. Plastering over such joints in a long masonry wall is a quick way to a call back!

Knowledge of structure and materials is something that is gained over time, and will need to be constantly updated. The tradesperson is forever a student in this regard.

ENVIRONMENTAL CONTROLS

Site drainage

Surface and subsoil drainage of new or existing buildings is a very important aspect of building construction. Some of the problems caused by poor site drainage are as follows:

- Boggy ground—This causes problems for vehicular and pedestrian access around the site, especially for trucks delivering materials. Workers will also create a hazard by walking in mud and then dragging it through the building, creating slippery conditions.
- Erosion—Soil erosion may be caused by water not being drained away adequately, which may lead to the footings being undermined and surface soil being washed into gutters and stormwater pipes and, finally, into our river systems.
- Electrical hazard—Where electrical equipment and portable powered tools are being used, leads may contact wet surfaces, either outside or inside (where water is trapped on concrete slabs during construction).
- Material damage—This may occur where materials are stored either outside or on concrete slabs. If water is not drained away, materials may be damaged either by direct wetting or by moist air.
- Insect and fungal damage—If excess water is not drained away from buildings and is allowed to soak in, especially around floor areas, a suitable

environment for cockroaches, subterranean termites and various types of fungi will be created.

Removal of surface water

Surface water may collect due to run-off from nearby land, washing down areas on-site, or from rainfall. It may be removed by several methods:

- Sweeping/mopping—Concrete slabs in the open tend to hold a certain amount of surface water. This may be removed by simply using a yard broom or squeegee to push or drag it to the edge. Where frames are already in place, small holes or notches may be made to allow the water to escape through the bottom plate. If this is not possible, a mop and bucket can be used to soak up the water and cart it away.
- Filling and grading—To prevent water ponding on the surface, all holes and depressions can be filled with crushed sandstone, road base or crushed brick and rock. They should be well compacted to allow the water to drain away rather than soak in. All areas around the building should have a suitable grade away from external walls so that water will not lie close to the work area (see Figure 6.50).
- Surface improvement—Where soils don't allow water to run off or drain through, they may become very boggy. This usually occurs with clay soils; therefore a surface improvement such as laying road base or crushed sandstone will improve conditions underfoot for the duration of the job. This is particularly important in driveway access areas, to prevent vehicles from becoming bogged and churning up the surface.
- Surface drains/gutters—Where it is difficult to grade the surface away from the building, an alternative would be to create surface dish drains or gutters to divert the water to a collection point or sump. These drains or gutters should be only very shallow, as deep drains will create a safety hazard (see Figures 6.47 and 6.48).
- Subsoil drains—These can be either temporary or permanent, and are used where surface grading is difficult or as a water collection and diversion

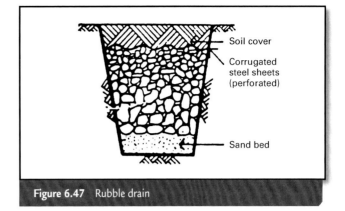

Figure 6.47 Rubble drain

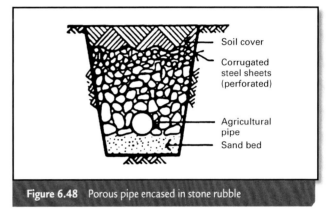

Figure 6.48 Porous pipe encased in stone rubble

line for surface grading. Where the water can't be directed to a stormwater line after collection in an agricultural line, it may be directed to an absorption trench or transpiration bed placed well away from the building. This allows the water to soak away through the subsoil via graded stone, broken bricks, tiles, concrete or gravels laid in the trench.

There are several types of agricultural pipes available, some of which are shown in Figure 6.49.

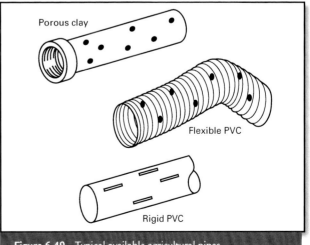

Figure 6.49 Typical available agricultural pipes

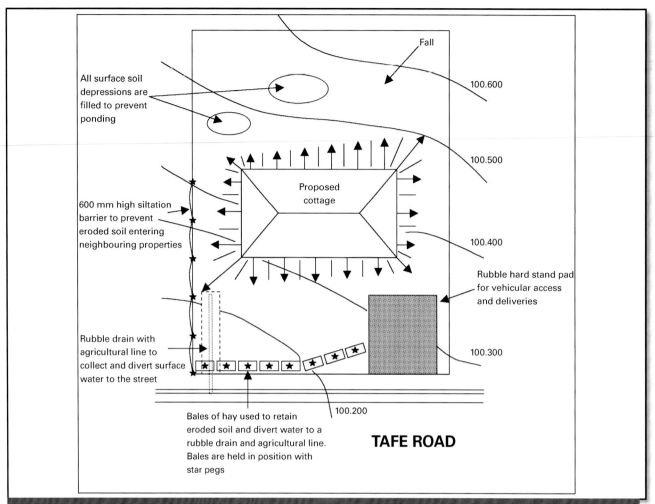

Figure 6.50 Building site describing surface water drainage and soil retention

SUGGESTED ACTIVITY one

Plan interpretation

1 Obtain a set of plans from the job you are currently working on now, or use a set from a previous job (ask your boss for a copy).

2 Identify the following details:

- The name shown in the title block to identify the owner or client.
- The scale of the site plan.
- The direction the front of the job faces and the street name it faces.
- Is the plan the original version or has it been amended?
- What access is provided from the roadway?
- List all the symbols and abbreviations, then identify the meaning of each.
- Do the plans state a reduced level (RL) measurement or a benchmark (BM) measurement? If so, what is it?

SUGGESTED ACTIVITY two

Formal drawing

Refer to the 'Practical drafting' section of *Building Materials and Hand Tools* produced by the Construction and Transport Division NSW.

Practice should be gained through scaled drawing exercises and suggested simple detailed drawings, such as:

- workshop practical projects, e.g. an oilstone case, timber-framing joints, carry-all toolbox, common joinery joints, laminated bread board, bench hook, shelving brackets, simple framed door;
- simple vertical sections;
- simple pictorial views;
- any other features.

Practice border and title-block preparation and set out, scaling, linework, lettering and interpretation of details to suit given performance criteria.

Worksheet 1

Student name: _____

Enrolment year: _____

Class code: _____

Competency name/Number: _____

To be completed by teachers:
Student competent ☐
Student not yet competent ☐

Task: Read through the sections beginning at *Key users of drawings* and up to *Specifications*, then complete the following.

Q. 1   What are drawings used for in the building industry?

_____

_____

_____

Q. 2   Name the key users of drawings:

1. _____

2. _____

3. _____

4. _____

5. _____

6. _____

7. _____

Q. 3   What are pictorial drawings, and why are they used?

_____

_____

Q. 4   Name two types of pictorial drawing:

1. _____

2. _____

Q. 5   What is an orthographic drawing?

_____

_____

_____

continued ▶

Q. 6 Name the five most common types of working drawings used in domestic construction:

1. _____
2. _____
3. _____
4. _____
5. _____

Q. 7 State the appropriate scales used to create working drawings in the building industry for the following views:

Site plans _____ Floor plans _____

Elevations _____ Sections _____

Q. 8 State the basic information found in a plan title block for a residential building:

_____
_____

Q. 9 What is the purpose of a specification?

_____
_____

Q. 10 How are the contents of a specification arranged to allow for easy use and reference?

_____
_____

Worksheet 2

Student name: _____

Enrolment year: _____

Class code: _____

Competency name/Number: _____

To be completed by teachers:
Student competent ☐
Student not yet competent ☐

Task: Read through the section on *Drawing symbols and abbreviations*, then, using the plan on p. 228, complete the following statements.

Q. 1   The name of the view shown on p. 228 is the _____

Q. 2   In the kitchen area, the abbreviation 'o' or 'w/o' identifies the position of the _____

Q. 3   The abbreviation 'DP' identifies the position and number of downpipes.

   The number of downpipes shown is _____

Q. 4   The garden tap is positioned outside the _____ (room)

Q. 5   The abbreviation 'HWS' stands for _____

Q. 6   The internal dimensions of the rumpus room are _____ × _____

Q. 7   The abbreviation 'm.h.' in the garage is to identify the position of the _____ in the garage _____

Q. 8   The abbreviation 'o/a' is referring to the _____ size of the entry door frame.

Q. 9   The provision of external ducting to the rangehood is located outside the _____ (room)

Q. 10 The width of the window to the dining room area is _____

continued ➤

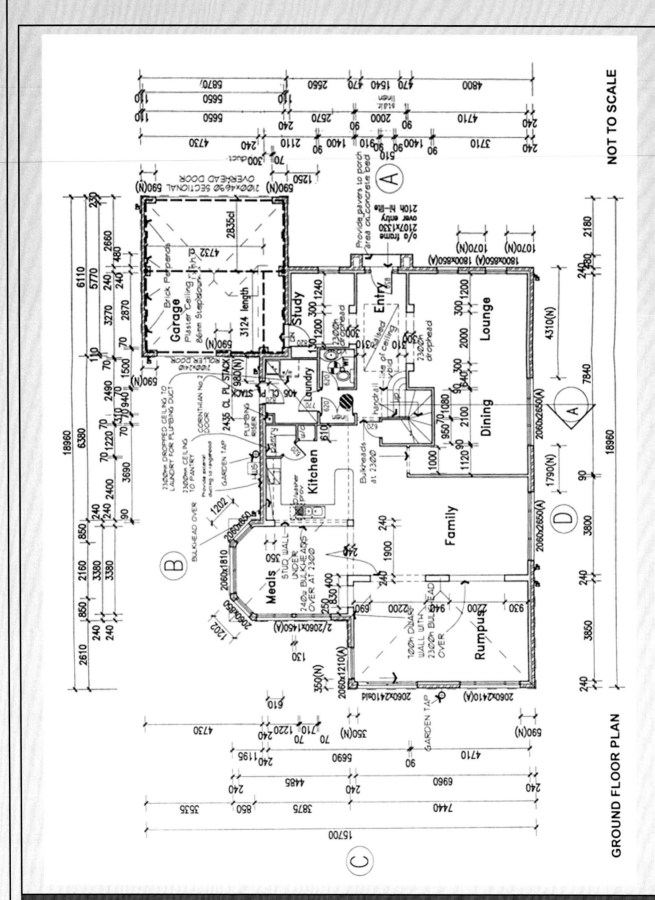

Figure 6.51 Plan of grand floor of building

Worksheet 3

Student name: _____

Enrolment year: _____

Class code: _____

Competency name/Number: _____

To be completed by teachers:

Student competent ☐

Student not yet competent ☐

Task: Read through the section on *Drawing symbols and abbrevations*, then, using the plan on p. 230, complete the following statements.

Q. 1   The name of the view shown on p. 230 is the _____

Q. 2   The boundary formed by survey pegs C and D is at the _____ side of the block (state orientation).

Q. 3   The abbreviation RL stands for _____

Q. 4   The council building line is set _____ m in from the street alignment.

Q. 5   The total area of the building block is _____ m².

Q. 6   The total number of trees to be removed is _____

Q. 7   The length of the west side boundary is _____ m.

Q. 8   The reduced level at the datum position is _____ m.

Q. 9   The approximate fall along the east side boundary is _____ m.

Q. 10 There are _____ (number) new trees to be planted on the block.

continued ➤

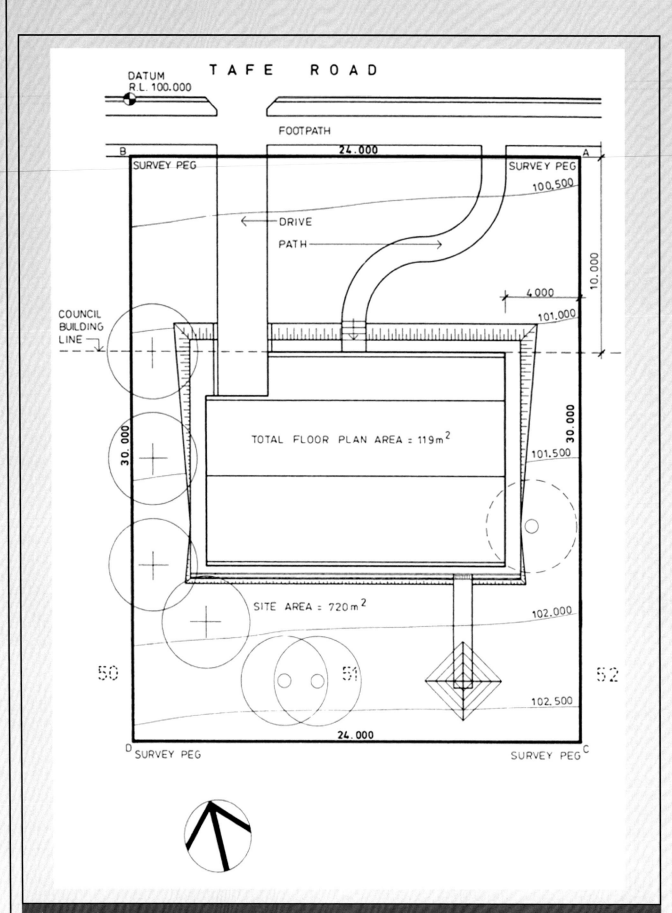

Figure 6.52 Plan

Worksheet 4

Student name: _____

Enrolment year: _____

Class code: _____

Competency name/Number: _____

To be completed by teachers:	
Student competent	☐
Student not yet competent	☐

Task: Read through the section beginning at *Issues for painters, tilers and plasterers* up to the end of *Residential building structure types*, then complete the following.

Q. 1 List the documents painters, tilers and plasterers should study to find the information they need.
_____
_____

Q. 2 What is a fold out?
_____
_____
_____

Q. 3 List and describe the steps in preparing a fold out.

1. _____
_____

2. _____
_____

3. _____
_____

Q. 4 Define the difference between the footing and the foundation.
_____
_____

Q. 5 Describe three common types of sub-floor construction systems.

1. _____ 2. _____
3. _____

Q. 6 Describe three common wall structures.

1. _____ 2. _____
3. _____

Q. 7 Why is it important for painters and plasterers to recognise and know about different wall structures?
_____
_____

Worksheet 5

Student name: _____

Enrolment year: _____

Class code: _____

Competency name/Number: _____

To be completed by teachers:	
Student competent	☐
Student not yet competent	☐

Task: Read through the sections beginning at *Paint, tile and plaster substrates* up to *Environmental controls* then complete the following.

Q. 1 What are substrates?

_____
_____

Q. 2 List all the types of substrates you have learned about.

_____
_____

Q. 3 Why is it important to be able to recognise a substrate?

_____
_____

Q. 4 List issues which can arise with MDF skirting, doors with plastic finishes or waterproof coatings, fibre cement sheets and plaster sheets.

_____
_____
_____
_____
_____
_____

Q. 5 List five problems which can be caused by poor site drainage.

1. _____ 2. _____
3. _____ 4. _____
5. _____

Q. 6 List five methods which can be used for removing surface water.

1. _____ 2. _____
3. _____ 4. _____
5. _____

REFERENCES AND FURTHER READING

Acknowledgment

Reproduction of the following *Resource List* references from *DET, TAFE NSW C&T Division (Karl Dunkel, Program Manager, Housing and Furniture) and the Product Advisory Committee* is acknowledged and appreciated.

Texts

Australian Vocational Training Scheme (1996), *Student notes Module 7974C—Introduction to plan reading and maintenance*, Australian Vocational Training Scheme

Liebing, R.W. (1990), *Architectural working drawings*, Wiley, New York

Major, S.P. (1995), *Architectural woodwork: Details for construction*, Van Nostrand Reinhold, New York

Noll, T. (1997), *The encyclopedia of joints and jointmaking*, RD Press, Sydney

Styles, K. (1986), *Working drawings handbook* (3rd edn), Architectural Press, UK

Web-based resources

Regulations/Codes/Laws

<www.epa.nsw.gov.au> NSW Environment Protection Authority

<www.standards.org.au> Standards Australia

Resource tools and VET links

<www.resourcegenerator.gov.au> ANTA resource generator

<www.ntis.gov.au> National Training Information Service

<www.hsc.csu.edu.au/construction> NSW HSC Online—Construction

Industry organisations sites

<www.aiqs.com.au> Australian Institute of Quantity Surveyors—Entry

Audiovisual resources

TAFE NSW, 2000, *Dress Timber*, Construction and Transport Division, Sydney

TAFE NSW, 2000, *Mortice and Tenon Joints*, Construction and Transport Division, Sydney

TAFE NSW, 2000, *Setting Out a Building Site*, Construction and Transport Division, Sydney

Australian Standards

AS 1100.301-1985 (plus supplement), 'Technical drawing—Architectural drawing', Standards Australia, Sydney

AS 1478.1-2000, 'Chemical admixtures for concrete, mortar and grout—Admixtures for concrete', Standards Australia, Sydney

AS 1684.2-1999, 'Residential timber-framed construction—Non-cyclonic areas', Standards Australia, Sydney

CHAPTER 7
LEVELLING PROCEDURES

Levelling, and the skilful use of levelling equipment, is a critical aspect of virtually any trade. Many people can 'see' if something is out of level or plumb without the use of elaborate levelling devices, and as often as not these people are your clients. So be warned: if you don't see it while you are on the job, then it is going to cost you time and money when your client does—after the work is done!

In addition, if something is out of level, aside from detracting from the overall finish, it may inhibit another trade from completing their work in a timely or appropriate manner. For example, a tiler's task is made unnecessarily complicated if a spa bath and its surrounding frame are not level and plumb. Likewise, bricklayers all too frequently have to 'gain' 10 or 20 millimetres because a concrete slab runs out of level.

The purpose of this chapter, therefore, is to explore the tools, skills and knowledge used in a variety of basic levelling **procedures** relevant to the 'wet' trades. Areas addressed from the unit of competency include:
- planning and preparing for levelling work;
- setting up and using levelling devices; and
- cleaning and maintaining levelling equipment.

Note: As this chapter covers a range of trades, some terms and/or procedures will not be relevant to everyone.

INTRODUCTION TO LEVELLING PROCEDURES

Some key terms and concepts

It has been said that a tradesperson may do little wrong if his or her work is square, straight, level and plumb. To this may be added 'wind' (pronounced as in 'wind up a rope') 'true' and perpendicular. The definition of these basic terms for construction and levelling is as follows:

- *Square*—The term **square** has at least two meanings:
  - at 90° to another surface or line as shown in Figure 7.1 (the word 'perpendicular', defined below, can also be used in this manner);
  - a rectangular or square shape (room, box, wall etc.) that has all its sides meeting at 90°. This is often checked by a two-step approach: first the sides are checked for parallel, then the diagonals are checked for being equal (see Figure 7.2).
- *Straight*—This can also have two meanings:
  - its common meaning: a straight line, or in a straight line;

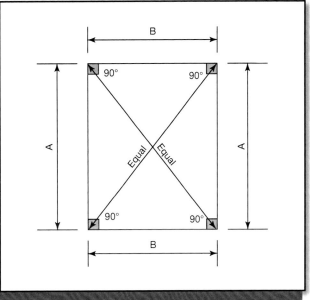

Figure 7.2 Checking for 'square'

  - that everything looks plumb and level in relation to everything else around it. For example, the tiles on one wall 'look' flat, level and 'plumb' (see below) in relation to a corner, door opening or some other vertical surface. In this context, 'straight' is usually used in conjunction with 'true' (see below), as in 'straight and true' (see Figure 7.3).

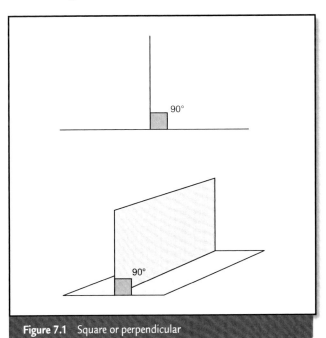

Figure 7.1 Square or perpendicular

Figure 7.3 A 'straight and true' corner

- *Level*—**Horizontal** (think of the horizon).
- *Plumb*—Vertical, straight up and down. The term **plumb** is derived from the tool known as a 'plumb' or 'plumb bob' (which in turn is derived from the Latin term for lead; see Figure 7.4). The word 'perpendicular' can also be used to mean plumb.

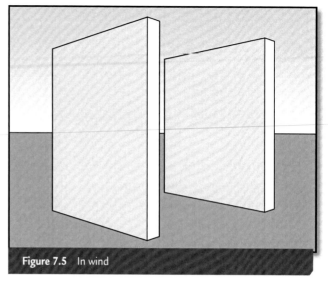

Figure 7.5 In wind

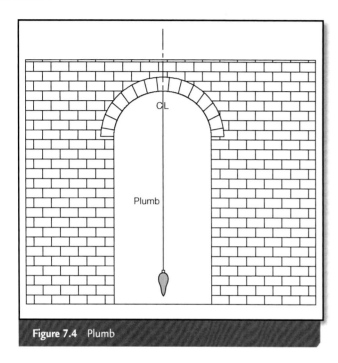

Figure 7.4 Plumb

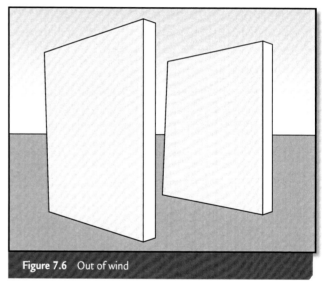

Figure 7.6 Out of wind

- *Wind* (sometimes spelt 'wynd', which is a Scottish Gaelic word for a winding narrow road or path)— 'Out of **wind**' means that that two edges of a surface, or two lines, are not in the same plane as each other. The brick walls on either side of a door opening, for example, when viewed as in Figure 7.5, should appear parallel if 'in wind'. They will appear 'out of wind' if one wall is just slightly out of plumb or vertical compared to the other (Figure 7.6).
- *True*—A broad term often used to encompass several of the terms above. A **true** line, for example, is one that follows the path that it should, without bumps or hollows. This line might be straight, or it might be forming a curve. A true wall of bricks would be one that is plumb, and on which all lines (joints, as well as the lay of each brick and the wall itself) follow as they should. Tiles that have been laid true would likewise offer a smooth finish, have even joints without lifted edges, and provide a surface that does not undulate (go up and down everywhere). The term is often used in conjunction with 'straight', as in 'straight and true' (see Figure 7.3).
- *Perpendicular*—Like 'square', **perpendicular** has a couple of key meanings:
    - vertical (see Figure 7.1)—it can be used as an alternative to 'plumb' in this sense;
    - at 90° to another surface or line (see Figure 7.2)—it can be used as an alternative to 'square'. Be wary of this usage, however, as it can lead to misunderstandings.

CHAPTER 7 LEVELLING PROCEDURES

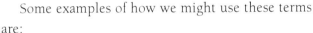
Figure 7.7 Checking tiles with entrance corner for wind

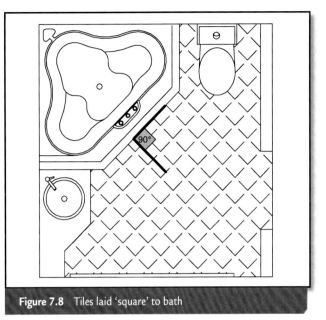
Figure 7.8 Tiles laid 'square' to bath

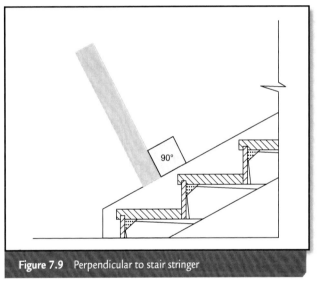
Figure 7.9 Perpendicular to stair stringer

Some examples of how we might use these terms are:
- 'Make sure the wall's plumb and true before you plaster it' (i.e. that the wall is vertical and straight in both directions, or, if a curved wall, that the curve has no flat spots in it).
- 'Check that tile inlay for wind with the entrance' (i.e. make sure that the tiles line up vertically with corner of the entrance (see Figure 7.7).
- 'Make sure those tiles run true to the wall' (i.e. that the tiles are running parallel with the wall).
- 'Run a level line a metre off the floor and paint in dark green up to that'.
- 'Run your floor tiles square to the bath' (i.e. lay the tiles so that the joints are at 90° to the bath wall or hob—see Figure 7.8).
- 'Paint a 100 mm wide line a metre long and perpendicular to the stair stringer' (meaning that a 100 mm wide line is to be painted at 90° to the stair member that runs down the wall—see Figure 7.9). As 'perpendicular' is more commonly used to mean 'vertical', this is the sort of usage that, while technically correct, can lead to mistakes.

A brief introductory word on safety

In general, levelling procedures are not particularly dangerous activities. However, it involves tools that contain lasers, and these come with the inherent, albeit low, risk of eye damage. This is particularly so because we are using our eyes frequently in levelling to line things up and make sure they are running true or parallel, for example. In doing so we can easily misjudge the location of the laser beam, particularly when using lasers for tasks such as levelling suspended ceilings.

This same misjudgment can lead to other risks. These would be comical if not for their consequences. Because you will be focused on looking intently at things, moving back and forward to get a 'good look' as it were, it is very easy to lose track of your surroundings. This means falling into, over, off or onto things, or getting hit by others or other equipment. It is therefore important to keep careful track of your surroundings, the position of laser beams, and those working with you.

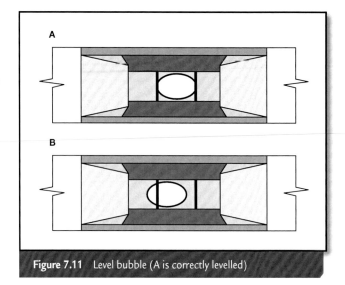

Figure 7.11 Level bubble (A is correctly levelled)

THE TOOL LIST

The following is a list of the most common (or becoming common) tools for basic levelling. Included is a short description of each item and its use, which will be expanded upon when we look at specific levelling procedures in more detail.

- *Spirit level*—These come in various lengths and are used for checking horizontal and vertical surfaces (Figure 7.10). They usually have two or more vials that are partially filled with a coloured fluid. Due to the shape of the vial, the bubble of air that remains is used to indicated level or plumb (see Figure 7.11). The origins of this tool date back to the mid 1600s, though the contemporary tool we use today came into being around 1920. These early levels generally had timber bodies and the position of the vial was adjustable. Most contemporary levels are sealed units and cannot be adjusted. It is important, therefore, to purchase a good-quality level and look after it well (don't use it as a hammer, and don't hammer it!).

- *Digital level*—This is an electronic version of the spirit level (Figure 7.12). The best of these can provide a highly accurate reading, presented either as a percentage of the fall or gradient, or as an angle in degrees. The advantage is that a digital level can be used to determine the angle of an existing surface, or to mark a required angle on a vertical wall face. In addition it can offer an auditory signal (beep) when level (or at a specific angle). The disadvantage is that digital levels can be too specific, as well as being slower to read and more sensitive to shock.

- *Line level*—This is a very small spirit level with hooks for hanging onto a string line (Figure 7.13). To have any chance of accuracy the level must be hung at the mid-span of the string. Even then, line levels are not a particularly accurate tool and are not as common on building sites as they once were. They are sometimes used for setting up footings and suspended ceilings, but only rarely nowadays unless the job is not critical and there is nothing better on hand.

Figure 7.10 Straight edge (with spirit level)

Figure 7.12 Digital level

- *String lines*—Contemporary string lines or 'brickies' line' (bricklayers' line) are made of plaited nylon and are brightly coloured for visibility (see Figures 7.13 and 7.14). These lines are strong, with a good degree of stretch, and are used for a variety of purposes, such as setting out building and boundary lines, brick courses, suspended ceiling set-out—indeed, setting out or lining up just about anything you can think of.

Figure 7.14 String lines

- *Straight edge*—Available in various lengths, this is a long timber, metal or plastic tool with parallel sides. Lightweight, it is used to aid in drawing long, straight lines (like an oversized ruler) and/or to increase the 'reach' of a spirit level (see Figure 7.10).

- *Chalk line*—This is a reel of string line in a container that holds powdered chalk (see Figure 7.15). This chalk is brightly coloured (generally blue or red) for visibility. When held tightly between two points and 'pinged' (pulled up and let go sharply in the middle), a neat line drawn in chalk is left on the surface. Chalk lines are used to mark slabs for tile positions, screws for plaster fixing in ceiling—just about anything that needs a temporary straight line over a long distance. Note, however, that the red chalk often has a mineral oxide (ochre) base, which is very hard to remove or paint over. Therefore, use blue chalk for anything that may be painted at a later date.

  Caution: The chalk is carcinogenic, and at the very least an irritant to the airways. Don't spread the chalk powder around needlessly or breathe it in.

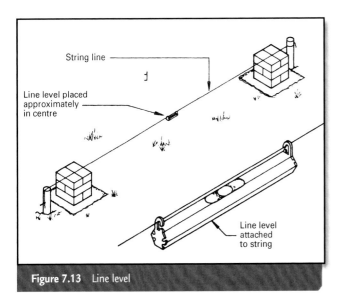

Figure 7.13 Line level

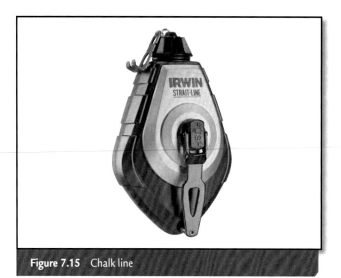

Figure 7.15 Chalk line

- *Plumb bob*—These come in many shapes and sizes and are really nothing more than a symmetrical weight on a string (symmetrical so that they hang straight; see Figure 7.4). When this weight is hanging uninterrupted, the string provides a highly accurate guide to plumb or vertical. Plumb bobs can be held 'free hand', in a frame (Figure 7.16) or from fixed point of a ceiling or wall. While a very old and basic tool (Egyptians are known to have used them as far back as 2600 BC) the humble plumb bob remains a useful piece of kit for any tradesperson.
- *Tripod*—Like plumb bobs, tripods have been around for quite some time. Today's tripods are generally aluminium, although timber units are still available. Indeed, for very sensitive instruments timber is preferred as it does not vibrate like metal and therefore settles down to a stable platform more quickly. Tripods for levelling instruments have adjustable legs with sharp points to allow them to be pushed firmly into position on the ground. The main distinction between contemporary tripods is the head. There are two basic types available: flat or domed (see Figures 7.17 and 7.18). Generally speaking, the flat head is the preferred platform for optical levels, and domed for self-levelling laser units.
- *Tape measures and rulers*—These are the all-important tools for measuring or assessing distances. The four-fold rule (Figure 7.19) unfolds

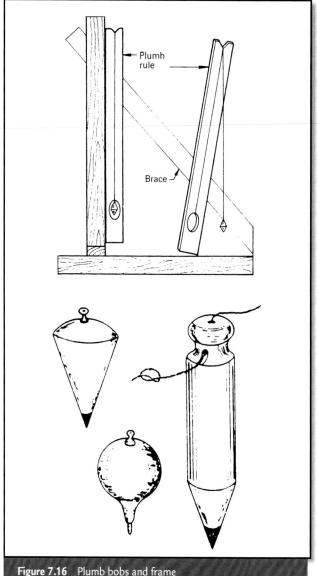

Figure 7.16 Plumb bobs and frame

Figure 7.17 Flat-headed tripod

Figure 7.18 Domed tripod

to give a full metre of measurement and should be used for the assessment of dimensions smaller than a metre. Retractable tape measures are used for longer distances and come as small belt clip units of up to 10 m in length (Figure 7.20). Longer tapes, generally 30 m to 100 m in length, are much larger and are wound up by hand (Figure 7.21). All of these tools are marked in increments of metres and millimetres.

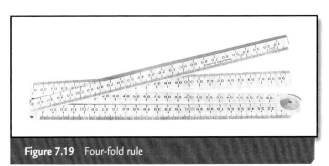

Figure 7.19 Four-fold rule

Figure 7.20 Retractable tape measure—8 m

Figure 7.21 Wind-up tape measure—60 m

- *Laser levels*—More properly called rotary laser levels, these levels cast a level line that rotates around the instrument's centre. They are used extensively in construction, including by bricklayers for footings and by plasterers for ceilings. Laser levels cast a invisible beam that is read by means of a small receiver. There are three types of rotary laser levels, of which only two are suitable for professional trade craft:
  - Manually levelled lasers—Originally the cheaper end of the market, manually levelled lasers rely upon the operator to set them up correctly using an inbuilt spirit level (Figure 7.22). Manual levelling is, however, also a feature of some of the contemporary high-end self-levelling units using remotely controlled servo motors. Manual units rely upon the operator for accuracy, and if they are moved without your being aware of it, you may find out only later that you are in error. Their advantage is that they can be set up at angles other than level or plumb (i.e. laid on their

side). At the time of writing there are very few purely manual units being marketed.
- Pendulum automatic level lasers (Figure 7.23)—Initially set up with the aid of an inbuilt bubble level, these level lasers use a pendulum system that ensures that the laser only casts a level beam or shuts off. This system compensates for up to about 3° of error. It also shuts off if knocked out of level while in operation. Automatic level lasers are reliable and relatively inexpensive to maintain or service. They are currently the mainstream tools of the domestic construction industry.
- Self-levelling lasers (Figure 7.24)—These are the higher end of laser levelling units. They use a set of servo motors to self-align to a level plane. They are a 'set and forget' sort of tool that you simply turn on and leave them to do the rest. They are, however, more expensive to service or recalibrate. The best of these units are used by professional surveyors for large

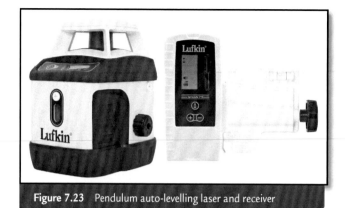

Figure 7.23 Pendulum auto-levelling laser and receiver

Figure 7.22 Manual laser

Figure 7.24 Self-levelling laser

building works, engineering, mining and the like. However, they are coming down in price and are now becoming more common in the domestic construction industry.
- *Laser line generators and dot lasers*—These tools are very different from rotary laser levels, and are used to cast a level or plumb line onto a surface or onto several surfaces at one time. They are available in various forms, from task-specific laser **squares** (projecting two fine beams at 90° to each other from

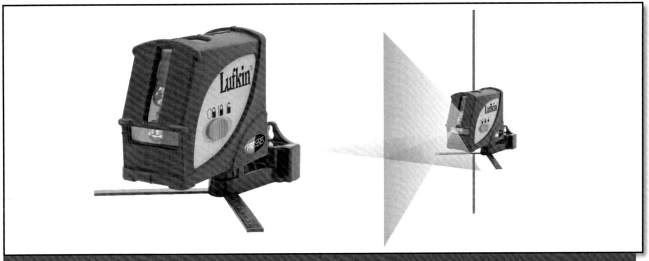

Figure 7.25 A multi-line laser line generator (laser array shown at right)

a corner) to multi-line generators that produce a wall of light to measure from (Figure 7.25). Unlike most rotary lasers, line generators produce a visible beam of light. They are powerful lasers that pose a greater risk to the eyes and skin but are a very quick and efficient tool for producing level or plumb lines, or dots on the ceiling and floor that are literally plumb in line with each other.

- *Optical automatic levels*—Frequently called 'dumpy levels', these are a self-levelling telescope with crosshairs for sighting (Figure 7.26). Very early dumpy levels were simple tubes without any lenses, and with a tribrach that actually had four, not three, levelling nuts. Much depended upon the skill of the user in setting up the instrument and in determining that the target was centred before taking a reading. Today, almost all optical levels purchased have good-quality optics (lenses) and are auto-levelled by means of a pendulum and prism system. These tools are very accurate when used and looked after correctly. Like any level, they must be constantly checked for accuracy prior to use.
- *Survey staff*—This is a very large, collapsible ruler for use with either laser or optical levels (Figures 7.26 and 7.27). When held correctly (plumb), the staff may be read by sighting through the optical level, or by positioning a laser receiver. Being able to read a staff accurately is an important part of many basic levelling procedures.
- *Water level*—This is little more than a clear flexible plastic hose or tube filled with water that can be used to transfer heights with great accuracy (Figure 7.28). The hose, when filled with water, must be free of all air bubbles and have about 300 mm of water-free length at either end. These ends may be made of glass tube to make reading easier; however, this is not essential as the tube itself must be of clear material to ensure that all air has been removed.

Figure 7.26 An automatic optical level or 'dumpy' (note staff in background)

BASIC LEVELLING PROCEDURES

Levelling and plumbing lines and surfaces

To draw a level or horizontal line we generally use a spirit level of the appropriate length. Usually this means the biggest one that will fit the area we are working on. And so we come to Rule One of levelling: never trust a level—always check it, or work with it so as to reduce the influence of error on your job.

Checking a spirit level for error

See Figures 7.29 to 7.31.
1 Place the level on a wall (vertically or horizontally, depending upon which bubble you are checking) and centre the bubble as accurately as possible.
2 Draw a line along its length.

Figure 7.27 Survey staff

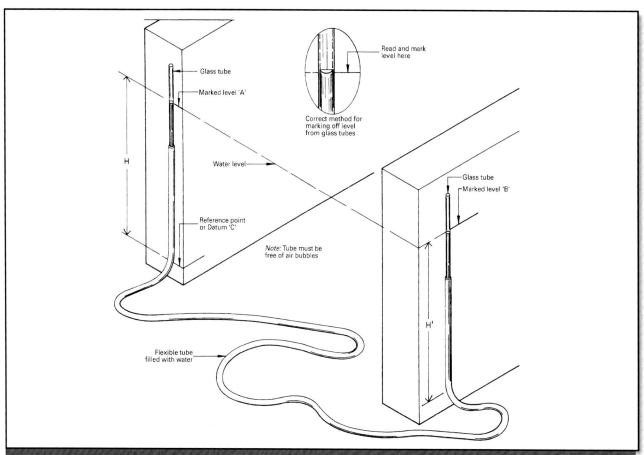

Figure 7.28 The water level in use

3 Turn the level end-for-end and place it back on the line.
4 Check the bubble and see if it is still exactly in the centre. If it is not then the level is 'out' (not giving a true reading of level).

The following procedure may also be used on any moderately level or plumb surface without drawing a line. It is how you check your level while it is in use:
1 Place the level on the surface.
2 Note the position of the bubble.
3 End-for-end the level and reposition it in the same place.
4 Check that the bubble is in the exactly the same position. If it is, the tool is accurate; if not, it is out.

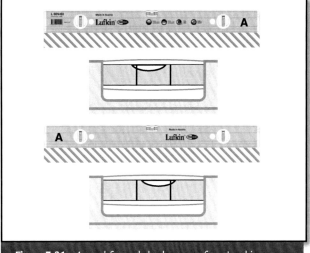

Figure 7.31 An end-for-end check on a surface: Level is out, surface is level

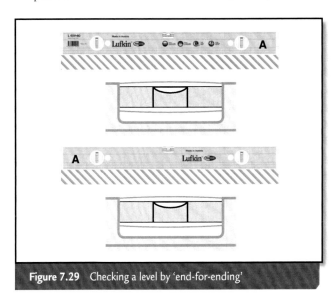

Figure 7.29 Checking a level by 'end-for-ending'

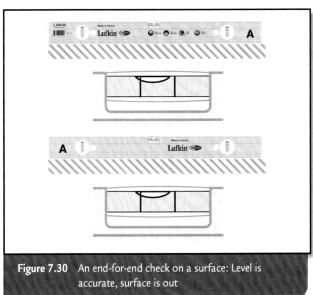

Figure 7.30 An end-for-end check on a surface: Level is accurate, surface is out

Marking the line

If the line has to go for longer than one length of the level, use the following procedure (see Figure 7.32):
1 Place the level at the required height on the surface or wall to be marked and centre the bubble exactly.
2 Using the level like a ruler, draw a line on the wall the full length of the level.
3 End-for-end the level (as if you were checking it for error).
4 Place the level on the surface or wall again at the end of the previous line and centre the bubble again. Draw the continuation of the line.
5 Repeat this 'end-for-ending' of the level until you have the line required.

In this way any small error in the level, or any tendency that you may have of reading the bubble just a little left (or right) of centre, will be reduced.
In the first direction it will go down (or up) and in the second it will go back up (or down) again, thereby eliminating the error.

Extending the reach of a level

End-for-ending, as described above, is acceptable up to a point. Wherever possible, though, it is better to use a straight edge to extend the length or reach of a level (see Figure 7.10 earlier). Even straight edges have their limits, however, so should you need to use

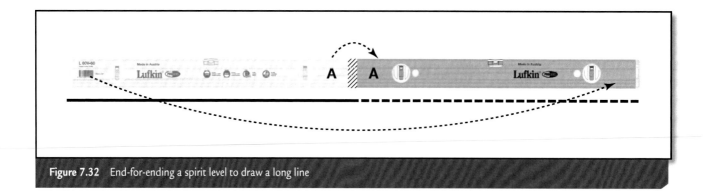

Figure 7.32 End-for-ending a spirit level to draw a long line

them over two or more lengths, then end-for-ending is the best way to reduce error.

Checking a surface for level

End-for-ending can also be used to check whether a surface is level. Place the level on the surface and check that the bubble is in the centre. Note carefully its position, then end-for-end the level and check the bubble position again. If it is exactly in the centre again (or if the level is slightly out, in the exact *opposite* position—see Figure 7.31) then the surface is level.

Plumbing a line

Using a spirit level

To check whether a line is plumb using a spirit level is much the same as levelling a line, as in the examples given above, only this time the level is held vertically. Again, it is good practice to end-for-end the level in order to check for any error in the level itself.

Using the plumb bob

Many plumbing procedures can be carried out using a plumb bob instead of a spirit level. The level is generally used because it is easier to hold still in most cases (e.g. against a wall). However if, for example, you wish to know that the centre of a plaster arch is in fact central to the hallway floor, then the plumb bob is a quick and sure way to check (see Figure 7.33).
1 Mark the centre line on the floor.
2 Hold the string at the apex of the arch and allow the plumb to swing freely, close to the floor.
3 When the plumb settles down, check that its position is over the line. If it is not, then you need to adjust your arch to suit.

Should you need to use the plumb bob outside in a light breeze, you can reduce its movement by having the weight sit in a bucket of water.

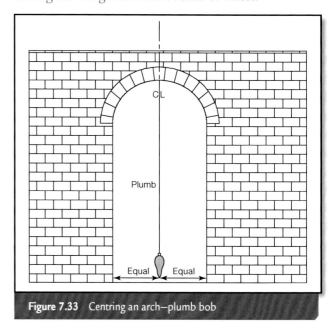

Figure 7.33 Centring an arch—plumb bob

Checking a surface for plumb

Using a spirit level

To check whether a surface is plumb using a spirit level is again very much like checking for level, as above, except that it is on a vertical surface instead. Again, end-for-end the level just to be sure.

Using a plumb bob

Figure 7.34 shows how this is done:
1 Hang the plumb bob a small distance away from the wall.

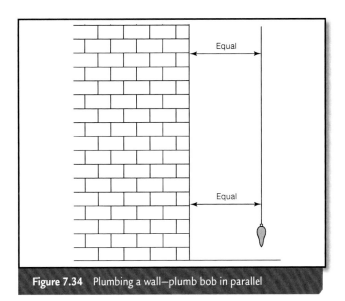
Figure 7.34 Plumbing a wall—plumb bob in parallel

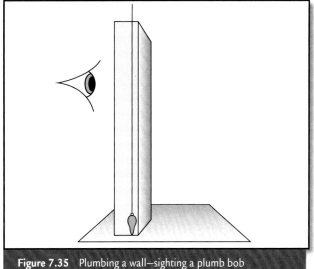
Figure 7.35 Plumbing a wall—sighting a plumb bob

2. Measure the distance between the wall and the string at both top and bottom.
3. If the measurements are the same, then the wall is plumb.

Alternatively, position the plumb bob so that the string and the wall are in line with each other and then sight to see if they are parallel (see Figure 7.35).

Finding the difference in height between two points

Finding the difference in height between two points is a basic form of 'traversing' whereby, using a spirit level, a series of level lines are used to 'step' up and or down from a known point to an unknown point. As demonstrated in Figure 7.36:

1. Locate pegs or marks at the furthest reach of the level, or level and straight edge combination.
2. Level out by using a rod, ruler or tape measure held vertically over each peg, and record the height difference between each point.
3. The sum of these heights is the difference in height between the first and last points.

This same approach can be used by marking along a wall instead of putting pegs in the ground.

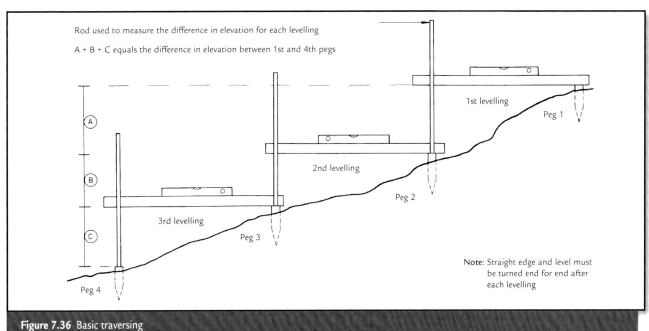
Figure 7.36 Basic traversing

Transferring a height

Using a spirit level
This is the same as marking a line, except that you don't draw the line. Likewise, it may or may not include a straight edge, depending upon the distance or location of the point you need to transfer to.

1 Identify the height you need to transfer.
2 Select the most appropriate (longest) spirit level available. If transferring across an empty space (a tile or paint line across a hall opening, or from one brick pier to another), you may need to use a straight edge. This way you can ensure that the level, or the straight edge, can be held firmly at either end.
3 Place one end of the level or straight edge on the height to be transferred and place a mark at the other end where required (see Figure 7.37).

Using a water level
Although water levels are seldom used today, there are still some instances in which they may come into their own. As shown in Figure 7.28 earlier, a water level can be used around walls and inside rooms when other levels would need the position to be constantly changed, possibly leading to error. This can be a good way to transfer the dado height for wall tiles or paint lines, for example.

LASER LINE GENERATORS

Lasers are fast becoming the tool of choice for many tradespeople because of their ease of use and one-person operation. In addition, they can be used to 'mark' a level or plumb line on a wall, or from ceiling to floor, without actually drawing anything on the wall at all. Many of the tasks dealt with above can be done using one of these instruments. Figures 7.38-7.40 show a simple and relatively

Figure 7.38 Laser line generator: Horizontal line at dado height for wall papering

Figure 7.37 Transferring heights with a spirit level and straight edge

inexpensive self-levelling laser line generator in action. Figure 7.38 shows a level line, while Figure 7.39 shows the crosshair for tiling set-out. Finally, Figure 7.40 shows the dot laser producing plumb points on a bulkhead and floor to help position an island bench.

Figure 7.39 Laser line generator: Crosshair mode

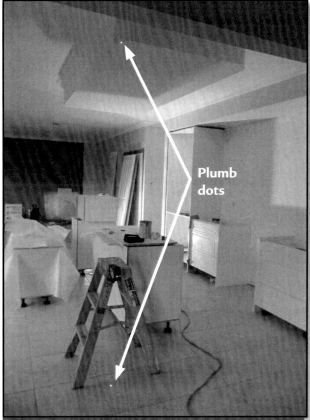

Figure 7.40 Plumbing a ceiling point to the floor (unit is magnetically attached to ladder)

The advantages and disadvantages of lasers

The advantages of lasers are that they:
- are quick to set up;
- are relatively inexpensive;
- are small;
- are light;
- require only one person to operate;
- can cast lines over long distances;
- are easy to see in dim light (such as when tiling in bathrooms).

The disadvantages of lasers are that they:
- are more easily knocked out of calibration than a spirit level (so they no longer produce a 'level' line);
- are harder to check for accuracy (which must be done frequently);
- have the potential to damage eyes and skin;
- are more dangerous to eyes in dim light (the pupil is more open in dim conditions, in order to let more light in, and so takes longer to close).

Safe use of laser line generators

Classes of lasers

Laser line generators generally use Class 2M or Class 3R lasers. Class 3B and Class 4 lasers are not permitted in the building industry. The higher the number, the greater the strength and visibility of the beam, the accuracy and the distance. However, the higher the number, the greater the risk too to the eyes and skin. Small rotary laser levels are often Class 1, Class 1M or Class 2 lasers, as these do not produce a visible beam.

All laser equipment must be labelled with a warning stating the class of laser (see Figure 7.41).

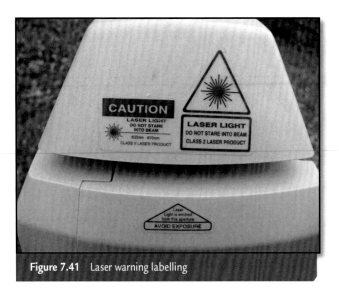

Figure 7.41 Laser warning labelling

PPE

For Classes 1, 1M and 2 lasers, no special PPE is required. For Class 2M and above, however, the operator should wear specially tinted glasses of the same colour as the laser beam itself—that is, red-tinted glasses for red beams, and green-tinted for green beams. These are generally provided by the manufacturer, but may need to be purchased separately.

Australian Standards

The manufacture and use of lasers comes under two Australian Standards:
- AS/NZS 2211.1:2004, 'Safety of laser products';
- AS 2397-1993, 'Safe use of lasers in the building and construction industry'.

These standards require the following with regards to each class:
- Class 2M lasers are low-powered and present a minor hazard. The normal blink and avoidance response to bright light is sufficient protection. Some administrative controls should be in place regarding training and use.
- Class 3R lasers produce a stronger light and require more stringent administrative precautions. It is particularly important to ensure that optical instruments (such as a dumpy level or theodolite) are not used in the immediate area as these could concentrate the beam into the eye.

Note: There is some confusion in laser classifications between the two standards. AS 2397-1993, the older but still operational standard, offers Classes 1, 2, 3A and 3B (Restricted) as being relevant to the building industry (followed by 3B and 4). AS/NZS 2211.1:2004 has the newer list of classes: 1, 1M, 2, 2M, 3R, 3B and 4. This chapter is based upon the new classifications—that is, Classes 3A and 3B (Restricted) have been replaced with 3R.

The standards also require that *all* operators of Class 2M and Class 3R lasers be trained in their use and have access to all operational manuals, and that a copy of AS 2397-1993 be kept on-site.

The operator (you) must ensure that the tool is used in accordance with the instructions in these manuals and the guidelines in the standards.

Because Class 2 and stronger lasers have the capacity to cast a line beyond the immediate area of use, a warning sign (Figure 7.42) must be in place at all points of access.

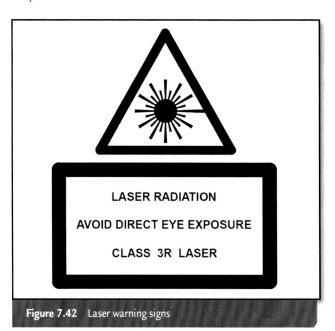

Figure 7.42 Laser warning signs

Laser safety officers

AS 2397-1993 requires that a laser safety officer be in place at all sites where Class 2 lasers or stronger are used. It is this person's responsibility to ensure that all tools are correctly maintained and that the beams of the more powerful line generator types are confined to the site of use (and not allowed to shoot across the street, for example). In small, one- or two-person businesses (e.g. tradesperson/employer and apprentice), this responsibility lands on the employer.

Lasers and the wet trades

It is important for workers in these fields to become very knowledgeable very quickly about laser levelling tools, as these are fast becoming the norm rather than the exception on-site. Tilers, for example, can now set up a level line accurately around a room with the flick of a switch, as can painters. Plasterers and bricklayers can check the centre of an arch with a laser plumb line generator instead of trying to find a way of pinning a string and weight to the ceiling, and almost all plasterers now use lasers for levelling suspended ceilings.

It is equally important, however, for tradespeople not to lose their 'eye' for true lines, be they plumb, level, straight or curved. And, as with spirit levels, don't simply rely on lasers being right: check them constantly (see two- and four-peg tests on page 269).

OTHER LEVELLING PROCEDURES

Levelling procedures that are considered 'basic' in the construction industry cover much more than the use of spirit levels, plumb bobs and laser line generators. Other types of procedures that you may be required to do using laser or optical levels are:
- Maintaining, marking or establishing a continuous height around various parts of a building or site.
- Determining the difference in height between one point and another.
- Determining the relative heights of various points using a datum.

In each of these cases, and the many variations of them that are possible, you may be required to use either an optical or rotary laser instrument. Before we can go further, however, there are some terms or words you need to become familiar with:

- *Datum*—The **datum** is the point to which all other heights are referenced. It is a beginning point. The datum may refer back to a mean sea level or Geoid height, the new Ellipsoid height (the Geocentric Datum of Australia or GDA) used by global positioning systems (GPS). On other occasions the datum may be a temporary mark made at a nominal height, usually 100 m (see Figure 7.43).

  The purpose of a datum is to show such things as the difference in height between the finished floor level of a building and the likely flood level, so that councils can ensure buildings are located appropriately. Alternatively the datum may simply provide a point of reference to ensure that differing floor levels, window heights, tile heights, ceiling heights, and so on, remain consistent.

  On a building site the datum is generally an easily identifiable peg or mark of a known or nominated height from which the tradesperson may work. This peg or mark is identified on a set of building plans as a TBM (temporary benchmark), BM (benchmark) or PM (permanent mark).

- *Benchmarks* (Figure 7.43)—**Benchmarks** are permanent or temporary marks made at a known height above the original datum (sometimes used as the datum with a nominated height).
  - *Permanent marks (PM)* are marks generally only to be found outside of the site boundary, set into a pavement or other fixed structure that can be accessed by the council. They are stamped with a reference number which is lodged at local, state and national offices. By referencing this number an accurate height above sea level (or the relevant ellipsoid or geodic datum) may be called up. This height is constantly being updated as the earth moves and the accuracy of technology improves.
  - *Temporary benchmarks (TBM)* are often no more than a nail in a fence, tree or pavement, or a pen mark on the same. At other times

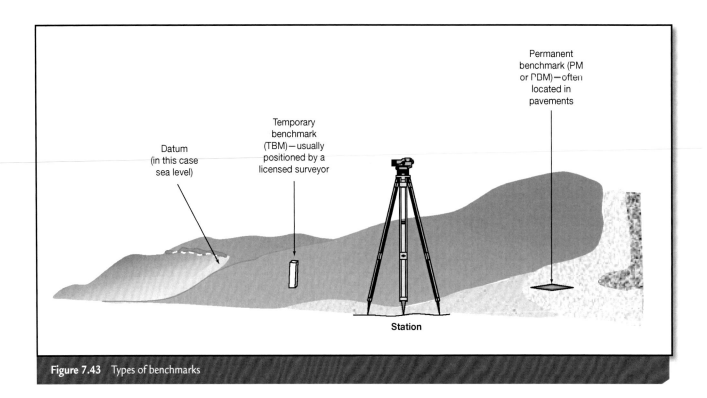

Figure 7.43 Types of benchmarks

they appear as a white painted peg placed out of harm's way on the site by a licensed surveyor, who has traversed this height from a nearby PM for which the height is already known.

- *Reduced level*—A **reduced level** (or **RL**) is the height given to a point relative to, or reduced from, the datum. The word 'reduced' in this context means 'to take from' or 'taken from'. So if our datum has a height of 100.00 m, and the point we are measuring is 1.000 m lower, then we would say that it has an RL of 99.00 m. Note that the datum itself is always expressed as an RL too, so our datum in this case is said to have an RL of 100.00 m.
- *Station* (Figure 7.43)—A **station** is the location of the levelling instrument. Depending on distance or undulations in the ground, there may be more than one station in a survey.
- *Stadia lines* (Figure 7.44)—**Stadia lines** are lines on the crosshairs of an optical level that can be used to determine distance. This is done by multiplying by 100 the number of millimetres between the top and bottom stadia as read on the

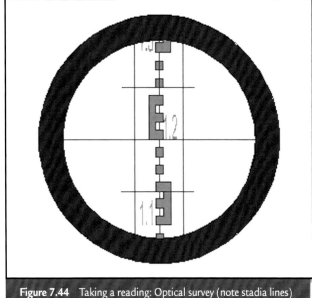

Figure 7.44 Taking a reading: Optical survey (note stadia lines)

staff through the instrument. The resultant figure is the distance from the instrument to the staff.

- *Line of collimation*—Technically, the **line of collimation** is the optical centre line of the telescope on a dumpy or automatic optical level (or any optical instrument for that matter). It can, however, also be used to refer to the line that a laser beam 'should' take if correctly calibrated (set by the manufacturer or repairer to produce a perfectly level beam of light). Note that the line of collimation passes along the centre of the main telescope optics, not those of the eyepiece (which may or may not be in the same line: see Figure 7.45).

Figure 7.46 Taking a reading: Laser receiver (note millimetre graduations on staff)

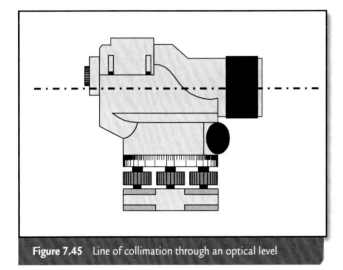

Figure 7.45 Line of collimation through an optical level

- *Readings*—These are the heights relative to the line of collimation extending from your instrument that you 'read' off the staff. In the case of a laser, this is the height or distance from the peg, ground or suspended ceiling batten that you measure when put your receiver in line with the laser beam (Figure 7.46). In optical instruments, it is this same distance or measurement that you can read off the staff or ruler when looking through the telescope (Figure 7.44).

Choosing the right tool

For the procedures listed at the beginning of this section, either a rotary laser or an optical instrument may be chosen and used successfully. However, for speed and accuracy, the laser is the better option for marking or establishing a given height or level over a site. An automatic optical or 'dumpy' level, on the other hand, is easier to use when trying to determine the rise and fall of the land, or simply the difference in height between two points. The downside of the optical level, however, is the need generally for a second set of hands to hold the staff.

The principle behind this form of levelling, be it by laser or optical instrument, is to cast a level line from which all other points of interest are measured. This is known as the line of collimation, as described earlier. Figure 7.47 shows this line being cast around the instrument a bit like a flat, perfectly level disc.

This is your level **plane** (a flat, horizontal surface passing through the air). With a laser this is exactly what's occurring and, in a foggy evening light, it is actually visible. With an optical

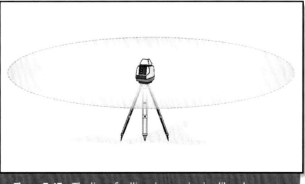

Figure 7.47 The line of collimation, projecting like a large, flat disc

instrument, this is the line of sight you will have as you turn the instrument through 360°.

When trying to find the height difference between just a couple of points, your task is fairly simple:

1 Measure from the 'disc' down to the peg or point you want (Peg C).
2 Now go and do the same for the other point(s) (Peg B).
3 Subtract one from the other and you have the difference in height between these two points.

Example 1
See Figure 7.48:

Peg A is measured at 0.600 m below the level plane or disc

Peg B is measured at 1.100 m below the level plane or disc

1.100 − 0.600 = 0.500 m

0.500 m is the difference in height between Pegs A and B.

If you need these heights referenced back to a datum (say Peg C) then:

1 Reference your disc over that point (i.e. find the height of the level plane above the datum point or peg).
2 Add this amount to the RL of the datum; this is the reference height of the level plane.
3 Take the remaining readings.
4 Take each reading away individually from the reference height to find the RL of each peg.

Example 2
Figure 7.48 shows a datum RL height of 100.00 m. The disc, reference plane or sight line is shown as being 0.800 m above this point. This means that the sight line is referenced at 100.800 m.

Peg A is measured at 0.600 m below the disc
Peg B is measured at 1.100 m below the disc
100.800 − 0.600 = 100.200 (Peg A)
100.800 − 1.100 = 99.700 (Peg B)

The RL for Peg A is 100.200 m and the RL for Peg B is 99.700 m.

This is a very basic way of using either an optical level or laser level. It is referred to as the 'height of instrument' or 'line of collimation' method and has

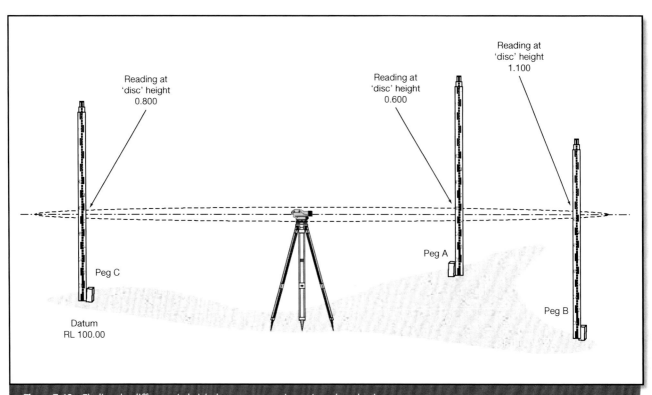

Figure 7.48 Finding the difference in height between two points using a laser level

been shown here for its instructional value. It clearly demonstrates the how and why behind more complex approaches. It also helps to explain what is happening in the following procedures.

However, you would only use this method for identifying one or two heights on any occasion. It is not the preferred approach for doing multiple sightings; for these you should use the rise and fall method of 'booking' readings. The rise and fall method is demonstrated towards the end of this section.

Maintaining, marking or establishing a continuous height around various parts of a building or site

Maintaining, marking or establishing a continuous height is a fairly simple levelling procedure that may be done by either a laser or an optical instrument, with the laser being the preferred tool due to its speed and one-person operation. Note, however, that the choice of instrument depends on the quality/accuracy of the laser and the distances involved. For shorter distances (a room for example) a laser line generator is equally good, if not better, for this purpose.

The following procedure (using the concepts outlined earlier) is used when boxing up for a concrete footing, setting up tile heights to a dado line, or installing a suspended ceiling for plastering:

If you already have a point set at the height to be transferred (Figure 7.49):

1 Following the manufacturer's instructions, set up the instrument as close to the centre of the job as possible (this reduces any minor error that may be in the instrument). As described earlier, lasers are usually set up above or below eye level so that you or other workers are less likely to stare accidentally into the beam.
2 Place your staff vertically upon this spot or mark.
3 Place the laser receiver on the staff, or even a simple stick of timber, and slide it up and down until it is correctly aligned with the laser. (A flat line usually shows on the screen. With the sound on, the beeps will stop and a steady tone will be emitted.)
4 Lock the receiver onto the staff or stick at this point.

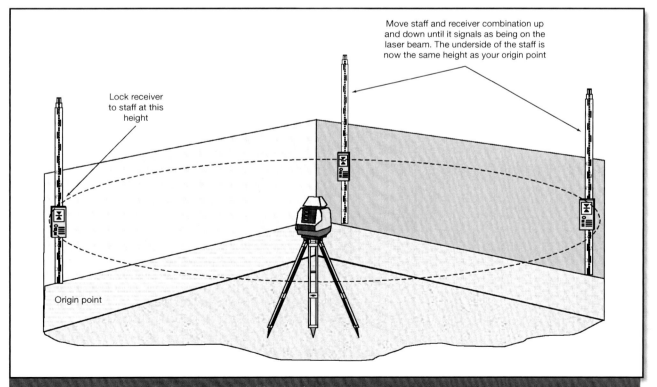

Figure 7.49 Maintaining or plotting a continuous height around a site

If you have to work above or below a datum point (Figure 7.50):

1 Following the manufacturer's instructions, set up the instrument as close to the centre of the job as possible.
2 Go to the datum/benchmark and set up your staff on this point.
3 Place the laser receiver on the staff and slide it up and down until it is correctly aligned with the laser. (A flat line usually shows on the screen. With the sound on, the beeps will stop and a steady tone will be emitted.)
4 Read off the measurement on the staff and then add or subtract the amount that you need to get to the desired height.
5 Lock the receiver onto the staff or stick at this new point.

Having followed either of the procedures above, you now can move the staff and receiver around the site wherever you choose. Whenever you align the receiver and staff combination with the laser beam, the base of the staff will be at the height you want.

This levelling procedure, employed upside down, is used to level suspended ceilings, as shown in Figures 7.51 and 7.52.

Figure 7.51 Ceiling being levelled in by laser

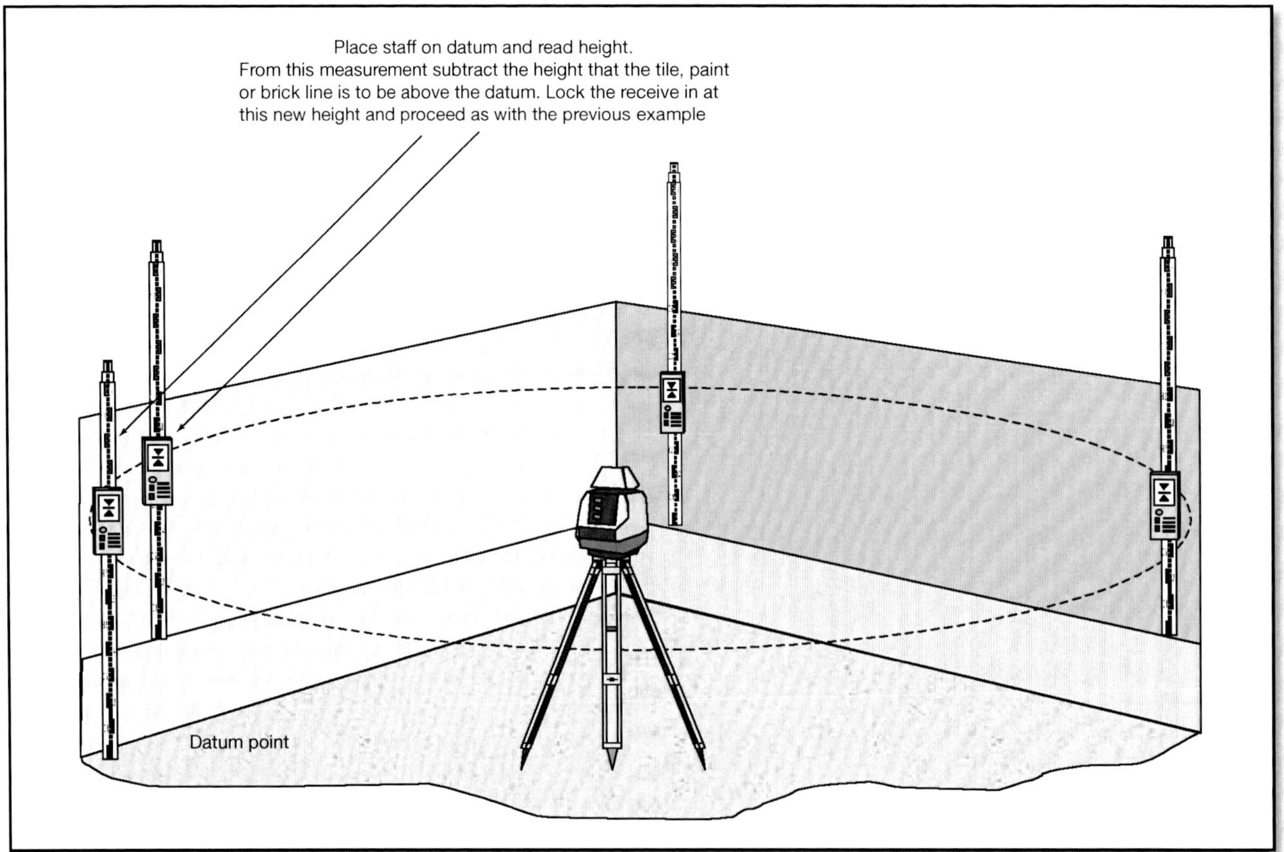

Place staff on datum and read height.
From this measurement subtract the height that the tile, paint or brick line is to be above the datum. Lock the receive in at this new height and proceed as with the previous example

Datum point

Figure 7.50 Maintaining or plotting a height relative to a datum

Figure 7.52 Taking a reading on channelling

Determining the relative heights of various points using a datum

This sort of work is seldom done by painters, tilers or plasterers, unless something very particular by way of curved ceiling work or sculpted tiling is being undertaken. It is, however, work that bricklayers may well have to do when setting up strip footings for walls and the like.

The basic method of determining the relative heights of various points, the 'height of instrument' method, has been described earlier. Using this method on more than three or four points, however, gets rather laborious and it is easy to make mistakes. The 'rise and fall' method (see page 260) was developed to reduce the chances of error, improve accuracy, and make the process faster.

This levelling procedure may be done with either laser or optical instruments, with optical instruments being preferred for accuracy over distances. In this example an automatic optical level will be described.

Setting up an optical level

Like laser levels, optical instruments are set up on a tripod with either a flat or domed head. The flat head is preferred because most automatic optical instruments have a flat-based **tribrach** (three-point levelling base). If you learn the following procedure properly, it will eventually take you no longer than 30 seconds to set up an optical level or manual laser level.

1 To set up the instrument you first need to securely position the tripod. This should be located out of harm's way, yet as central to the task as possible. In setting up the tripod, establish it so that the head is approximately level (use your eye and 'sight' it in to anything in the neighbourhood that is approximately level, such as brickwork, gutters or windows).

2 Now place the dumpy carefully onto the head and screw it into position. To level the instrument, there are three thumb screws available for adjustment (see Figure 7.53) and a blister or bubble level (Figure 7.54). Adjust these screws using the following technique:

   a Position yourself so that you can look directly down onto the bubble level, or so you can see the bubble clearly using the mirror (Figure 7.54).

   b Picture in your mind the concept described in Figure 7.55. In this diagram the screws each sit under one arm of a 'T'. Hold two of the screws between fingers and thumbs as shown in Figure 7.56. These are the screws sitting under the head of the 'T' (screws A and B in Figure 7.56).

Figure 7.53 The tribach screws

Figure 7.54 Blister level and mirror

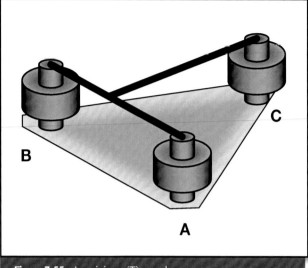

Figure 7.55 Imagining a 'T' over the screws

c You begin by adjusting these two screws simultaneously. You will do so by moving your thumbs either towards each other, or away from each other as shown in Figure 7.56.
As you do so watch the bubble in the blister level. Your aim is to get it 'centred' left to right. You are not trying for absolute centre at this point, just left and right.

3 Next you go to the third screw (screw C in Figure 7.57). Without touching the other screws, turn this screw until the bubble is moved to the absolute centre. (If the bubble remains slightly to the left or right of centre, go back to the first two screws and adjust them simultaneously again—never one at a time.)

Parallax error

The basic optical train of the telescope in an automatic level is shown in Figure 7.58 (the prism compensating system has been excluded for clarity). This demonstrates the way the main lenses and the eyepiece work together to form an image. When you focus a telescope you bring the image as developed by the main lenses to a point focused upon by the eyepiece. This is their 'mutual point of focus'.

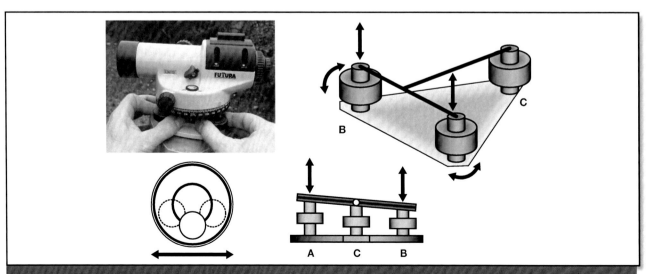

Figure 7.56 Adjusting the first two screws for level—move thumbs towards or away from each other

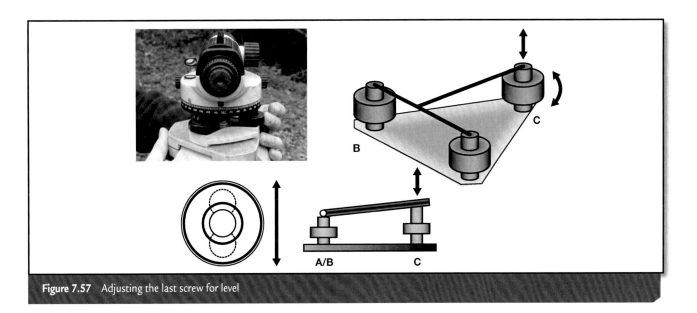

Figure 7.57 Adjusting the last screw for level

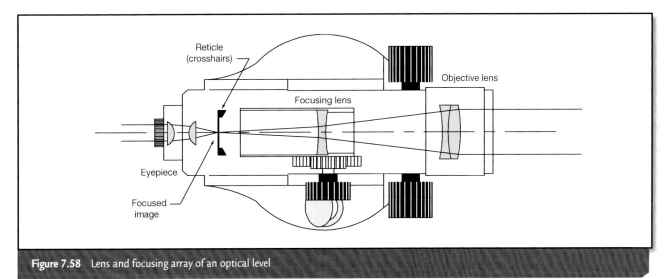

Figure 7.58 Lens and focusing array of an optical level

Now note the position of the **reticle** (the frame that holds the crosshairs and stadia lines that you sight with). Set up correctly, the image created by the telescope objective and negative lenses will come into focus at this point, and the eyepiece will be focused on it—that is, their mutual point of focus is on the crosshairs.

When this *is* the case no movement of the image in relation to the crosshairs will be perceptible when you move your head up or down while looking through the telescope. When this *is not* the case then the mutual point of focus is in front or behind the reticle (Figures 7.59 and 7.60). This means that movement between the object and the crosshairs is

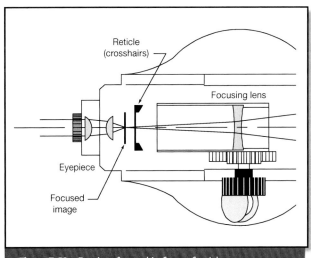

Figure 7.59 Eyepiece focused in front of reticle

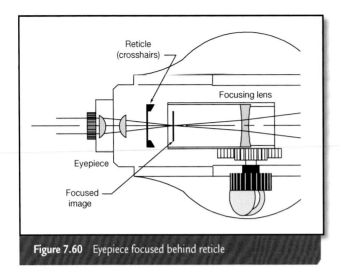

Figure 7.60 Eyepiece focused behind reticle

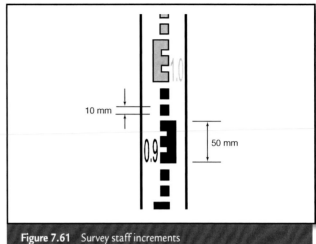

Figure 7.61 Survey staff increments

discernable as you move and so it is possible to read the staff incorrectly (i.e. the staff seems to move up and down behind the crosshairs, allowing you read it at varying points). This is called **parallax error**.

To remove parallax error:

1. Hold a blank piece of paper (the back of a business card is fine) about 100 mm in front of the telescope.
2. Without adjusting the main focusing knob, simply look through the scope with a relaxed eye. Try not to focus on anything (hence the blank white card).
3. Turn the eyepiece-focusing ring slowly until the crosshairs become sharp (focused). Remember to keep your eye relaxed. Focus the crosshairs to your eye, not your eye to the crosshairs. Now when you focus the telescope on an object (a staff for example) using the main focusing knob, the image will become sharp and clear when on the crosshairs; that is, you have made the crosshairs the mutual point of focus (as in Figure 7.58)

Every person has a different eye focus (eye shape) and therefore this procedure must be repeated whenever the person doing the levelling is replaced by someone else.

Taking a reading (optical levels)

The surveyor's staff (Figure 7.61) is marked in solid block increments of 10 mm. When reading any measuring tool, the most accurate you can be is one half of one graduation. With staffs of this type, therefore, you should work only to 5 mm increments.

Trying to gain a more accurate reading is generally pointless as you are using the staff on rough ground or rough-sawn pegs, so the smallest stone is going to shift your reading by two or three millimetres anyway. When you need a more accurate measurement (over shorter distances) you would use a staff graduated like a standard ruler, as shown in Figure 7.46 on p. 253.

When looking through an optical level you will see the image as shown in Figure 7.62. You will note that, aside from the crosshairs passing through the centre, there are two other lines. These are stadia lines (see 'Using stadia lines' on page 268) and you must be wary of them. It is easy to accidentally take your reading from one of these lines, rather than from the central crosshair. Figure 7.62 shows a reading of 1.0 m.

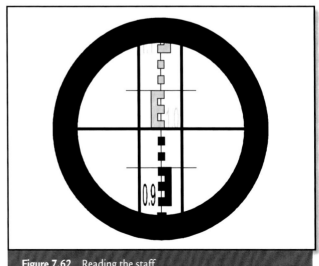

Figure 7.62 Reading the staff

The 'rise and fall' method

This is a system for booking (recording) the readings taken from a survey staff. The booking sheet looks something like the one shown in Figure 7.63.

Readings are booked, or written down, in the order that they are taken. The value of this system is that it has a built-in system for checking that you have added up your values correctly, and at the same time it provides you with the reduced levels (RL) or height in relation to your datum point.

The term 'rise and fall' refers to the relationship between one reading and another, one peg or staff position being seen as either higher or lower than the previous peg or the one to be read next. Figure 7.64 shows how this works: Peg B records a rise from Peg A, while Peg C records a fall from Peg B.

Sighting terms

This brings us to a few more terms. Figures 7.65 and 7.66 help to explain these concepts.

- *Backsight*—Oddly enough, this is the term for the *first* sighting of any **survey** (Figure 7.65). The **backsight** is also the first sighting taken when you change the position of the instrument (i.e. you change station) and begin taking readings again. It refers to looking back at something for which you already have the RL or that is nominally your starting height (100 m for example).
  Memory tool—Think of looking to the past.
- *Foresight*—The **foresight** is the last sighting taken before moving your instrument (changing station) or ending the survey (Figure 7.65).
  Memory tool—Think of looking to the future.
- *Intermediate sight*—The **intermediate sight** is all sightings taken in between the backsight and the foresight (Figure 7.66).
- *Rise*—The **rise** is an increase in height between a reading and the one taken immediately before it.
- *Fall*—The **fall** is a decrease in height between a reading and the one taken immediately before it.

Backsight	Intermediate sight	Foresight	Rise	Fall	Reduced level	Notes

Figure 7.63 Rise and fall booking sheet (column headings)

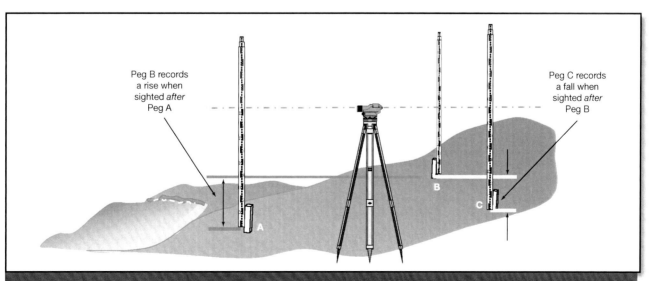

Figure 7.64 Rise and fall

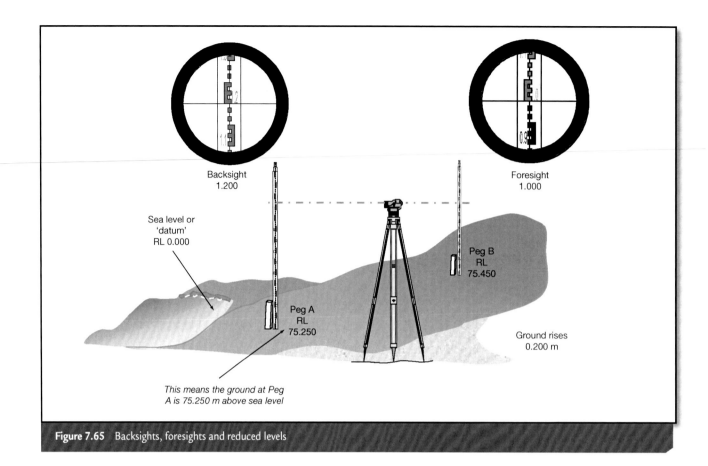

Figure 7.65 Backsights, foresights and reduced levels

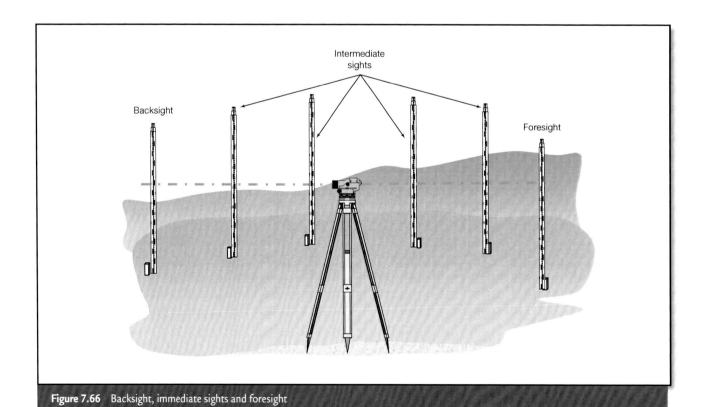

Figure 7.66 Backsight, immediate sights and foresight

Booking a basic site survey

The following example is for booking a survey for a strip footing for bricks on a sloping block. As shown in Figure 7.67, the land is sloping too much for the level to sight a staff at all the points from one position: the line of sight either runs into the ground or is too high for you to read the TBM. This means that two station positions will be required.

Figure 7.68 shows the layout of the survey needing be done including the location of each station.

1 Figure 7.69 shows the level set up at Station 1 so that a sighting can be taken 'back' to the temporary benchmark (TBM). This first reading is a backsight, and is 'booked' in the backsight column as shown in Figure 7.70. Note that included in this entry is

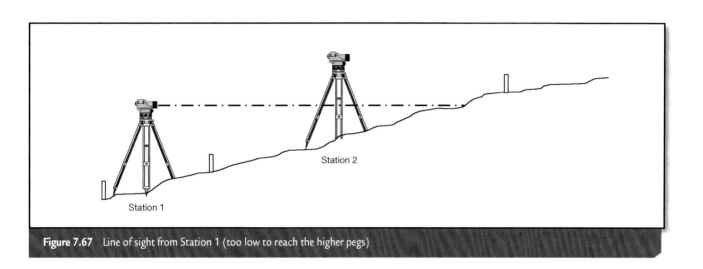

Figure 7.67 Line of sight from Station 1 (too low to reach the higher pegs)

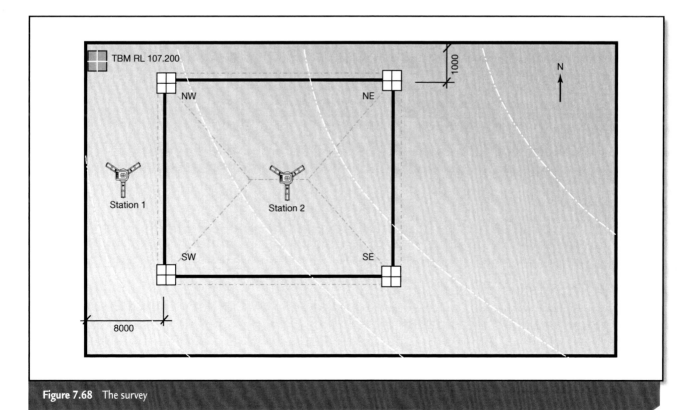

Figure 7.68 The survey

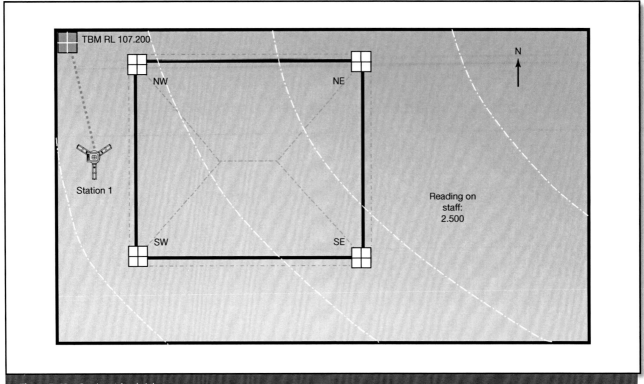

Figure 7.69 Station 1 backsight

Backsight	Intermediate sight	Foresight	Rise	Fall	Reduced level	Notes
2.500					107.200	TBM

Figure 7.70 Booking the backsight

the known reduced level or RL for the TBM, and that the name of the peg (TBM) is entered in the notes column.

2 The next sighting (Figure 7.71) is a foresight to Peg NW—the last sighting taken from this station before the instrument is moved.

3 This information is then booked as shown in Figure 7.72. Note that the change point (CP—the point at which the level is removed to a new location) is listed in the 'Notes' column beside the name of the peg; in this case 'NW'. The next sighting will look 'back' to this peg.

4 The instrument is now moved to Station 2 and a backsight is taken 'back' to Peg NW (Figure 7.73). As there is only one 'Peg NW', the backsight is always recorded on the same line as the previous foresight to the same peg (Fig 7.74).

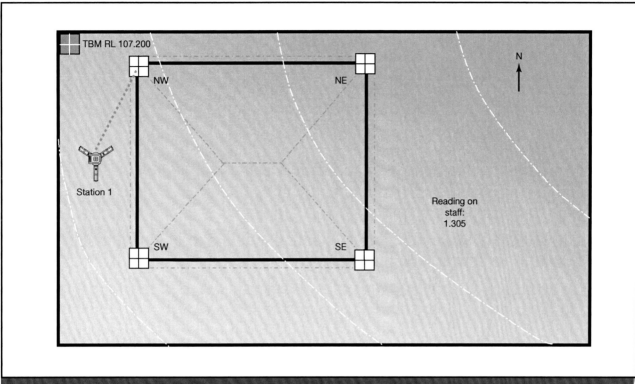

Figure 7.71 Station 1 foresight

Backsight	Intermediate sight	Foresight	Rise	Fall	Reduced level	Notes
2.500					107.200	TBM
		1.305				NW−CP

Figure 7.72 Rise and fall booking sheet (column headings)

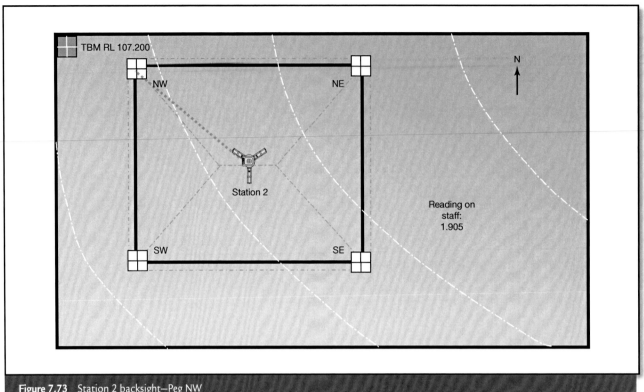

Figure 7.73 Station 2 backsight—Peg NW

Backsight	Intermediate sight	Foresight	Rise	Fall	Reduced level	Notes
2.500					107.200	TBM
1.905		1.305				NW–CP

Figure 7.74 Entering the backsight from Station 2

5 Sightings are now taken to all the remaining pegs. These will be 'intermediate sights' except for the last one, which will be a foresight (see Figure 7.75).

It is important that sightings are recorded *as they are taken, in the order that they are taken,* and *in the correct columns* (Figure 7.76). It is also important that all peg names or locations are correctly identified in the notes.

6 The relative rise and fall for each location is now calculated by subtracting each sighting from the one preceding it. This is done in the order that they were taken, and within the readings taken from a single station (see the loops in Figure 7.77). If the result is 'positive' it is recorded as a rise. If 'negative', it is recorded in the 'Fall' column. For example:

Peg NW: 2.500 – 1.305 = 1.195 (rise)
Peg NE: 1.905 – 0.910 = 1.040 (rise)
Peg SE: 0.910 – 1.255 = –0.345 (fall)
Peg SW: 1.255 – 2.365 = –1.110 (fall)

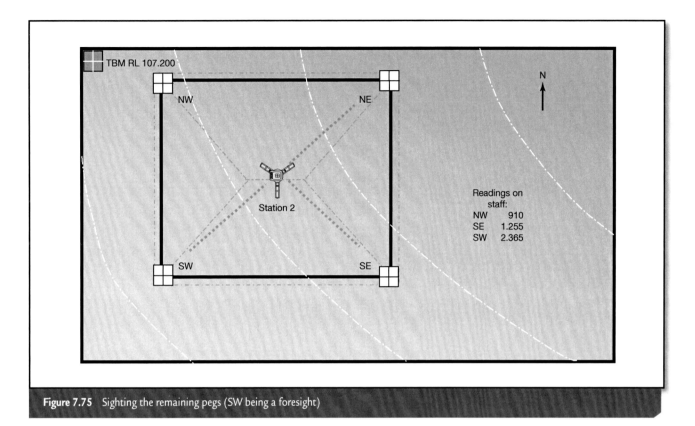

Figure 7.75 Sighting the remaining pegs (SW being a foresight)

Backsight	Intermediate sight	Foresight	Rise	Fall	Reduced level	Notes
2.500					107.200	TBM
1.905		1.305				NW–CP
	0.910					NE
	1.255					SE
		2.365				SW

Figure 7.76 Entering the remaining sightings

7   The reduced levels (RL) for each location are now found by adding the fall to, or subtracting the rise from, to the RL of the peg preceding it. In this example:
Peg NW: 107.200 + 1.195 (rise) = 108.395
Peg NE: 108.395 + 1.040 (rise) = 109.435
Peg SE: 109.435 − 0.345 (fall) = 109.090
Peg SW: 109.090 − 1.110 (fall) = 107.980

These values are entered into the 'Reduced level' column as shown in Figure 7.78.

Checking the booking

We can now run a series of checks on the booking that will determine if we have carried out all of the preceding actions correctly. Because each RL is calculated from the one prior to it, the potential for

268 PAINTING AND DECORATING, AND MORTAR TRADES

Backsight	Intermediate sight	Foresight	Rise	Fall	Reduced level	Notes
2.500					107.200	TBM
1.905		1.305	1.195			NW–CP
	0.910		1.040			NE
	1.255			0.345		SE
		2.365		1.110		SW

Figure 7.77 Entering the rise and fall developed from each reading

Backsight	Intermediate sight	Foresight	Rise	Fall	Reduced level	Notes
2.500					107.200	TBM
1.905		1.305	1.195		108.395	NW–CP
	0.910		1.040		109.435	NE
	1.255			0.345	109.090	SE
		2.365		1.110	107.980	SW

Figure 7.78 Entering the calculated reduced level for each peg

what might be called 'cumulative error' has been deliberately built in. This means that an error in any one RL calculation will flow on, and will be reflected in the final RL. The steps for checking are as follows (see Figure 7.79):

1. a Add all the backsights together.
   b Add all the foresights together.
   c Subtract the foresight total from the backsight total.
2. a Add all the rises together.
   b Add all the falls together.
   c Subtract the falls total from the rises total.
3. a Subtract the initial RL from the last RL.
   b Check that the totals calculated in each of these steps are the same.

If these figures are not the same, then there is a booking or calculation error that must be found. You will then need to go back over each of the rise and fall and RL calculations, fix the mistake and then redo the check above.

Using stadia lines

The purpose of stadia lines (Figure 7.80) is to give the surveyor a means of estimating distance with reasonable accuracy. All that is required is to multiply the difference between the readings taken from the top and bottom stadia lines by 100.

Example
See Figure 7.80:
Top stadia reading: 1.050 m
Bottom stadia reading: 0.950 m
Difference: 1.050 – 0.950 = 0.100 m
Distance: 0.1 × 100 = 10.00 m

Backsight	Intermediate sight	Foresight	Rise	Fall	Reduced level	Notes
2.500					107.200	TBM
1.905		1.305	1.195	+	108.395	NW–CP
+	0.910	+	1.040		109.435	NE
	1.255		+	0.345	109.090	SE
		2.365		1.110	107.980	SW
4.450	–	3.670	2.235	1.455	–	
= 0.780		= 0.780			= 0.780	
Step 1		Step 2			Step 3	

Figure 7.79 Checking the booked results

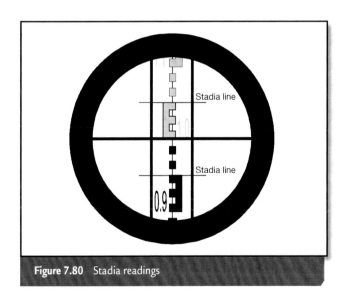

Figure 7.80 Stadia readings

CHECKING YOUR LEVELLING EQUIPMENT

As stated earlier, it is critical that you do not take levelling instruments for granted. Such tools, lasers and optical levels are easily knocked out of calibration and this can lead to significant, time-consuming (and therefore expensive) error.

The 'two-peg test' is a simple and quick test that should be used frequently on all optical instruments, and particularly before any major work over long distances. It is a test that can also be used on laser line generators, although these tools generally come with a basic testing procedure described in the operator's manual.

The 'four-peg test' is designed for rotary laser levels as these must be tested over two axes, not one.

The two-peg test

The steps for a two-peg test are as follows:
1 (See Figure 7.81)
    a Establish the instrument on reasonably flat ground, giving clear sighting for 15–20 m in each direction if possible.
    b Locate the first peg so that a staff can be easily read, but far enough away to make any error in the instrument noticeable.
    c Locate the second peg at exactly the same distance from the centre of the instrument, on the opposite side.
    d Where possible, make the two pegs and the instrument in line with each other.

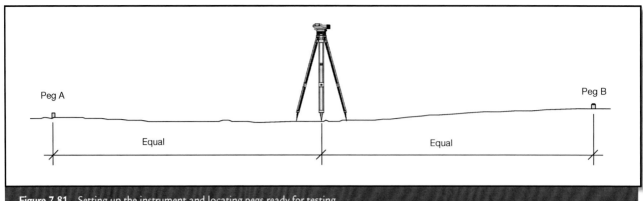

Figure 7.81 Setting up the instrument and locating pegs ready for testing

2 (See Figure 7.82)
   a Take and record a reading at Peg A.
   b Take and record a reading at Peg B.
     (For this example these readings are 1.200 m and 1.000 m respectively.)
   c Calculate the difference between these two readings.

	Station 1	Station 2
Peg A	1.200	
Peg B	1.000	
Difference	0.200	

Figure 7.82 Establishing the difference in height between the two pegs

In carrying out these two steps correctly you will have established a known level line height that runs between the two pegs. As shown in Figure 7.83, if the instrument is correctly calibrated then this level line passes through the optical centre of the telescope. It is on the line of collimation, even though we won't know this until the next steps are taken.

However, should the instrument be out and actually be reading slightly uphill, then a known level line height has still been established—except that this line is actually *above* the centre of the instrument. Figure 7.84 demonstrates this point.

Should the instrument be out and sighting slightly downhill then once more a known level line height has been established, only this time it is *below* the centre of the instrument (see Figure 7.85).

3 The instrument is now moved to a second station approximately 5–10 m beyond, but in line with, Pegs A and B.

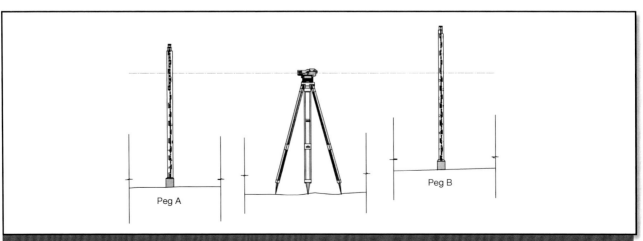

Figure 7.83 A correctly calibrated instrument establishing a level line height

CHAPTER 7 LEVELLING PROCEDURES 271

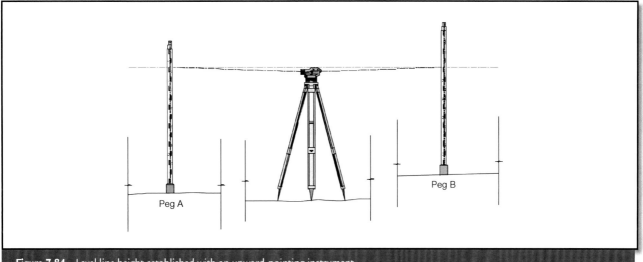

Figure 7.84 Level line height established with an upward-pointing instrument

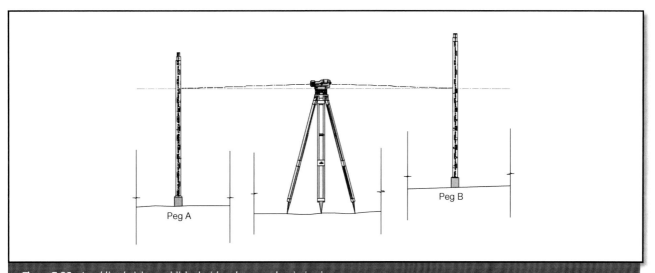

Figure 7.85 Level line height established with a downward-pointing instrument

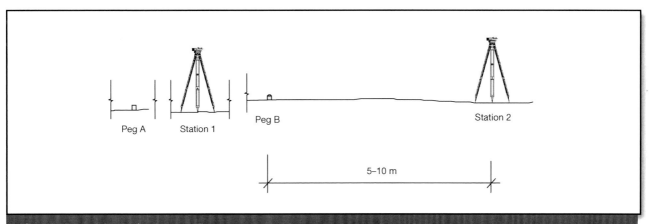

Figure 7.86 Locating Station 2

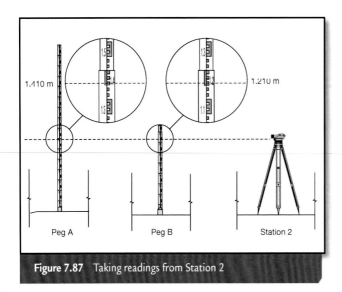

Figure 7.87 Taking readings from Station 2

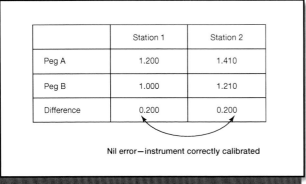

Figure 7.88 The difference between Station 1 and Station 2 readings

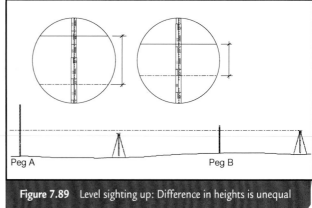

Figure 7.89 Level sighting up: Difference in heights is unequal

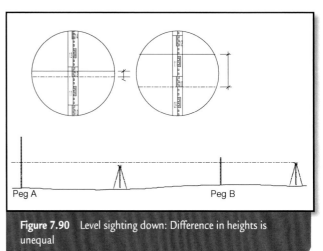

Figure 7.90 Level sighting down: Difference in heights is unequal

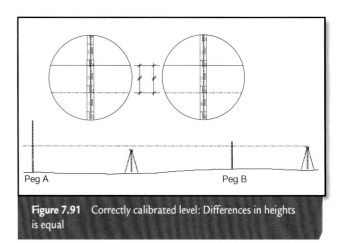

Figure 7.91 Correctly calibrated level: Differences in heights is equal

4 Readings are now taken back over the two pegs and recorded as shown in Figures 7.87 and 7.88. Once again the difference between the two sighting is calculated. This result is then compared to the difference between the previous two sightings. If there is a difference of greater than 3 mm over the distances described in this example, then the instrument is suspect and needs professional servicing.

To understand how this test works, study Figures 7.89–7.91. In Figure 7.89 the instrument is out and sighting uphill. The result is that the readings 'climb' away from the instrument and lead to a greater difference in height at Peg A than at Peg B.

Figure 7.90 shows the opposite instance: the instrument sights downhill and so the reading shows a lesser difference in height at Peg A than at Peg B.

Figure 7.91 then shows an instrument that is correctly calibrated. In this instance the difference in the height at each peg is equal; that is, both lines are level and running parallel with each other.

Rotating laser levels: The four-peg test

For a four-peg test the same approach is used as for the test above, except that four pegs are used instead of one. This is because the instrument needs to be checked in two directions (at 90° to each other) across the horizontal plane. In effect this is the two-peg test performed twice: north–south and then east–west. A 'plan' view of the layout of pegs and stations is shown in Figure 7.92.

Figure 7.93, using example data, demonstrates the recording and calculating required. Note the you are looking for discrepancies across each axis only, not between axes. In the example the result is nil error and so the tool is correctly calibrated. Once again, if the error is greater than 3 mm across the distances described in this example (or greater than the manufacturer's stated tolerances—see the specifications section of the operator's guide), then the equipment needs to be professionally serviced.

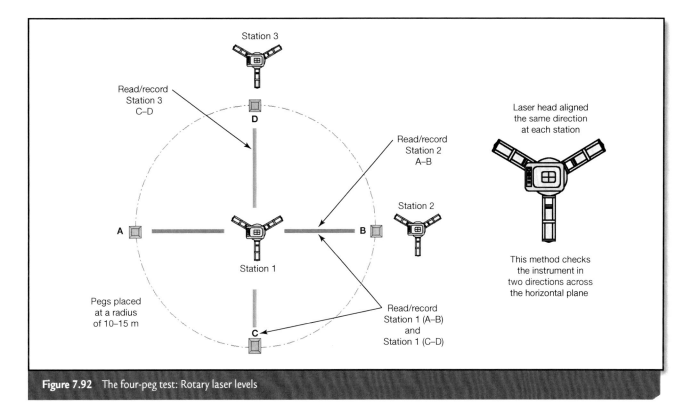

Figure 7.92 The four-peg test: Rotary laser levels

	Station 1 (A–B)	Station 1 (C–D)	Station 2	Station 3
Peg A	1.020		1.220	
Peg B	0.900		1.100	
Peg C		0.750		0.570
Peg D		0.890		0.710
Difference	0.120	0.140	0.120	0.140

Difference between Station 1 (A–B) and Station 2, and between Station 1 (C–D) and Station 3, should be less than 3 mm

Figure 7.93 Booking the results

CLEANING AND STORAGE

The 'wet' trades are, surprisingly enough, just that: wet. This means that levels and the associated levelling equipment can easily get coated in adhesives, plaster, paint, render or mortar. A spirit level that has lumps of dried mortar or tile adhesive stuck to its milled edges is next to useless. Similarly, such material over the lens of either an optical or a laser level will seriously impair your capacity to use these tools successfully. It is critical for anyone wishing to become proficient in the wet trades to understand the importance of cleaning and storing their tools correctly. Learning how not to get the tools coated
in such materials in the first place is better still.

It is good practice to:
- wipe/wash your hands before even getting a laser or optical level out for set up;
- set up such instruments away from areas that are likely get materials such as plaster, tile adhesive or paint dropping onto them;
- brush, wipe and/or wash down spirit levels and straight edges that, by necessity, have come in contact with wet materials. Do this before such materials have had a chance to dry, and be aware that mortar and some tile adhesives are corrosive to metals, particularly aluminium;
- keep string lines out of wet materials; if they fall in the mortar or concrete, wash them down quickly.

Having cleaned your tools, now put them away safely. Don't just throw them into the back of the ute: put them in a toolbox and store them in such a way that other equipment cannot fall on them, or so that they will not get thrown across the tray of the ute at the first roundabout. Laser equipment is extremely sensitive and must be stored and carried with great care—bouncing around is not its favourite past-time!

It is also good practice to:
- always store laser equipment in the cushioned case supplied for it by the manufacturer;
- place such tools carefully in the toolbox or carry them in the cab of the ute on the floor (so they can't fall any further);
- place spirit levels in a padded carry bag, or at least place them carefully in a secure position in the toolbox;
- position straight edges such that they cannot slide around or have other things slide into them;
- store all your levelling equipment away from containers of wet or powdered materials.

A CLOSING WORD

As stated at the outset of this chapter, levelling is a critical aspect of all trades, wet or otherwise. The tools and procedures discussed are aids in achieving that all-important goal of square, straight, level and plumb. They are only tools however, and the result is totally dependent upon the skill of their user. In addition to using the right tool for the right job, you should never forget to develop skill in the use of the most important tools of all, your eyes and hands. Don't just trust technology, test it. Test it with your eyes and hands: if it doesn't look level, in wind or plumb, or if it doesn't feel flat and true, then it probably is not.

Worksheet 1

Student name: _____

Enrolment year: _____

Class code: _____

Competency name/Number: _____

To be completed by teachers:	
Student competent	☐
Student not yet competent	☐

Task: Read through the section *Introduction to levelling procedures*, then complete the following.

Q. 1 A tradesperson is said to do little wrong if the following four elements are achieved:

1. _____
2. _____
3. _____
4. _____

Q. 2 What are the steps for checking a rectangular layout or set-out for 'square'?

1. _____
2. _____

Q. 3 What is the deference between 'plumb' and 'perpendicular'?

_____
_____

Q. 4 What is meant by 'straight and true'?

_____
_____
_____

Q. 5 Name two hazards that may be encountered while carrying out levelling operations:

1. _____
2. _____

Worksheet 2

Student name: _____

Enrolment year: _____

Class code: _____

Competency name/Number: _____

To be completed by teachers:
Student competent ☐
Student not yet competent ☐

Task: Read through the section on *The tool list*, then complete the following.

Q. 1 List two advantages of a spirit level over a digital level:

1. _____
2. _____

Q. 2 List two advantages of a digital level over a spirit level:

1. _____
2. _____

Q. 3 Why are 'line levels' seldom used anymore?

_____
_____

Q. 4 Name two important characteristics of a straight edge:

1. _____
2. _____

Q. 5 When shouldn't you use red oxide on a chalk line?

_____
_____

Q. 6 What is the one main advantage of using a manual laser level?

_____
_____

Q. 7 How does a pendulum rotary laser automatically establish a level beam?

_____
_____
_____

Q. 8 How does a self-levelling rotary laser automatically establish a level beam?

_____
_____
_____

Q. 9 What is a laser line generator?

_____
_____

Q. 10 List two advantages of the laser level over an optical level:

1. _____

2. _____

Q. 11 List two advantages of the optical level over laser levels:

1. _____

2. _____

Q. 12 The 'E' on a survey staff represents:
- a 500 mm
- b 0.500 m
- c 5 mm
- d 0.050 m.

Worksheet 3

Student name: _____

Enrolment year: _____

Class code: _____

Competency name/Number: _____

To be completed by teachers:
Student competent ☐
Student not yet competent ☐

Task: Read through the section on *Basic levelling procedures*, then complete the following.

Q. 1 Describe the steps for checking a spirit level for accuracy:

1. _____

2. _____

3. _____

4. _____

Q. 2 When marking out a line that is two or more level lengths long, we:
- a end-for-end the level
- b use a straight edge that will cover the distance
- c make sure we keep the level the same way round
- d follow option **a** if option **b** is not available.

Q. 3 Describe two methods for checking plumb:

1. _____

_____

2. _____

_____

_____

Q. 4 On the diagram below, sketch the procedure for finding the height difference between one point and another using a spirit level and a straight edge:

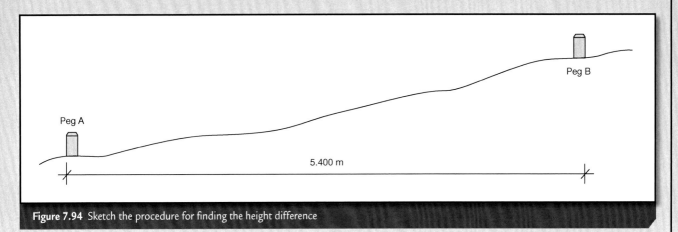

Figure 7.94 Sketch the procedure for finding the height difference

Worksheet 4

Student name: _____

Enrolment year: _____

Class code: _____

Competency name/Number: _____

To be completed by teachers:
Student competent ☐
Student not yet competent ☐

Task: Read through the section on *Laser line generators*, then complete the following.

Q. 1 What Australian standards apply to lasers in the construction industry?

_____
_____

Q. 2 From what class or strength of laser upwards do you need to have a laser safety officer at the workplace?

_____

Q. 3 What is the most powerful laser allowed in the construction industry?

_____

Q. 4 What is the purpose of the tinted glasses supplied with a laser unit?

_____
_____

Q. 5 What colour glasses should be used with a red laser?

_____

Q. 6 The class of laser being used is identified by:
   a looking closely at the beam to see how bright it is
   b looking for the required label on the unit
   c looking at the type of batteries used
   d looking at the distance that the laser can cast its beam.

Q. 7 What is the most significant OH&S risk of using laser levels?

_____
_____

Q. 8 According to the Australian standards there are several measures required to be taken to reduce the risks associated with using laser levels. List four:

1. _____

2. _____

3. _____

4. _____

Q. 9 Describe the difference between a laser level and a laser line generator.

_____

_____

Worksheet 5

Student name: _____

Enrolment year: _____

Class code: _____

Competency name/Number: _____

To be completed by teachers:	
Student competent	☐
Student not yet competent	☐

Task: Read through the section on *Other levelling procedures,* then complete the following.

Q. 1 The line of collimation:
    a is always the centre line of the eyepiece on an optical level
    b is the central vertical line of the crosshairs
    c runs horizontally through the centre or main axis of the telescope
    d can also be referred to as a 'stadia line'.

Q. 2 In levelling, a benchmark is:
    a a standard to which you will do your work
    b a temporary mark giving a known or nominal height above sea level
    c a permanent mark giving a known or nominal height above sea level
    d both **b** and **c**.

Q. 3 On a plan, a PM is shown just outside the boundary with an RL of 112.0. This means that there is a:
    a chalk or paint mark 112 mm outside the boundary
    b chalk or paint mark that has a known height of 112 m above sea level
    c fixed government survey pin with a known height above sea level of 112 m
    d fixed government survey pin with a nominal height above sea level of 112 m.

Q. 4 Parallax error is said to have been eliminated when
    a the reticle (crosshairs) are plumb and horizontal
    b the eyepiece and the main lenses focus upon the same point (mutual point of focus)
    c the mutual point of focus is behind the reticle
    d the mutual point of focus of the eyepiece and the main lenses is the reticle.

Q. 5 The following set of stadia readings have been recorded: 2.150 and 2.730. The distance from the staff to the levelling instrument is therefore:
    a 0.580 m
    b 58.0 m
    c 580.0 m
    d 5.8 m.

Q. 6 Using Figure 7.95, determine the following:

1. The height difference between:

   Peg A and Peg B _____

   Peg B and Peg C _____

2. The reduced levels of:

   Peg A _____

   Peg B _____

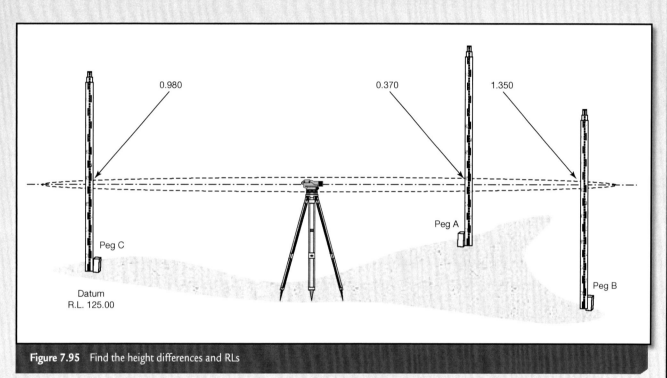

Figure 7.95 Find the height differences and RLs

Q. 7 An optical level is checked for level by means of the two-peg test. Why does a laser level require a four-peg test?

_____

_____

Q. 8 Use the booking sheet below to determine the rise and fall and the reduced levels for the survey shown in Figure 7.96. Carry out all checks.

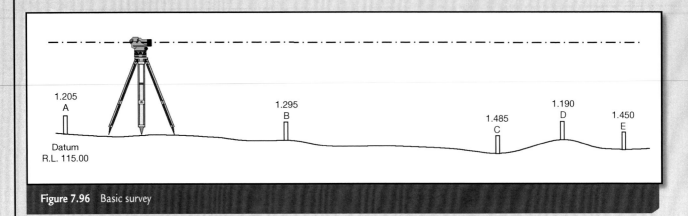

Figure 7.96 Basic survey

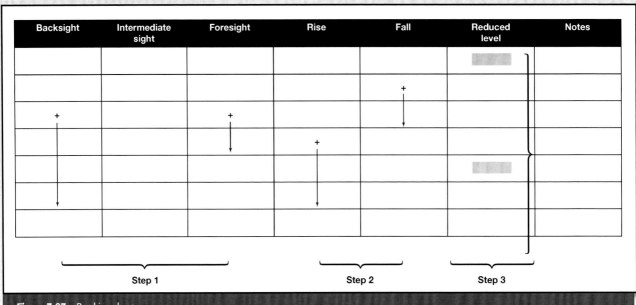

Figure 7.97 Booking sheet

Worksheet 6

Student name: _____

Enrolment year: _____

Class code: _____

Competency name/Number: _____

To be completed by teachers:	
Student competent	☐
Student not yet competent	☐

Task: Read through the section beginning at *Checking your levelling equipment* up to the end of that section, then complete the following.

Q. 1 Why should levelling equipment be tested before use?

_____

_____

_____

_____

_____

_____

Q. 2 List the two different types of test which can be used and the equipment on which they can be used.

1. _____

_____

2. _____

_____

Q. 3 What should you do if there is a difference of over 3 mm in either of the types of test?

_____

_____

_____

_____

_____

Worksheet 7

Student name: _____

Enrolment year: _____

Class code: _____

Competency name/Number: _____

To be completed by teachers:
Student competent ☐
Student not yet competent ☐

Task: Read through the section beginning at *Cleaning and storage* up to the end of *A closing word*, then complete the following.

Q. 1 What can happened to levels and other levelling equipment?

_____
_____

Q. 2 Why can this be a problem? Give examples

_____
_____
_____
_____
_____

Q. 3 What are first four elements of good practice for tools?

1. _____
2. _____
3. _____
4. _____

Q. 4 How should tools be stored?

1. _____
2. _____
3. _____
4. _____
5. _____

Q. 5 Should you always trust technology? If not, what should you do?

_____
_____

REFERENCES AND FURTHER READING

Texts

Australian Building Codes Board (2006), *Building Code of Australia*, Australian Government, Canberra

Laws, A. (2009), *Site Establishment, Formwork and Framing*, Pearson Education Australia, Sydney

Web-based resources

<http://hsc.csu.edu.au/construction/other_units/compulsory/levelling/3349/frontpage.html>

<www.levelling.uhi.ac.uk>

PowerPoints

Costin, G.P. (2009), 'PowerPoint to accompany Chapter 1 Levelling & Setting Out: Using booking sheets', in A. Laws, *Site Establishment, Formwork and Framing*, Pearson Australia, Sydney

Costin, G.P. (2009), 'PowerPoint to accompany Chapter 1 Levelling & Setting Out: The two peg test', in A. Laws, *Site Establishment, Formwork and Framing*, Pearson Education Australia, Sydney

Australian Standards

AS/NZS 2211.1:2004, Safety of laser products

AS/NZS 3690:2009, Approval and test specification – residual current devices (current-operated earth-leakage devices)

AS 2397-1993, Safe use of lasers in the building and construction industry

CHAPTER 8
HAND TOOLS, PLANT AND EQUIPMENT

The purpose of this chapter is to help you identify, use safely and maintain the hand and power tools that you are most likely to encounter in your daily work. Some of these tools could be called 'generic' in that they are used by just about everyone, irrespective of their role in the construction industry. On the other hand, the trades of plastering, bricklaying, tiling and painting have become very specialised over the years, and hence so have some of their tools, plant and equipment.

This being the case, the chapter begins with an overview of the most common of these general hand tools, after which the hand tools specific to each field of work are dealt with trade by trade. Power tools are dealt with similarly, as is plant and equipment. The chapter closes with a look at the common issue of cleaning and maintenance. Areas addressed from the unit of competency include:
- planning and preparation for using tools and equipment;
- identifying and selecting hand, power and pneumatic tools;
- using tools safely;
- identifying, selecting and using plant and equipment; and
- cleaning up after using tools and equipment.

INTRODUCTION TO TOOLS

Human beings are tools users. Arguably, this is how today's civilisation developed in the manner, and to the level, it has. Whether we continue to develop sustainably, or degrade our civilisation into some future archaeological digging, depends much upon our tool and material choices, as these influence what society consumes or excretes. Humble as it may be, you have a role in developing a sustainable culture, for how sustainable *you* are partly depends on the choices you make in the purchase, use, maintenance and ultimate disposal of plant, tools and equipment.

Safety too, lies as much in the correct choice of tool as it does in its correct application. Once again this choice begins at purchase. Cheap tools not only lead to cheap results and greater use of energy and resources, they can also lead to expensive medical bills.

'Which is the best …?' is the perennial question asked by apprentices. It is often followed by the frustratingly vague, 'Well it depends …'. The following is a list of criteria for choosing the right tool. Your choice depends on:
- the task: what you intend to use the tool for;
- the task: what the tool is designed to be used for;
- the material: what you intend to use the tool on;
- the material: what the tool is designed to be used on;
- the build quality of the tool;
- the reliability of the tool;
- **ergonomics** (ease of use, vibrations);
- the tool's safety features;
- dust, particle, or spray, generation (including methods for limiting, containing or removing this waste);
- ease of adjustment or set-up;
- disposal or recycling options;
- consumables (availability, quality, and options such as sanding pads and grades);
- the back-up and service offered by the manufacturer;
- the warranty or guarantee available.

Quality tools cost money, accept it. This doesn't mean you can't strive for a bargain, but it does mean you should be wary of 'cheap' tools, even those that offer a generous guarantee. Guarantees are useless when you're two hours from the nearest store and that cheap cordless drill strips its plastic gears. Buying quality doesn't guarantee that things won't break, but it does reduce the probability of them doing so.

Tradesperson or *bricoleur*

At the beginning of the above list, you will have noticed that materials and usage are mentioned twice: once for *your* intent; once for *the designer/manufacturer's* intent. In France this difference in intent is captured in the concept of *bricolage* or the **bricoleur**. Historically, the *bricoleur* is a handyman or woman who has a little bit of the unknown about themselves, their methods and their work, and who is therefore sometimes considered untrustworthy. Generally, a tradesperson will have a large collection of tools, each with a fixed purpose in their design and intended use. The *bricoleur*, on the other hand, being more of a traveller, has a much smaller collection of multi-purpose tools. These may be exactly the same as some of the tradesperson's tools, but they are used with less restraint—that is, the *bricoleur* will use anything, anyhow, to get the job done.

This is also the risk for novices. It is important that you learn to use the right tool for the right job, make your choices based upon that knowledge and, when you don't know something, make it your business to find out. In time you *may* develop (not everyone does) a practical expertise that allows you to use tools skilfully in a manner other than for their designed purpose. However, being able to make these judgments appropriately, based on safety, efficiency and quality, requires experience and knowledge. Leaping down this path without such expertise is dangerous both to you and those around you.

With this in mind, the next section outlines the basic hand tools and their intended use, followed by trade specific-tools and, finally, other plant and equipment.

BASIC HAND TOOLS

Tools can be clustered into groups with a designed purpose, such as: setting out; hitting; levering; cutting, scraping and spreading; screwing; cramping or holding; and, finally, cleaning. Those designed for levelling, plumbing, measuring or giving a straight line were dealt with in Chapter 7, 'Levelling procedures'.

Before looking at these groups specifically, however, a word about holding tools generally is appropriate. It is important that you understand from the outset that for the proficient user, the tool is an extension of themselves. It is not the brush that puts on the paint, it is the painter with the brush. Tools should be held lightly—securely, firmly and with control, but lightly. Develop a sense of feel 'through' the tool, to the point of application, cutting, striking or whatever: this is critical to skilful use. The tool is to be part of you, not apart from you. Such skill can be gained only through repetition and practice. It may take longer to achieve for some than for than others, but it is a journey that all must undertake.

The following are some guidelines for holding tools correctly:

- Watch carefully how the most skilful tradespeople you work with hold a tool (they often can't tell you; you need to watch and model, then adapt, their approach).
- Use an 'open' hand when holding tools, as this helps set up a connection between you and the tool both visually and physically (see Figure 8.1).
- Relax your body. Breathe out, not in, when striking an energetic blow; continue to breathe when trying to be steady, don't hold your breath.
- Use your body. Body position and stance are critical. Figure 8.2 gives the appropriate stance for most actions (left-handers would use the opposite stance).
- Be conscious of where your strength is coming from and how to maintain your balance should the piece you are working on or your support move unexpectedly. Note that in Figure 8.2 the knee never goes beyond the toe.
- Use energy, not force. Using force can lead to mistakes or becoming unbalanced, or cause injury to you, others or the work piece.
- Be conscious of where the tool will end up should it slip, jump or otherwise move unexpectedly. Make sure no part of your body is in line with that movement (see Figures 8.3 and 8.4).

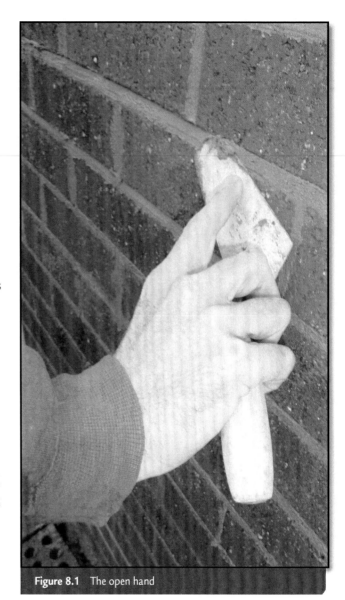

Figure 8.1 The open hand

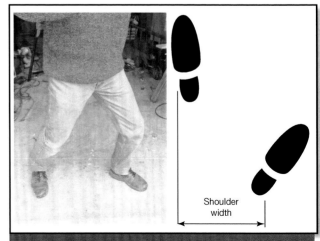

Figure 8.2 Good stance: Width depends upon energy required

CHAPTER 8 HAND TOOLS, PLANT AND EQUIPMENT

Figure 8.3 Good hand positions

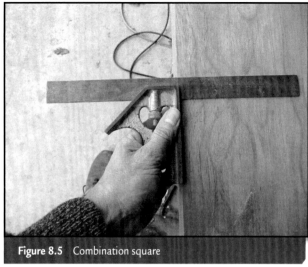

Figure 8.5 Combination square

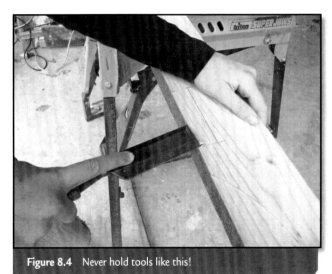

Figure 8.4 Never hold tools like this!

- *Steel square*—a larger, fixed-metal square, useful in setting out or checking the accuracy of walls and the like (Figure 8.6).

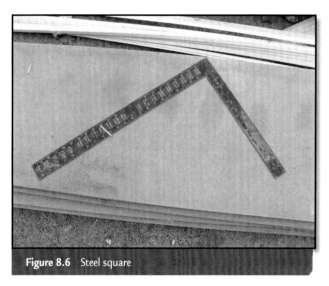

Figure 8.6 Steel square

Tools for setting out

Many of these tools were dealt with in Chapter 7; however, there are a few which need mentioning here.

Squares

Some squares are trade-specific and so will be dealt with later; others are more general in their use, as follows.

- *Combination square*—a common tool used to produce square or 45° angled lines. The blade of the tool can be moved so as to scribe a line a set distance from the material's edge (see Figure 8.5). Included is a small level which can be handy in tight spaces, although is not to be overly relied upon.

Compasses

These are tools for describing arcs, circles, finding centres or developing other geometric lines, as follows.

- *Common compass*—come in various forms. Most have fixed metal spikes at both ends, while some have a receiver for a pen or pencil.
- *Trammel heads*—metal heads that can clamp onto a straight length of metal or timber. Trammels can hold either a metal scribing spike or a pen/pencil. Their advantage over the common compass is that they can be attached to any length of material to produce an arc (Figure 8.7).

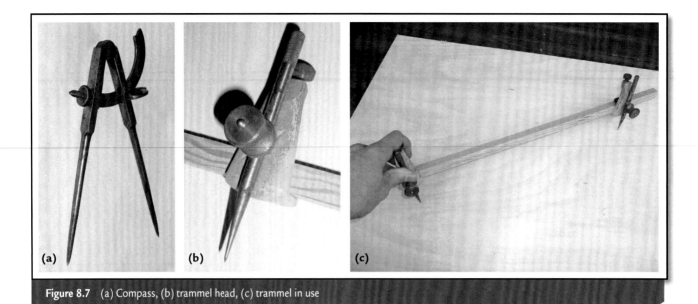

Figure 8.7 (a) Compass, (b) trammel head, (c) trammel in use

Using squares and compasses

Figure 8.5 shows the right way to hold a square onto an item. This method allows you to feel if the square is being held true to the surface or not. As when using a common compass, the correct method lets you feel if the compass radius alters while it is describing the arc.

Tools for hitting

Mallets

Mallets come in a variety of forms, sizes and materials (Figure 8.8). Their purpose is usually to give a more 'weighted' (stronger), yet easily controlled, blow. Often they are used in conjunction with another tool or to position specific materials, as described below.

- *Timber*—generally used for striking chisels or other tools with breakable handles.
- *Plastic*—used for striking metal without marking (such as tile trims, plaster corners or suspended ceiling components).
- *Rubber*—used for tapping brick or block work, tiles and pavers.
- *Metal*—included in this group are lump or 'mash' hammers, used for providing a firm, controlled blow to bolsters and cold chisels. They are also used for putting in timber pegs, star pickets and the like.

Hammers

Hammers are generally used for tapping, driving or breaking. With the exception of lump or 'mash' hammers (see above), a hammer is generally identified by a smaller head in relation to its handle size. Mallets give a broad surface blow, hammers a more point-specific strike.

- *Claw hammer*—a common hammer used by the contemporary tradesperson in various fields of work (Figure 8.9). They are used for driving in or taking out nails, and on punches, bolsters and cold chisels for light work. (They should be used with caution on chisels: turn the head flat to increase the strike area and soften the blow.)
- *Warrington, cross pein or brad hammer*—a small hammer used to tap in panel pins and brads. This is an important hammer for painters when refixing glazing beads, for example (Figure 8.10).
- *Sledge-hammer*—a large metal hammer for demolition work or driving in larger pickets or pegs (Figure 8.11).
- *Punch*—used in conjunction with a hammer to drive the head of a nail or spike below the surface (Figure 8.12).

Holding and using hammers

How you hold a hammer depends upon the type or weight of the hammer and the blow required.

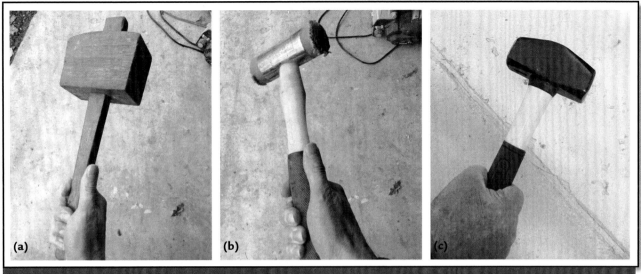

Figure 8.8 Mallets: (a) Wooden, (b) plastic tipped, (c) mash

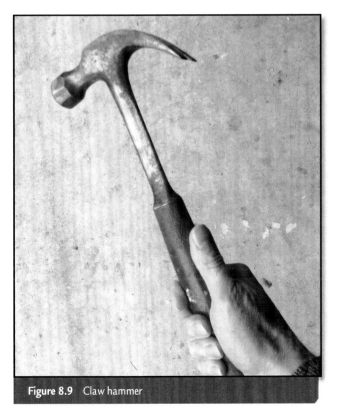

Figure 8.9 Claw hammer

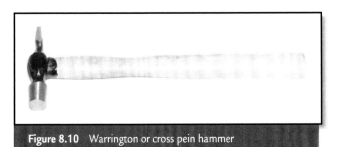

Figure 8.10 Warrington or cross pein hammer

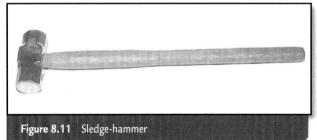

Figure 8.11 Sledge-hammer

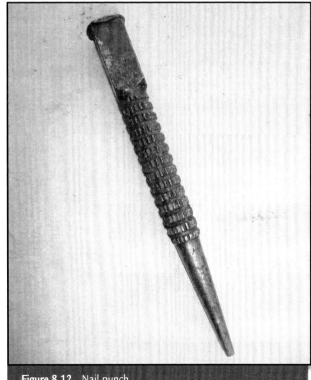

Figure 8.12 Nail punch

Figures 8.8 and 8.9 show the correct grip for the common types of hammer. Ensure your grip is firm but not tight or locked, and that the muscles of your arm are relaxed.

As you swing a hammer, your wrist and elbow should be unlocked. Your mind and eye should be on the target, not the hammer head. Strike firmly but make use of the rebound so that you use less energy bringing the hammer back up again. For heavy, demolition-type blows, swing 'through' the object but maintain your sense of balance so that you are not pulled forward by the hammer's weight.

Tools for levering

This group includes tools such as crowbars, pinch (or jemmy) bars, spanners and sockets of various types. Many tools may be used as levers (chisels for example); however, those listed here are purpose-built for the job.

- *Crowbar*—a long metal bar with a round, square or hexagonal section (see Figure 8.13). Generally crowbars have a flat, chisel-like point at one end and at the other, either a spear point or a flat head for striking or compacting. They are used for breaking up concrete or stone, digging holes in hard ground, or levering. When working in the sun, never leave a crowbar lying on the ground; thrust it into the ground so that it stands upright. This will prevent the crowbar from getting hot and make it less effort to pick up next time.
- *Pinch or jemmy bar*—a steel bar of a hooked or bent shape (see Figure 8.14), generally with a claw-like end for removing nails. These are used for levering, or prying loose material during demolition.

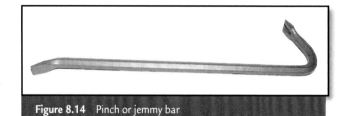

Figure 8.14 Pinch or jemmy bar

- *Spanner*—a tool for tightening or loosening nuts and bolts, generally made of metal (sometimes plastic). There is an almost infinite array of spanner sizes and types to suit different situations, nut or bolts sizes, required leverages or available space. Adjustable spanners or 'shifters' are useful if used carefully when a specific spanner is unavailable (Figure 8.15).

Figure 8.13 Crowbar

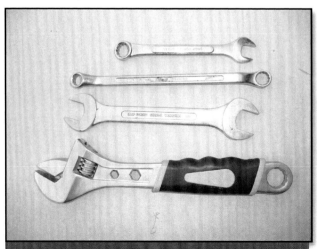

Figure 8.15 Spanners (shifter at bottom)

- *Sockets*—These are metal tools that fit over nuts or bolts (or in some cases inside them, as in torque screws or bolts), allowing an arm or lever to be attached and the nut or bolt turned. This lever may be a straight, long bar or a 'ratchet' arm which can be set to turn clockwise or anticlockwise on demand (Figure 8.16).

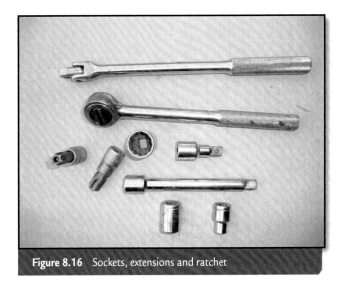

Figure 8.16 Sockets, extensions and ratchet

Using levers

Watch your hands and fingers! When levering it is very easy to jamb your fingers between the tool and the work piece or surrounding objects. This is particularly so if the tool slips or the levered object lets go suddenly. Should there be any chance of this happening, use your palms with the fingers spread wide and clear of the tool.

Tools for cutting, scraping and spreading

The list of tools for cutting, scraping and spreading is a long one, particularly as it covers four rather diverse trades. They can, however, be clustered into the following categories: saws, knives, chisels, scissors, abrasives, hole saws and drill bits.

Saws

Saws are generally metal blades with a row of sharp teeth that, when dragged back and forth over a material, will make a clean cut. The shape, size, direction, and 'set' of the teeth on saws varies depending upon the intended use and material to be cut. Most handsaws (Figure 8.17) purchased today have hardened teeth that cannot be sharpened and are considered disposable. Almost criminally, no supplier or manufacturer of these saws has seen fit to provide a recycling point, which means that most of the metal from these saws ends up as land fill.

Other common saws include the hacksaw (for metal cutting; see Figure 8.18) and keyhole saws (for cutting small openings in plaster and other sheet material; see Figure 8.19). Note that the forefinger position on a hacksaw is outside the frame, or within

Figure 8.17 Handsaw

Figure 8.18 Hacksaw

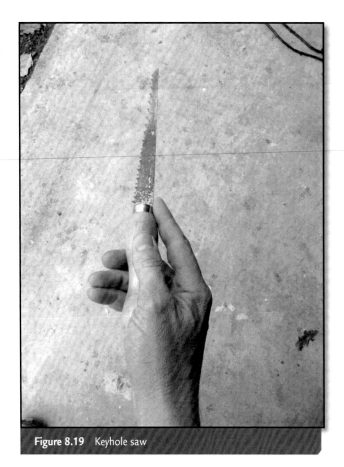

Figure 8.19 Keyhole saw

Figure 8.20 Utility knife

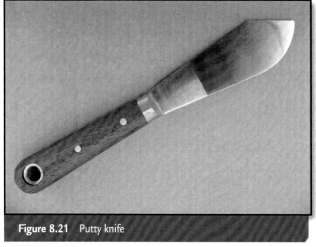

Figure 8.21 Putty knife

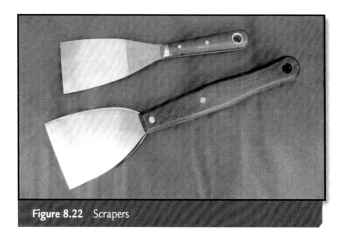

Figure 8.22 Scrapers

the handgrip. If it is elsewhere and the blade breaks, then the blade can fold back and amputate your finger.

Knives

The most common knives used by tradespeople are known as utility knives (Figure 8.20), which are used to cut all sorts of materials from paper to plastics and even light metals. Knives are also useful for positioning items such as masking tapes, trimming paints and fillers. The safer of these tools have retractable or folding blades that lock into position when opened. Never use a non-locking pocket knife for trade work; these can fold back suddenly and sever a finger.

Other tools that fall into the knife category are actually tools for applying, spreading or scraping off materials. The putty knife (Figure 8.21), for example, is a common resident in the tool boxes of many trades, as are paint scrapers and jointers of various widths and forms (Figures 8.22 and 8.23).

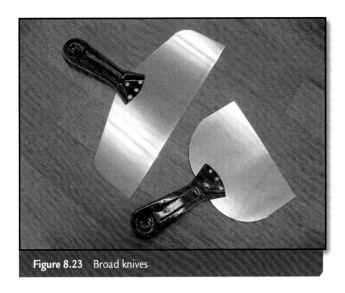

Figure 8.23 Broad knives

Chisels

Chisels are sharp metal implements for cutting or paring timber, plastics and, in some instances, light metals. Although wood chisels (Figure 8.24) are used mainly by the woodworking trades (e.g. carpenters), painters and plasterers may also sometimes find them useful for preparation work.

A very different form of chisel is the cold chisel, which is used by bricklayers for chipping into bricks and pavers, and is also handy for shearing metal rods. However, the main cutting tool of the bricklayer is the bolster (Figure 8.25). This is a much broader, flatter form of cold chisel that is used to cut bricks accurately.

Figure 8.25 Bolster

Scissors and shears

Most readers will be familiar with common paper scissors, and scissors in this form are used by painters. Other forms, however, may be less familiar: tin snips or 'aviation' snips (Figure 8.26), used for cutting light metals and some plastics; wire cutters, used for the same purpose and also for plastic trims;

Figure 8.24 Wood chisel

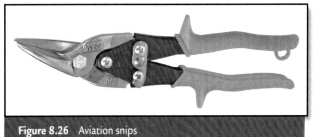

Figure 8.26 Aviation snips

bolt cutters (Figure 8.27), used for heavier wire or metal rod; and fibre-cement sheet (fibro) cutters, a type of shears used for cutting cement sheet products (Figure 8.28).

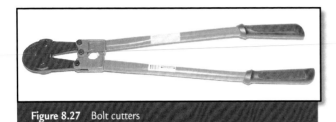

Figure 8.27 Bolt cutters

Figure 8.28 Fibro cutters

Abrasives

Abrasives take many forms, from the common 'sandpaper', to metal files and abrasive stones.

Papers

Papers are commonly referred by the misnomer, 'sandpaper'. Papers are actually a sheet material (cloth, paper or fibre-reinforced paper) that has an abrasive material bonded to it. This material is rarely if ever 'sand', though it may be a naturally occurring material sourced from mineral sands. The most common papers available are listed below.

- Naturally occurring grits:
  - flint (light to dark grey);
  - emery (very dark grey to black);
  - garnet (pink/orange to red).
- Manufactured grits:
  - silicon carbide (dark grey to black);
  - aluminium oxide (white to red/orange brown);
  - glass (a yellow-brown 'sandy' colour)

Papers are graded by the size of the grit, and the space (open or closed) between the each grain. Open papers are best for paint removal as they tend not to clog so easily. Closed papers are best for finishing work as they cut more evenly, leaving the surface smoother and the cut lines of the grit less visible.

When using papers, remember the following points:

- If a surface is rough, start with a coarse paper first to get rid of the humps and hollows (fine paper will just follow them). Then follow with progressively finer grits.
- Be dust conscious. Dusts can be a nuisance or highly toxic. Over time even nuisance dusts can lead to industrial asthma (a debilitating breathing complaint). Wear a P1 dust mask as a minimum safety precaution.
- Use a cork sanding block wherever possible. When this is not practicable, fold the paper into three layers (not two), as this stops the paper from slipping against itself.

Files

Files come in various forms, from open cut 'rasps' to fine cutting 'mill' files. Made of hardened metal, files are designed for shaping wood, metal, plastic and even stones. Although they are not common tools for bricklayers or painters, files are starting to be used more frequently by tile layers and plasterers, who must work with metal trims and framing components.

The following points should be remembered when using files:

- Files should not be rubbed back and forth over a surface. The cutting action is on the forward stroke only. The return stroke should be unweighted (slightly lifted), otherwise the file becomes blunt more quickly.
- Rubbing chalk into the face of a file can help to reduce clogging by rejecting the swarf (filings).

Abrasive stones

Abrasive stones are used for wearing down very hard surfaces in order to remove sharp edges, or to straighten or sharpen them. Most hand stones are either silicon carbide or aluminium oxide, though the use of industrial diamonds is becoming more popular.

Hole saws and drill bits

This is a many and varied array of tools. Augers, hole saws, jobber bits: the list is endless (see Figure 8.29). In fact, there is a drill bit for just about any occasion. The list below includes only the most common tools, and only those that fit a standard power drill.

- *Hole saw*—used to cut simple holes in a variety of materials, including timber and plaster sheet. There are diamond versions for cutting tiles and concrete, although the latter is a more specialist tool.
- *Spade bit*—a flat-bladed drill bit for rapid cutting into timber, plaster and similar soft materials, including plastic (at lower speeds).
- *Auger bit*—a more accurate and cleaner-cutting bit than the spade bit, but more prone to blocking and overheating.
- *Jobber bit*—the main stream drill bit for mild steel, timber, plastic and plaster.
- *Masonry bit*—a tungsten carbide tipped drill bit designed for drilling holes in stone and masonry.
- *Diamond drill*—this bit looks very different from a standard drill bit in that it has no twisted shank; rather, it is cylindrical and has a flat end. Diamond drills are used for cutting holes in porcelain tiles. They must be used carefully with a constant (though only light) supply of water to keep them from overheating.

Using cutting and scraping tools

Cutting and scraping tools are sharp, or they should be… Although this seems obvious, it still comes as a surprise to some. They are in fact more dangerous when blunt, as blunt tools require more force. Figures 8.3 and 8.4 showed some dos and don'ts for this category in relation to hand position. Watch your stance in particular, as this will help you guide your tool with control and retained balance, rather than using force. (For drill bits and hole saws, see the section on power drills.)

Tools for screwing

This is a broad range of tools used for driving an endless array of screw types.

Although cordless drivers are increasingly common, a collection of humble screwdrivers is still essential to almost all trades, as there is invariably something that will need to be fixed or removed with a screw of some form. Figure 8.30 shows the main types of drivers in use today, including flat, Phillips and torque screwdrivers and Allen keys.

When using screwdrivers, it is again important to maintain your balance and focus upon what is happening at the tip. This is particularly so with slotted drivers that can easily slip and penetrate either the work piece or your hand if it is in the way.

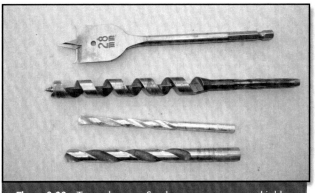

Figure 8.29 Top to bottom: Spade, auger, masonry and jobber

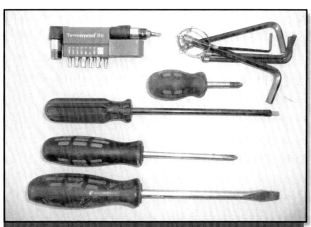

Figure 8.30 Top to bottom: Torque screwdriver, Allen key, Phillips (butt) screwdriver, flat (electrical) screwdriver, Phillips screwdriver, flat screwdriver

Tools for holding or cramping

This cluster is many and varied, encompassing basic pliers to sash cramps. The purpose of these tools is to either hold something as an extension of your hands, or hold things together so that other work may take place or an adhesive may set.

- *Pliers*—each trade tends to have some form of preferred pliers, but Figure 8.31 shows some of the most common, which include:
    - side cutters (which can cut as well as hold);
    - 'multi-grips'—these can be adjusted to hold large or small items;
    - locking pliers—these are adjustable and can be used to cramp pieces together and lock;
    - needle nose pliers—used for getting into tight spaces, or holding very small items.

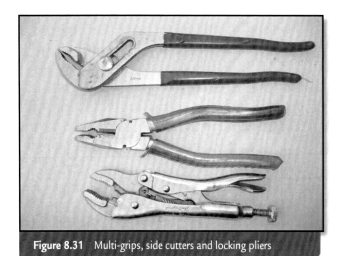

Figure 8.31 Multi-grips, side cutters and locking pliers

- *Cramps*—almost a must in every trade. Using cramps is our main method of temporarily holding items together while work is being done or glue is drying, or simply so that we can see how the finished item might look. The main forms are:
    - 'G' cramps;
    - sliding cramps;
    - quick action cramps;
    - sash cramps.

Using cramps and pliers

There are two dangers with using this cluster of tools: the risk to the work piece and the risk to self

Figure 8.32 Sliding cramps in use (note card to protect fascia)

and others. The first instance comes about through over-tightening, or by not having something protecting the finished material from the jaws of the tool (Figure 8.32). Pliers in particular can leave grooves and gouges from their serrated faces.

The second instance comes about through placing too much faith in cramps, especially sliding and quick-action cramps. It is dangerous to hold in place overhead items, or planking upon which you will stand, with nothing more than a sliding cramp, yet it is done all too regularly.

Tools for cleaning

Cleaning is an ongoing exercise for all tradespeople: it comes with the territory, as it were. Sometimes, particularly in painting, cleaning is part of the preparation. The following are the basic tools, most of which (it would be hoped!) will already be familiar.

- *Brooms*—for general sweeping (Figure 8.33).
- *Brushes*—for small sweeping jobs.

Figure 8.33 (a) Straw broom, (b) yard broom, (c) broad floor broom

Figure 8.34 Cleaning bin and brush

- *Sponges and sponge cloths*—for wiping off excess material and cleaning.
- *Scourers*—for scrubbing down. Care must be taken in their selection and use as scourers can mark some surfaces such as acrylic baths and stainless steel fittings.
- *Scrubbing brushes*—used for tool cleaning, for example on shovels, knives and trowels (Figure 8.34).
- *Wire brushes*—these are particularly useful for cleaning metal prior to painting (Figure 8.35).
- *Buckets*—for carrying or mixing cleaning fluids, or carrying out waste (Figure 8.34).
- *Vacuum cleaners*—most trades would use a wet/dry unit (see Figure 8.107 on page 328).
- *Shovels and pans*—for scooping up rubbish (Figure 8.36).
- *Hoses*—for washing down larger areas, and supplying water for mixing various materials (Figure 8.37).
- *Site skips*—these are large bins placed on-site for waste materials (Figure 8.38). Large sites may have multiple skips for separating recyclables.

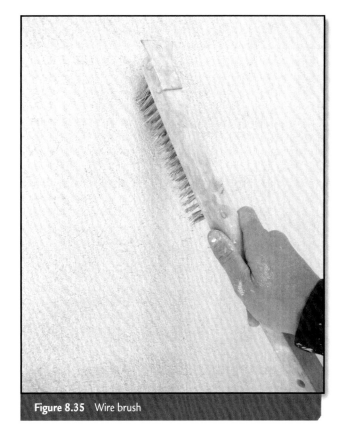

Figure 8.35 Wire brush

Figure 8.36 Shovels: (a) Square mouth, (b) long-handled square mouth, (c) round mouth, (d) long-handled round mouth

Figure 8.37 Hose and nozzle

Figure 8.38 Site skip

TRADE-SPECIFIC HAND TOOLS

All of the tools listed so far are common and may be encountered or required in almost any trade area. Those that follow, though not confined to the trade under which they are offered, are more trade-specific. In some cases—hawks for instance—the tool will be mentioned under one trade only, although it may be used by several. Conversely, trowels will be listed several times due the significant differences in shape and usage for each trade.

Plastering and rendering tools

The most common tools specific to plastering are those for spreading and scraping or sanding.

- *Trowels*—these are many and varied in size and shape. Most tradespeople tend to use only a limited array, favouring one or two particularly. The common plastering/rendering trowel is a flat, rectangular tool with an offset handle (Figure 8.39). Some have a slightly curved edge which tends to hold the plaster centrally to the trowel while feathering (thinning) the lay of plaster at the edges.
- *Corner trowel*—internal and external corner trowels are available and are sometimes used for final coat applications. Some are adjustable (see Figure 8.40).
- *Hawk*—a hand-held mortarboard that holds plaster or mortar mix ready for use (see Figure 8.41).
- *Jointing knife*—this is similar to those mentioned previously (under general tools), except that it often has a slight curve in the edge of the blade.

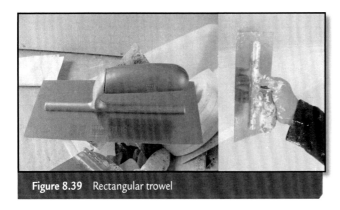

Figure 8.39 Rectangular trowel

Figure 8.40 Adjustable corner trowel

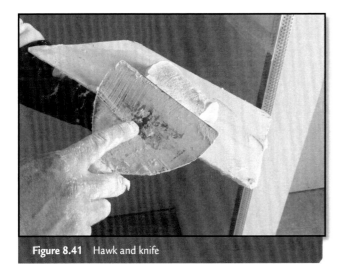

Figure 8.41 Hawk and knife

This allows for a minor build-up of material over jointing tape.
- *Taping knife*—a broad, rectangular knife used to bed in the tape on joints (Figure 8.42).

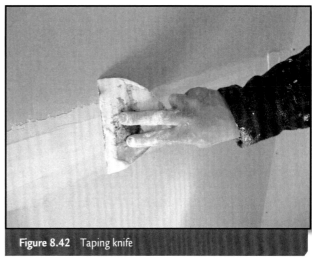

Figure 8.42 Taping knife

- *Floor scraper*—used for scraping up the waste left or dropped on concrete floors. The long handle means it can be used a bit like a pushed broom.
- *Plasterer's hammer* (also know as dry wall or lath hammer)—this was once the hammer of choice for plasterers, as it has a chisel-like blade at one end that may be used for scoring plaster sheet. Many now have just a claw hammer.
- *Circle cutters*—a form of compass used to scribe (cut out) a circle in plasterboard materials.
- *Putty or adhesive knife*—plasterers prefer long-bladed putty knives to apply plasterboard adhesives. This allows you to reach further up the wall, and more easily into the bucket without getting adhesive on your hands (Figure 8.43).

Figure 8.43 Adhesive knife

- *Small tool*—a specific plasterer's tool for getting into intricate joints, such as on ornate cornice (Figure 8.44).
- *Float*—similar to a trowel but with a thicker base. Often used in rendering, it offers a different finish and generally a thicker coat. Floats may be of different materials, the choice being based upon the finish required. There are literally dozens of types and sizes to choose from (Figure 8.45)

Figure 8.44 Small tool

Figure 8.46 Loading a tape box

Figure 8.45 Trowels and floats on display

- *Automatic taping box*—used for the automatic laying of tape and bedding compounds (Figure 8.46). Used wisely, automatic taping boxes improve jointing quality and speed. Though less common on the domestic scene, automatic corner taping boxes are also available.
- *Jointing box*—a box that rapidly and evenly applies various widths of joint compound to ceilings and walls (see Figures 8.47 and 8.48). This is a

Figure 8.47 Filling a jointing box

very efficient tool that, when used correctly, can significantly improve timelines and reduce waste.
- *Sanding blocks and floats*—various forms and qualities are available, from plastic to timber or metal (Figure 8.49). Sheets of sandpaper or sanding gauze are fitted to the blocks (some use a

Figure 8.48 Using a jointing box

Figure 8.50 Plasterer's square

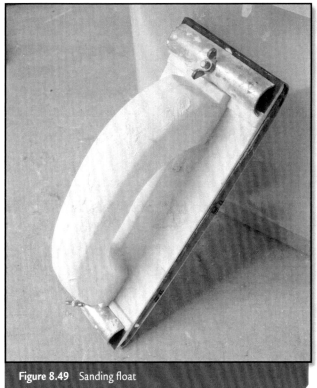

Figure 8.49 Sanding float

self-adhesive or Velcro; others are simply clamped into place).
- *Plasterer's square*—a large (usually aluminium) T-square used for marking out or directly cutting plaster sheet (Figure 8.50).

Bricklaying tools

As with plastering, the trowel is the dominant specialist tool of the bricklayer, and therefore it will be addressed first. There are, of course, many other tools that need a mention, and these will also be discussed below.

Trowels

Unlike the plasterer's trowel, bricklayers' trowels have the handle to the rear, and the blade is diamond or teardrop in shape. Yet even here there are preferred forms. Depending upon the task, these differences may be little more than personal preference, whereas on other occasions they may make an otherwise difficult task remarkably easy (the pointing trowel for example).

- *London-pattern trowel*—this is the traditional trowel of Australian bricklayers and is still the preferred pattern of many (Figure 8.51). It is claimed that the long, narrow tip allows a cleaner, neater job than the American pattern.
- *American-pattern trowel*—this has become more popular in recent years. Those who favour it claim that it makes bricklaying quicker as it can carry

more mortar. Others suggest that, being shorter, it is more balanced and less tiring on the hand and wrist (Figure 8.52).

- *Pointing and gauge trowels*—these are small tools that allow mortar to be pushed off the trowel into joints, or to create struck joints. Gauge trowels (Figure 8.53) have a rounded tip, whereas pointing trowels have a pointed end.

Figure 8.51 London-pattern trowel

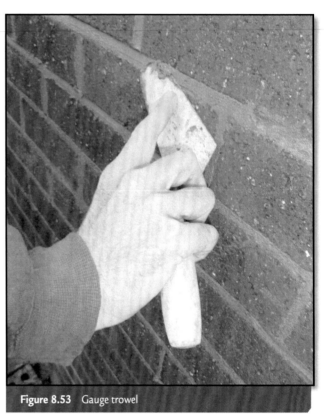

Figure 8.53 Gauge trowel

Other bricklaying tools

Other tools used by bricklayers include:
- *Brick hammer*—this tool is similar to a geologist's hammer: one face of the head is like a carpenter's hammer, while the other has a flat, pick-like blade. The tool is designed for chipping pieces off bricks or pavers (Figure 8.54).
- *Plugging chisel*—a long, slim chisel used for cleaning out mortar from perps or bed joints (Figure 8.55). A necessary tool when replacing a brick in an existing wall.
- *Corner block*—usually a plastic tool that sits on the corner of brickwork or a profile and holds a string line in place (Figure 8.56).
- *Line pin*—driven into the mortar to hold string lines; often used in conjunction with a corner block (examples of pins are shown in Figure 8.57).

Figure 8.52 American-pattern trowel

Figure 8.54 Brick hammer

Figure 8.56 Corner block and profile

Figure 8.57 Example of line or mortar pins

Figure 8.55 Plugging chisel

- *Mortarboard*—a flat board, typically a piece of plywood, used to lay mortar on ready for use.
- *Dutch pin*—driven into mortar beds to hold profiles in place (Figure 8.58)
- *Larry or mortar hoe*—used to aid the mixing of mortar in a wheelbarrow or box.
- *Jointer*—a round sectioned rod, or 'u' shaped channel, with a handle or simply bent so as to offer a handle like grip at one end. Used for rounding

Figure 8.58 Dutch pin

or smoothing out joints to an even finish (Figure 8.59). Flat jointers are also available for backfilling joints as necessary.

Figure 8.59 Round jointer

- *Raker*—a tool used to clean out fresh mortar from joints to a specific depth (Figure 8.60).
- *Brick* **profile**—generally a vertical, square section post that is pinned or clamped in place at the corners or ends of brickwork. It provides a reference plane, brick course heights and a location for corner blocks (Figure 8.61).
- *Profile clamp*—a very specific form of clamp that holds profiles to timber or metal frames. Generally screwed into position (Figure 8.62).
- *Brick carrier*—a form of cramp that allows multiple bricks to be carried easily at one time (Figure 8.63).

Figure 8.61 Profile

Figure 8.60 Raker

Figure 8.62 Profile clamp

Figure 8.63 Brick carrier

Painting and decorating tools

The brush or roller is to a painter what a trowel is to a bricklayer: the dominant tool of the trade. While a lot can, and should, be done with just the house brush, unlike trowels there are dozens of brushes and rollers to choose from. Surfaces are not painted by brushes and rollers alone, however: spray equipment of various forms is taking a greater share of the trade. So, aside from brushes and rollers, there are several other important tools to be considered.

Brushes

Figure 8.64 shows an array of just some of the brushes available: there are many! Choosing a brush depends on the task and the context: the task will influence the size, shape and width of the brush, while the context (e.g. temperature, humidity and surface) will influence the choice of bristle as well as the other variables. In almost all cases, quality brushes will serve you better than poor ones.

The important aspects of a quality brush are bristle type, the manner in which the bristles are set (Do they come out easily and get left in the paint? Do they retain their shape?), the ferrule (Will it rust?) and the handle. Apart from size, they may be grouped by bristle and handle shape as follows:

- *Wall brushes*—these have large stubby handles and are square-cut and 'full' brushes, designed to hold a large quantity of paint (Figure 8.65). In a skilled hand, a good-quality wall brush can still be used to cut in neatly to windows, architraves and other trims.
- *Sash cutters*—these are generally narrower and slimmer than the fat and wide wall brush. Their main characteristics are a long, narrow, round handle and proportionally longer bristles (Figure 8.66). Sash cutter bristles can be square or angle cut.

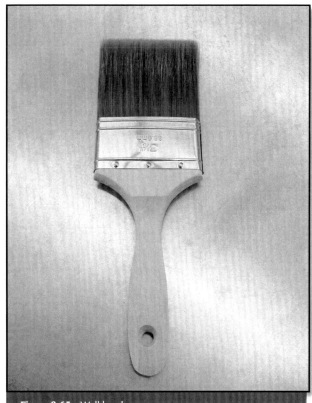

Figure 8.65 Wall brush

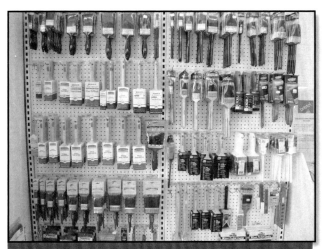

Figure 8.64 Just one rack of the many brushes available

Figure 8.66 Sash cutter (note the drop sheet behind)

Figure 8.68 Oval wall brush

- *Rat-tailed brushes*—similar to sash cutters but with a flat rather than round handle (Figure 8.67).
- *Oval brushes*—these can be found as wall or sash cutter types but with a distinctive oval body of bristles allowing for a fuller load of paint (Figure 8.68).
- *Round brushes*—generally found as a form of sash cutter, or much smaller as a detail brush.

- **Fitches** *(trimming or detail brushes)*—often rat-tailed in handle, these brushes can be flat, slim and square-cut; round and pointed; or round and stumpy (Figure 8.69). Fitches are designed for smaller trims and fine detail.

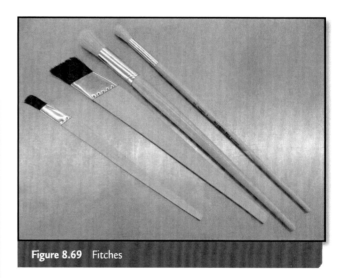

Figure 8.69 Fitches

- *Block brushes*—large, square-set brushes for painting bricks or concrete block work.
- *Stencil brushes*—S-stumpy, round, square-cut brushes for dappling over stencils, or certain 'dry brush' dappling techniques.
- *Foam brushes*—foam applicators on a paint brush handle. In small areas these can offer a finish that replicates a roller better than a bristle brush might.
- *Paint pads*—these come in variety of sizes and effectively 'smear' the paint onto a surface. They can be useful for ceiling work when used with an extension pole.

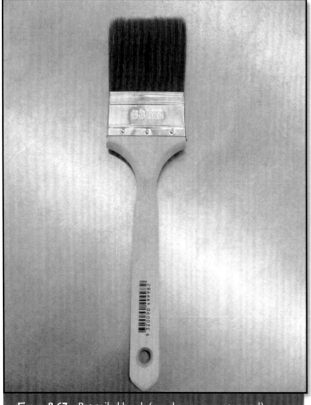

Figure 8.67 Rat-tailed brush (may be square or tapered)

Using brushes

Using a brush is a skill developed over time, in context, using a variety of mediums and on a variety of surfaces. It is also best developed under the guidance of a skilled professional. This being the case, only a few basic pointers are relevant here:

- Hold the brush as shown in Figure 8.66. The exact hold will influenced by the angle of the surface to be painted, the surface material, the paint and the brush size.
- Don't overload the brush; dip the brush only about a third to half of the bristle depth.
- Tap the brush gently on the container edge; reduce wiping the brush on the edge of the can to a minimum.
- Brush the paint out fully, using strokes in a variety of directions to ensure good coverage (particularly on porous or grained surfaces), but generally finish with strokes in one direction (this will depend on the finish or texture desired).
- Avoid brushing back over earlier work, which may already be partly dried.
- Work methodically over a surface (have a plan of where you will start and finish).
- Rather than cleaning brushes during minor breaks, they can be kept wet by a variety of methods; for example, oil-based brushes can be placed in water, and water-based brushes wrapped in cling film.
- Clean your brushes promptly and thoroughly; where possible keep them moist between jobs, or dry them and hang them so that the bristles remain straight.

Rollers and trays

Like brushes, many types of rollers are available (such as the detail roller shown in Figure 8.70). In the main they vary in size, texture, material and quality. Cheap rollers tend to leave more fibres and other impurities in the finish, even when washed or used several times. Your choice of roller will depend on the desired finish and the type of paint you are putting on: oil-based paints cannot be used on foam rollers, for example. Paint trays (see Figure 8.71)

Figure 8.70 Detail roller

Figure 8.71 Paint tray and roller

may be of plastic or metal, and either form can offer a choice of good-quality or cheap units.

When using rollers it is important to remember the following points:

- 'Roll in' a new roller on a surface that is not important, or on a piece of sheet that can be discarded. Despite cleaning a new roller, and

wiping down the nap or pile (the outer surface material) to remove any loose lint, some may still remain and come off on your finished surface.
- If you want a quality finish, buy a quality roller. A cheap roller will cost you in time-consuming repair work.
- Having washed a roller, wring it out thoroughly. If any free water or solvent remains, it will thin the paint and make it run or bubble. It is best to roll it out on an unimportant surface (as if new) until the paint is flowing correctly.

Other painting and decorating tools

Other tools used for painting and decorating include:
- *Extension handles*—primarily used on rollers but also available for paint brushes and sanders (Figure 8.72). These extendable rods save a lot of stretching and unnecessary scaffolding.
- *Mahlsticks (also maulstick)*—supports or props for the brush hand. These are used in detail work, such as sign writing, to keep the hand from spoiling earlier work, and consist of little more than a dowel with a ball or wad of chamois or cloth to stop it from slipping. Mahlsticks are usually made by the painter to suit themselves (though they can be purchased).
- *Scrapers*—various forms are available, from wide blade scrapers to small units with replaceable blades. Used for removing wallpaper and other old surface coatings (Figure 8.73).
- *Paper hanger brushes*—broad stiff brushes without a handle, used for laying on wallpaper.
- *Drop sheets*—canvas or plastic sheets used to protect existing surfaces, furniture or fixtures during the painting process (see the drop sheet in the background of Figure 8.66).
- *Paint pots, buckets or 'kettles'*—paint from larger cans or drums is decanted into these small, lighter, containers for immediate use. This allows the unneeded paint to be sealed and stops it from drying out unnecessarily. It also reduces the physical stress on the painter and the size of accidental spills.

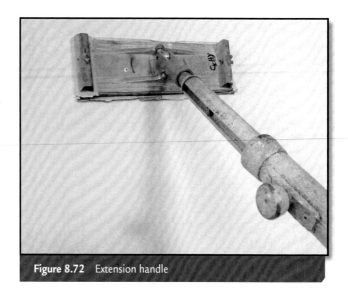

Figure 8.72 Extension handle

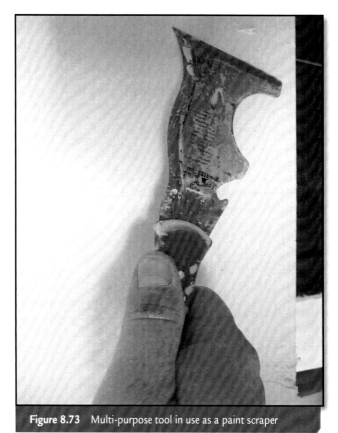

Figure 8.73 Multi-purpose tool in use as a paint scraper

- *Masking tools*—masking tape is more a consumable material than a tool; however, there are various means of applying it (and various qualities of tapes). Some machines will dispense not only the tape, but also a wide, clear plastic film as well. These are particularly useful around window and door trims (Figure 8.74).

Figure 8.74 Masking tool

(sometimes diamond), they make a groove which will become a break once pressure is applied with the snapping arm (Figures 8.75 and 8.76). On large tiles (600 mm) it is sometimes best to remove the tile after scoring and snap it over a long, straight edge of metal or timber (this can also work for some porcelain tiles, saving wheel cutting).

Figure 8.75 Tile cutter

Tile laying tools

The contemporary tile layer requires some very specialised tools, not only to be competitive, but also simply to get the job done. Porcelain tiles are becoming more common, and the only accurate way to cut these cleanly, or to drill or otherwise cut holes in them, is by diamond wheel. In addition, bathrooms and ensuites have become more demanding to install, with styles that include metal trims and very tight joint demands. Furthermore, waterproofing codes have tightened and the price of failure in this regard is high—for all parties. Ceramic tiles still have the greatest market share however, and as such the common hand tools listed below apply in the main to this form of tile. Most forms of diamond cutters will come under the heading of plant and equipment later in the chapter.

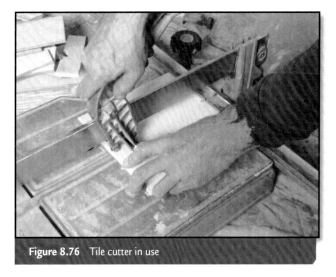

Figure 8.76 Tile cutter in use

Tile cutters

Large and small professional-quality tile cutters are available; the choice depends upon the size of the tiles being cut. Fitted with tungsten carbide scoring wheels

Nippers

Nippers are plier-like tools with tungsten carbide (or otherwise hardened) edges for chipping or cutting tiles to shape. Several types are available, and the choice of which to use depends on the task and the material being worked.

- *Straight nipper*—used for straight cuts (Figure 8.77).

Figure 8.77 Straight nipper

- *Curved (mosaic) nipper*—this allows for more detailed work, cutting smaller sections off at a time, and also enables curved work. It is used to make mosaics, as the pieces don't come off hard and straight.
- *Quarry nipper*—a larger nipper for quarry tiles, which are thicker and harder to cut. The jaw opens wider and the handles are longer for greater leverage.
- *Parrot nipper*—shaped like a parrot's beak, this nipper allows holes to be shaped or enlarged in tiles once an initial penetration is made with a drill or angle grinder (Figure 8.78).
- *Trim or PVC nipper*—used for cutting plastic and light metal tile trims.

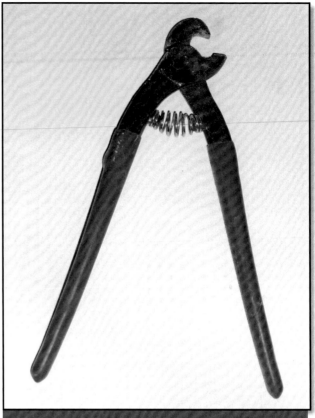

Figure 8.78 Parrot nippers

Trowels

Three forms need to be considered here: notched, plain and gauge.

- *Plain trowels*—used for screeding in much the same way a concreter might. These are generally rectangular, and most tilers will just have the one size that they will make fit all purposes (similar to plasterers' trowels; see Figure 8.39 on page 303).
- *Notched trowels*—these are for spreading adhesives and vary in notch size depending upon the bed depth required (Figures 8.79 and 8.80). They can be purchased as either left- or right-handed (depending on which side the notches are located) and with notches of 4 mm, 6 mm, 8 mm, 10 mm, 12 mm and 15 mm. It is not unknown for tilers to make their own notched trowels if the desired size is unavailable.

Figure 8.79 Applying adhesive with a notched trowel

- *Gauge trowels*—these are the same as the bricklayer's version, and are used for measuring in materials for mixing and for applying adhesives and other mixtures to trowels (see Figure 8.80).

Figure 8.81 Grout float and scourer

Figure 8.80 Notched and gauge trowels

Grouters

Grouting tools come in various forms:
- *Squeegee*—a simple rubber strip embedded into a timber or metal handle. Squeegees are also used for cleaning.
- *Float*—a more trowel-like form of grouter, and favoured by most tradespeople. The base is rubber or rubber-coated, which allows for more grout to be applied and more pressure to squeeze the grout into the joint (Figure 8.81). Floats can be pointed (known as 'boats') or rectangular.
- *Grout sponge*—a large sponge on a float. The sponge is segmented, trapping the excess grout. This is then deposited into a bucket by passing the float over a pair of rollers (Figure 8.82).
- *Grout saws and rakes*—small tools used to remove grout and adhesive from between joints.

Figure 8.82 Grout sponge (note cuts in foam showing under fingers)

Other tiling tools

Other tools used for laying tiles include:
- *Scriber*—a pencil-like tool with a tungsten carbide or diamond tip (or wheel) for scoring tiles prior to snapping them.
- *Spacers and wedges*—various sizes of these are available depending on the tile used or the joint width desired. Spacers are of a fixed size, whereas wedges allow you to vary the joint as needed.
Note: Don't depend upon spacers alone to give you the right gap: use your eye, straight edge or level to ensure that the tiles are running true.
- *Suction grips*—used to help lift tiles that are down a little and causing a lip (Figure 8.83).

Figure 8.83 Suction grips

- *Caulking gun*—a tool common to all trades that doesn't really fit any category. It is an adhesive, caulk and sealant delivery system that has become almost indispensable (Figure 8.84). Various models are available, but they all do effectively the same thing: squeeze fluids from a tube. Learning to use a caulking gun efficiently is important. When applying silicone or caulking to visible corners or edges, a neat finish is achieved by running the back of an empty tube (or a cut water bottle) up the joint. This will produce a clean curve and scoop up any residue without spreading excess everywhere (keep your fingers out of it!).

PLANT AND EQUIPMENT

This section is broken up into two parts: power and powered hand tools, followed by other plant and equipment. Both sections are dealt with from a general perspective rather than being trade-specific.

Power and powered hand tools

This group covers both battery- and mains-powered equipment. Indeed, many of the tools mentioned are available in either form. However, as 240-volt power, and the associated leads and safety practices, are still necessary on-site, these factors will be dealt with first.

Power supplies

The building process still requires 240-volt power, and this will probably remain the case for some time. Getting this power to a new site used to be achieved by using a temporary power pole, but this form of supply is now mainly confined to larger commercial sites (see below). Contemporary domestic building practice allows for the early installation of the house switch board (Figure 8.85), so most builders use a portable generator till then. This also reduces costs: there are no electrician and power authority fees for the installation and removal of the power pole; the power is cheaper, as domestic rates are lower

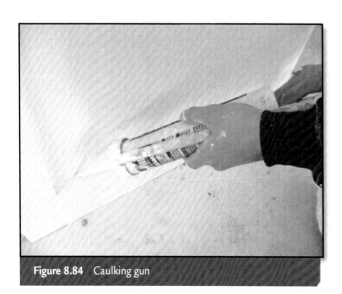

Figure 8.84 Caulking gun

Figure 8.85 Accessing domestic power boards

than the industry rates charged for power from a temporary pole; and the power lead runs are shorter (and safer).

Temporary power pole supply

The tradesperson will still come across temporary power pole supply on larger commercial sites (Figure 8.86). It is important to be aware that the power supplied to these poles may include 415 or 'three-phase' power for use by heavy equipment. In addition, power may be reticulated around large sites to multiple locations and include sub-boards.

Only electricians may connect power to temporary supplies, and even then only with authorisation from the mains supply authority. However, it is often up to the builder to supply and install the pole itself (which may or may not have an existing board on it). The location of temporary power is important, and the person choosing where to put it should consider:

- its accessibility for end users;
- the accessibility of the nearest mains supply;
- placing it so that water will flow away from the location of the pole.

In addition, the board must be weatherproof, lockable and placed at a convenient height, and preferably have weather-protected penetrations to allow leads to exit the box without the lid remaining open.

Portable generators

Portable generators are petrol- or diesel-powered units that can produce 240-volt power. Although considered a two-person lift, they are still small and lightweight for what they produce (Figure 8.87). Diesel units are more expensive but can maintain a continuous load and are more fuel efficient. Petrol units are more common, as they are cheaper to purchase and because most construction work requires only short bursts of power rather than continuous use.

Figure 8.86 Typical builder's temporary power pole and board

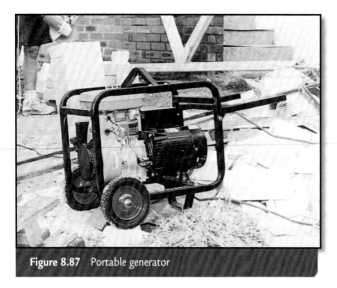

Figure 8.87 Portable generator

The maximum capacity of portable generators is as follows:
- 2 kVa: 1600 w (6 amps)—enough for two power tools;
- 5 kVa: 4000 w (16 amps)—for five power tools;
- 8 kVa: 6800 w (26 amps)—for eight power tools or light machinery use.

Maintenance of temporary and portable power

Take nothing for granted with power supplies: carry out checks regularly. For temporary power poles, check:
- the stability and security of the board;
- for vandalism or tampering;
- the level of weather protection.

For portable generators, check:
- the oil levels (four-stroke units);
- the petrol level (and petrol type, clearly indicated on or near the filler)
- the spark plugs and air filters;
- the pull-start cords for fraying;
- the power outlets.

Power and safety precautions

Irrespective of whether it comes from an existing mains power supply, a temporary power pole or a portable generator, 240-volt power is dangerous and can kill. In addition, the wet trades are so called for a reason: they involve water. Water and electricity are not a good mix; rather, they are a deadly combination that can lead to the sudden grounding of current, with you in the circuit. Your first defence, in practice and by Regulation, is an **earth leakage device** or **RCD unit (residual current device)** that complies with AS 3190-2009.

All power to *all* power tools, including battery chargers for cordless tools, *must* be routed through an RCD unit first. Do not assume that an existing house or building power point is protected: plug in your own unit.

The RCD unit (Figure 8.88) is designed to detect earth leakage (whether through you or incidentally) and cut power supply instantaneously. That means you live to exclaim, 'Hey! What happened to the power?'

Figure 8.88 RCD unit

Testing and tagging

All electrical equipment used in a workplace must be tested and tagged as prescribed by the relevant state's OH&S Regulating Authority (WorkCover NSW or WorkSafe Victoria, for example). This procedure must be carried out by a licensed person (generally an electrician), and a coloured tag indicative of the month of testing must be applied to the lead (Figure 8.89). Codes of Practice outlining this procedure may be found on the websites of each state or territory Regulating Authority.

Extension leads

Extension leads are still a necessary evil on work sites and must be treated with care. The length of a lead, and the power drain imposed, determines the

CHAPTER 8 HAND TOOLS, PLANT AND EQUIPMENT 319

Figure 8.89 Tested and tagged lead

Table 8.1 Maximum length of extension leads

Amp rating	Conductor area (mm²)	Maximum lead length (m)
10	1.0	25
	1.5	32
15	1.5	25
	2.5	40
20	2.5	32
	4.0	40

Figure 8.90 Lead stand

sectional size of the conductor (wires within the lead). Table 8.1 outlines the length of lead allowed given the amperage and conductor size.

In addition, the following precautions should be undertaken:

- All leads must be fully unrolled before use. This reduces heat build-up, which can cause damage to the lead or be dangerous.
- Leads must be suspended off the ground or work platform to a height of 2.1 m so as prevent them being stood upon or driven over, and to avoid any damage to the insulating shroud (Figure 8.90).
- When rolling up a lead, first go to the switch board/power board and switch off and unplug the lead. Then roll the lead up from the pin end. Tradespeople have died rolling the lead up the other way before disconnecting.
- Never patch or bandage a lead—aside from being dangerous, it is in breach of Regulations to do so.
- Never allow a power lead to lie in water or on wet ground.

Power boards

Note now and remember: **No** *double adaptors allowed*. Only power boards complying with AS 3105 are to be used on a work site. This means that each outlet has its own switch, and the unit is protected against excessive current drain (too many power tools on the one board) and is robust. The unit can be an RCD in its own right (see Figure 8.88) or it can be connected to an RCD (the RCD is always plugged in first).

Mains-powered tools

There are too many tools that fit in this category, particularly given the diversity of the four trades in question. There are, however, a few tools that are common across the trades, albeit handled slightly differently by each trade and used on different materials. Most of the safety factors will be addressed with each tool; however, a note on PPE and clothing is appropriate here.

Power tools, PPE and clothing

Power tools generate waste more rapidly than hand tools, and spread that waste further. The waste is also much finer and so more easily enters and remains in the lungs. It must therefore be understood that wearing PPE (personal protective equipment) appropriate to the tool and material is critical. In particular, wearing a high-quality, approved dust mask is paramount when grinding or cutting materials such as tiles, concrete or bricks. This also applies to those who are around you while you are doing so.

As for general clothing, loose apparel such as baggy T-shirts or shirts left hanging out is not safe. As this author has witnessed, a humble drill can cause a severe injury to the stomach when the wind suddenly blows a T-shirt into the drill. The drill and bit are instantly drawn into the body. So think ahead and wear tidy clothing that suits the material and the task; angle grinding metal while wearing shorts is not a great fashion statement.

Angle grinders

Angle grinders are a multi-purpose tool used to cut bricks, pavers, stone, ceramic and porcelain tiles as well as metal. The tool is available in a variety of sizes based upon blade diameter, ranging from 100 mm to 300 mm. Tilers will generally work with small-diameter tools (100–125 mm; see Figure 8.91), while bricklayers tend to favour a 9-inch or 230 mm unit (Figure 8.92).

Figure 8.92 Angle grinder—9 inch (230 mm)

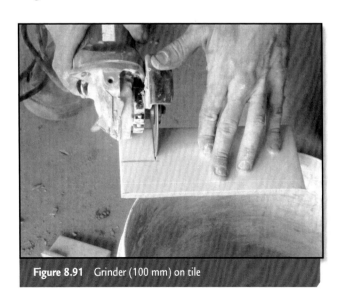

Figure 8.91 Grinder (100 mm) on tile

There are many types of blades available for angle grinders; the following is a list of only the main groups.

- *Diamond blade*—there are various cutting patterns available, some which cut faster and others cleaner. Some blades are specifically designed for cutting porcelain, for example. Diamond blades are colour-coded for their speed and the coarseness of their cut.
- *Metal cutting blade*—standard cutting blades are about 2 mm thick; however, very fine blades are now available which cut cooler and quicker, but are more fragile. These thin blades are excellent for cutting aluminium trim.
- *Metal grinding blade*—a thicker blade designed for grinding rather than cutting. Never grind with a cutting blade; it is too thin and may break.
- *Stone or concrete blades*—these are similar to metal cutting and grinding blades but made of a different compound. Don't try to use them on metal, as they can overheat and shatter.

When using an angle grinder, it is important to remember:

- PPE—safety goggles and hearing protection are essential;
- to use guards—these are moveable, but must always be in place when being used;
- to watch out for flying debris—angle grinders throw out a lot of material fast and hot (particularly with metals), so you must ensure that no-one,

including you, is in line with the direction of waste disposal;
- to use both hands—these tools 'jump' or 'run' easily if you do not hold them correctly or work the material correctly.

Note: Angle grinders cannot be used in an open (unenclosed) space during a total fire ban day.

Circular saws

Circular saws (Figure 8.93), or power saws, are used by many trades, but in the wet trades this tool takes the form of a wet saw (Figure 8.94). This means that the blade is fed water during the cutting process to reduce heat stress. However, the addition of water brings the risk of electrical shock if the tool is not handled correctly. In light of this risk, these saws are now available as cordless units (some with their own water bottle rather than a hose), while other manufacturers have added an inbuilt RCD unit to the tool.

The size of the tool depends on the material worked or the task you need to perform. Saws come in sizes from 125 mm up to 260 mm; generally you would choose one that can penetrate completely through the material with at least 12 mm to spare.

When using a circular saw, remember that all power saws will kick backwards towards the user should they jam or be used incorrectly. Wet saws also have water to contend with, making both the tool and the surface more slippery. Some basic safety guidelines are as follows:

- Check the guard, making sure it returns without jamming.
- Use the saw at full depth unless making a groove or cutting thin sheet material (see next point). With all saws this reduces the chance of kickback due to there being less blade contact at the point of incision. In addition the cutting action is upwards so that the reaction is downwards instead of backwards.
- When cutting thin material, such as plywood or fibre-cement sheeting, allow the blade to penetrate only 25 mm through the material. This limits the pressure of the cut side of the sheet as it bends down on the blade (a common cause of kickback).
- For wet saws, using the full depth reduces heat in the blade for the same reasons mentioned in the second point above.
- Always cramp your material, don't hold it in your hands.
- Always position your body beside a saw, never stand or let others stand behind it.
- Never put your hand or feet behind the saw.
- For wet saws, use only the minimum necessary water supply.

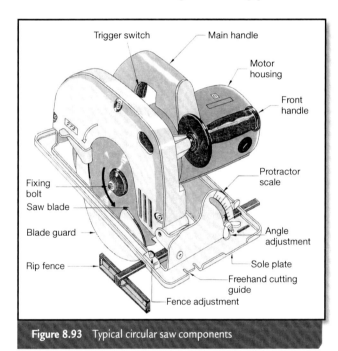

Figure 8.93 Typical circular saw components

Figure 8.94 Bench-mounted wet saw

- Should the saw begin to jam or give an indication that it might kick back, don't panic! Take your finger off the trigger and hold down firmly on the saw.
- Never start the saw in the material; remove it and start again.
- For drop-in cuts (cuts in the middle of the material), hold the nose of the saw firmly on the material. While moving the saw slowly forward, hinge the saw down into the material.
- Never move a saw backwards in the material with the blade turning. This increases the chance of kickback.
- Keep an eye on the power lead. *Do not* put the lead over your shoulder. With wet saws this increases the chance of electric shock, as water can flow into the plug. With ordinary saws, the lead can slide off your shoulder and swing into the blade.
- After cutting, hold the saw away from your body and check that the guard has returned before putting the tool down.

Drills (cordless and mains powered)

Used for drilling large or small holes, or as a screwdriver, the drill is part of every tradesperson's kit; even painters tend to have one, if only for mixing purposes. Most drills on the contemporary work site are cordless (Figure 8.95), with mains-powered units being used only for larger needs such as mixing plasters, tile adhesives and the like (Figure 8.96).

Figure 8.96 Large drill used for mixing

Except for some of the larger units, the 'chuck' of most drills today is keyless. Keyed chucks (Figure 8.97) use a tooth and cog arrangement to tighten the bit into the tool. Some large drills used for masonry, demolition and concrete work use a special internal lock system called an SDS chuck. This allows for a more fluid hammer action.

Whether fast or slow, drills turn the bit as a means of cutting. Although with low-powered drills this is not much of an issue, with high powered drills this can lead to serious injury if the bit becomes locked or jammed in the material and the drill body moves instead. A high-quality, high-torque, 14-volt drill can snap the wrist of an unwary person who is using it one-handed. This author has seen a friend break his jaw with a medium-sized 240-volt drill using a 19 mm spade bit. Some do's and don'ts when using a drill are as follows.

- Use two hands when drilling with large bits, hole saws or anything that might grab.

Figure 8.95 Cordless drill

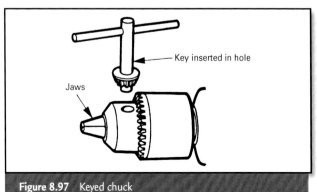

Figure 8.97 Keyed chuck

- With corded tools for mixing, keep an eye on the lead and ensure it stays out of water.
- Stand so that you do not rely on the material to hold you up.
- Stand so you can maintain your balance if something moves unexpectedly.
- Use the right speed for the right bit and material: fast for small bits, slow for large bits.
- Drive screws with the slower speed.
- Cramp material down; don't hold sheet material with your hands or feet.

Screw guns

A specialist plasterer's tool, this is effectively a drill with a fancy belt feed and screwdriver attached (Figure 8.98). Screw guns have been rapidly adopted across the trade, and they more than double the speed at which sheet materials such as plasterboard can be fixed. Designed as a one-handed tool, they are cordless and have a clutch depth limiting system that prevents overdriving.

When using screw guns, remember to:
- maintain them;
- clean them;
- not drop them;
- never put your hand in front and pull the trigger!

If you follow these guidelines, the tool will not cause you injury and will last for years.

Figure 8.98 Screw gun

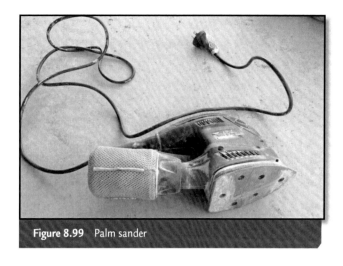

Figure 8.99 Palm sander

today are belt sands, orbital and reciprocating sheet sanders, and palm sanders (Figure 8.99). The latter is becoming increasingly popular due to its size and adaptability.

When using sanders, the safety points below should be followed:
- Where possible, attach the sander to a vacuum unit.
- Wear a dust mask (whether attached to a vacuum unit or not).
- Take a break: the vibration from orbital sanders can cause injuries to fingers, wrists and arms.
- Disconnect the sander from the power before changing belts or pads.
- Use reciprocating sanders where possible for timber work where the grain is to be seen.
- Clean the tool regularly to stop the build-up of fine dust in the air vents.
- Fit sanding belts with the arrow in the right direction, or the belt may break.
- With belt sanders, make sure the material is securely held or it may fly backwards.

Other plant and equipment

This section is a mixed bag of equipment of which perhaps only ladders, trestles and elevated work platforms are common across all trades.

Portable compressors

A portable compressor is a pump that compresses air into a cylinder for controlled use on spray

Sanders

Powered sanders are frequently used for preparing a surface for painting. The forms mostly used equipment, air chisels, sand blasting and other pneumatic tools (see Figure 8.100). The size of these compressors can vary considerably, from very small hobby-based units to larger, trailer-mounted affairs. Your choice will be based upon the supply of air you will need in litres per minute (L/m) or litres per second (L/s), although most are marketed by horse power (hp) and cylinder volume. A further choice will be of power supply: electric (240-volt), petrol, diesel, or a combination.

Most compressors on domestic sites have cylinders of approximately 60 litres, 2–3 hp motors, and deliver around 200 L/m.

Figure 8.100 Portable compressor

Using compressors and compressed air
If a petrol- or diesel-driven unit, follow the procedure for small motors by checking the oil and fuel, and then check the compressor. Ensure that the bleed/drain valve is closed and that all hose connections are tight.

Compressed air is dangerous, and in foolish or unskilful hands, deadly. Some basic safety practices should be followed:
- Never 'play' with compressed air.
- Never point or spray people with compressed air.
- Never force compressed air directly onto a person's skin.
- Wear safety glasses around compressed air tools and hoses.
- Switch off the power, and open and bleed down the cylinder before transporting the compressor.
- Check all hoses and clamps at least weekly.
- Securely tie down compressors when transporting them.

Spray guns

Painting by brush, and even roller, is becoming a minor part of the trade compared to spray work. Even architraves and skirting in domestic homes are being painted in this manner. At some point, therefore, you will be required to be come familiar with this technology.

Very little contemporary spray equipment requires an independent compressor, although many of the better spray units come with their own compressors built in, while others use a pump and operate 'airless'. Each has advantages and disadvantages, as outlined below.

Airless spray guns
The advantages of airless spray guns are that:
- they are compact;
- they can pump straight from the can, making for less cleanup;
- there tends to be less bounce-back of material from the surface when the pressure is set correctly;
- they have high rates of fluid delivery;
- there is good penetration to recessed areas;
- they give good coverage;
- less, or no, thinning is required.

The disadvantages of airless spray guns are that:
- there is potentially a higher risk of injection injury (penetration of the skin by paint materials);
- there is no control over paint volumes (the spray is all on or all off);
- it is difficult to feather paints;
- the initial purchase costs are higher;
- it is difficult to achieve very high quality finishes (although high-end modern units can do so);
- as there is a greater potential for static build-up (sparks and fire), units must be grounded.

Air spray guns
Air spray guns (see Figure 8.101) consist of high-pressure (independent compressor) and high-volume low-pressure (HVLP) units.

The advantages of air spray guns are that:
- they are more easily regulated and therefore allow for less overspray when working in sensitive areas;
- the paint volume can be reduced to allow for feathering;
- the set-up costs are lower;
- they are generally safer to operate,

The disadvantages of air spray guns are that:
- they require more cleanup;
- paints must be thinned;
- more coats are required;
- they have more components.

Air and airless units have very different spray guns and tips. The tips of airless guns have to be much harder (sometimes made of tungsten carbide) to cope with the highly abrasive nature of the technique. Air guns require two feeds: for air at the back of the gun and for paint at the front (Figure 8.102).

Learning to spray takes time, and experience with different types of spray guns, pumps and air supplies is critical to this skill development. Given the safety hazards, the toxic wastes and the cost of this equipment, it is a skill that must be developed only under experienced supervision.

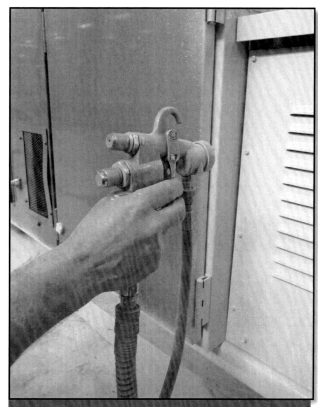

Figure 8.102 Air spray gun with two feeds: Air and paint

Concrete mixers

Mixers come in many shapes and forms, from the well-known concrete trucks with rotating drums of over 3 m³ in capacity to the small tilting mixers seen on most building sites. It is these smaller mixers that will be addressed here.

Tilting drum mixers

A tilting drum mixer is a drum rotated by an electric, petrol or diesel motor (Figure 8.103). Mounted on a small, steel frame with large wheels, the drum is tilted by means of a lever that pivots the drum around its drive shaft (Figure 8.104). The motors and gearing for these units are based on low speed and high torque. This means that should your fingers or arm get caught in the machine, it will not stop.

Figure 8.101 High-pressure low-volume air spray unit (note pressured paint pot at front left)

Figure 8.103 Concrete mixer

Figure 8.104 Concrete mixer and wheelbarrow

As with any tool, hands-on training guided by an experienced tutor or fellow tradesperson is better than words in a book. There is an art to getting the right mix, be it plaster, mortar or concrete, and achieving this mix depends on an awareness of the moisture content of the materials you are working with. Aside from this skill, which can only be taught to you in the doing, there are some basic points to remember when using tilting drum mixers:

- Never place your hands or tools (shovels for example) in the mixer while it is turning.
- Keep the inside of a mixer clean (spotless, if possible).
- Use a wash-down area: never allow sludge from a mixer to get into drains/waterways.
- Don't let the inside of a drum rust between uses; a light oil should be applied.
- Switch off the tool before cleaning.
- Switch off and remove the power leads before washing down the mixer.
- Never spray water around the outside of the mixer while it is connected to the mains power.
- Always wet down the drum, or add at least a small amount of water to it, before adding the dry ingredients (generally almost the full water complement will go in first).
- Check the drive belt at least every six months for wear.
- Check electric motors for weather and water protection.
- On hot days, clean the drum out between mixes to prevent drying and build-up.

Wheelbarrows and brick trolleys

These are the tradesperson's basic carry-all. The main difference between a garden wheelbarrow and the trade version is its size and structural stability. Many barrows today have plastic bins, which are tough yet light. Metal barrows are still available, however, and are preferred by some despite being significantly heavier (Figure 8.104).

Brick barrows or trolleys (Figure 8.105) are two-wheeled derivatives that look more like a small cart. They are designed specifically to carry bound brick

Figure 8.105 Brick trolley

blades from a pack. Both tools have large pneumatic tyres to help get across rough or boggy ground.

Some hints for using barrows and trolleys are as follows:
- Bend your knees when lifting, and lift up straight.
- Check and constantly monitor your balance, as barrows are easy to tip over in inexperienced hands.
- If the barrow starts to tip, lower it; don't continue lifting.
- Fill a barrow evenly if putting in tools or dry materials such as bricks or tiles.
- Spread the load of the barrow to the front so that it is not back-heavy.
- Tip the barrow gradually, as tipping too fast can lead to it overbalancing.
- Make ramps and boardwalks the width of the bin; single planks are dangerous and easy to slip off.
- When using a brick barrow or trolley, be sure to separate each blade of bricks before lifting. Excessive pressure when lifting can snap the plastic strap.
- Clean barrows regularly when carrying wet materials, particularly in hot weather.

Brick saws

Used to cut bricks and pavers, this tool is common to all bricklayers. It's the red, dusty looking thing on the back of the ute (Figure 8.106). A type of wet saw, brick saws are water-fed by a hose that puts water directly to the point of incision. Unlike other saws, these saws can use a pedal to raise the blade, allowing both hands to be kept on the material being cut.

Figure 8.106 Brick saw

At first use, even the most experienced tradesperson finds this a frightening tool. Its blade size, action and close proximity to fingers are daunting. Then it becomes all too easy and the fingers creep a bit closer. That's how people get caught by this machine: they relax too much.

Hands-on teaching is required for this tool, and should be done on a one-to-one basis by an experienced tradesperson. Some basic points to remember when using a brick saw are listed below.
- Ensure adequate but not excessive water supply. If it is too dry the blade will overheat, become excessively dusty and be more likely to catch and drag the brick or paver. If it is too wet you won't be able to see the material properly, aside from flooding the site.
- Set up the saw on or near a wash-down area or soak. Never allow the sludge water to enter stormwater drains or waterways.
- Work the material into the wheel gradually.

- Do not let the material be 'behind' the blade, or the wheel to come over the top of the material: either action will cause the material to lift and fly backwards.
- Never allow any part of your hands to be in line with the blade.
- Wear tight sleeves or, better, sleeves rolled up. Loose clothing can be drawn into the saw by the breeze made by the blade.
- Hearing and eye protection are critical. Breathing protection is also wise, as there is a large amount of dust.
- Do not wear gloves; these make your hands larger and if caught by the blade, will drag your hands in.
- Clean the tool regularly (ensuring that the power is disconnected before doing so).
- Check the rubber mat and replace it when necessary. This is vital to prevent the material from slipping on the wet metal tray.

Vacuum cleaners

Depending upon the requirements of the trade, vacuum cleaners (Figure 8.107) can be simple domestic units or large industrial types. Industrial vacuum cleaners differ from domestic units in having wet and dry usage, a large reusable canister (some have large disposable bags), larger motors and greater suction power.

Caution: Any vacuum cleaner used to pick up hazardous materials such as asbestos or silica-based dusts (cement products) must be emptied, and the contents disposed of according to strict guidelines. These are determined by the relevant state or territory OH&S Regulating Authority (WorkCover in NSW, for example), including wearing protective clothing and respirators. Bag removal must take place within the contaminated area.

When using industrial vacuum cleaners:
- check the leads;
- never use extension leads in wet areas;
- always empty the bag after use;
- Check and clean the filters regularly—if asbestos has been vacuumed, these filters will need to be disposed of in the same manner as the waste material.

Trestles, planks and ladders

Falling from heights remains one of the major causes of workplace injuries in Australia. In recognition of this, state and territory OH&S Regulating Authorities have the right to enforce strict regulations regarding the use ladders, planks and trestles.

Trestle and plank platforms

Trestle ladders must comply with the Australian Standard AS 1892, 'Portable ladders'. Trestle and plank platforms (see Figure 8.108) also must comply with the OH&S Regulations pertaining to scaffolding (AS/NZS 1576, 'Scaffolding'). This means the following:

- Platforms must be at least 450 mm wide (two planks).
- The trestles and platform must be established level and stable.
- The trestles must be able to accommodate toe-boards, mid-rails and hand-rails if the fall is greater than 2 metres, or less if the fall could be onto dangerous objects such as starter bars in concrete.

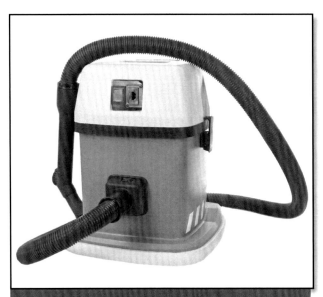

Figure 8.107 Industrial vacuum cleaner

Figure 8.108 Trestles, planks and trailer

- There must be easy access by way of an affixed ladder or steps.
- Planks must comply with AS 1577, 'Scaffold planks', by being at least 220 mm wide and 32 mm thick if hardwood, or 38 mm thick if Oregon.

Aluminium folding trestles

Aluminium folding trestles differ from ladder trestles in that they are a solid plank with folding legs of a fixed height (Figure 8.109) and are not required to be two-planks wide. They have become very common on both domestic and commercial building sites. The main considerations when using these trestles are as follows.

- Set the trestle up on a level, stable surface only; the trestle must not be able to rock.
- Check the load rating and do not exceed it.
- Do not jump from trestles (either when getting off it or when moving from one to another).
- Stay focused when on a trestle; it is all too easy to walk off it.

Trailers

The well-equipped, well-laid-out trailer is one of the most important pieces of equipment you can have. Figure 8.108 shows how easy access to trestles, planks and profiles has been maintained when in many trailers these items would end up at the bottom.

Ladders

There are basically two forms of ladder to be found on building sites: extension ladders and step ladders.

Figure 8.109 Aluminium trestle

Both may be made of timber, aluminium, fibreglass or carbon fibre reinforced plastics. Aluminium is still the most favoured, due to its light weight and low cost. Timber ladders must not be painted so that faults are not hidden. All ladders on building sites must carry the manufacturer's mark, including the load rating in kilograms.

Extension ladders (Figure 8.110) come in various lengths, the most common being 3.0 m, 3.3 m and 3.6 m. The purpose of an extension ladder is to gain access to higher levels; they are not a work platform, as three points of contact must be made at all times (two hands and one foot, or two feet and one hand). Ladders must be established at a base to height ratio of 1:4 (see Figure 8.111) and must either be tied to the structure they are resting on or held to prohibit sideways movement. Other factors include:

- Only one person may be on the ladder at a time.
- Do not place ladders against windows or doors, or in front of doors.

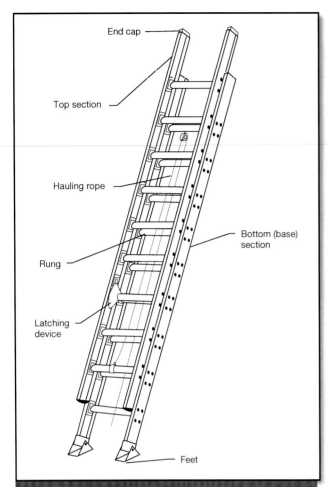

Figure 8.110 Typical extension ladder components

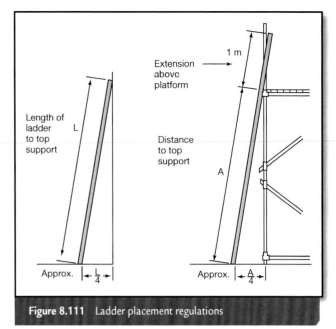

Figure 8.111 Ladder placement regulations

Figure 8.112 Stepladder

- Never join ladders together to gain extra length.
- Be aware of electrical lines and keep ladders clear of them. The distance required depends on the voltages involved; check with the relevant authorities before positioning the ladder.
- Make sure that the foot of the ladder is on stable level ground or is chocked to make it so.
- Do not link extension ladders with planks.
- Always clean mud or dirt from ladders before storage.

Stepladders (Figure 8.112) are perhaps the most misused ladder in the industry. Take the following precautions:

- Always fully extend the frame before climbing.
- Never stand higher than the third-last rung.
- Never leave tools or materials (paint for example) on the top of the ladder: these will fall when the ladder is moved.
- Do not use stepladders as trestles for planks.
- Ensure the ladder is stable on all four legs before climbing.

- Do not overreach from a stepladder, as it will topple.
- Always clean ladders before storage.

Elevated work platforms

Known as **EWPs**, **elevated work platforms** are common to all sites and are becoming more popular given the greater stringency on safe working at heights. EWPs come in a variety of forms, from 'cherry pickers' (articulated knuckle booms or trailer boom lifts) and scissor lifts (Figure 8.113) to vertical towers and rough terrain mobile lifts. Training for this form of equipment is regulated and should be conducted by experienced professionals.

Figure 8.113 Scissor lift

Some basic factors to keep in mind, should you be required to work in or around an EWP with a ticketed operator, are listed below:
- EWPs must be the following distances from power lines (as prescribed by AS 2550):
  - normal power lines up to and including 133 kv (most street power on poles): 6.4 m, or 3.0 m with a qualified 'spotter';
  - main transmission lines greater than 133 kv (usually on towers): 10 m, or 8 m with a qualified 'spotter'.
- Safety harnesses must be worn and connected correctly to the EWP basket at all times (except for scissor lifts).
- Should any movement occur in the outriggers, or if you notice that the unit is no longer sitting level, lower the platform immediately.
- If you see hydraulic fluid leaking from any of the outriggers or arms, lower the platform immediately
- Do not work in high winds (generally winds of more than 12 miles per second, or otherwise as stated in the manufacturer's instructions).

YOUR HANDS, EYES AND EARS: THE FORGOTTEN TOOLS

All of the above technology—indeed all the technology in the world—will fail you if you forget the most important tools you have: your hands, eyes, ears and the rest of the body, with everything functioning in unison. Not only do you need to practice using your body and each of its parts as you would any other tool, but you must also look after it as you would any other tool.

Learn to use and trust your eyes in levelling and sighting straight lines or true curves. Look for light and shade that will help you to discern that a plaster wall is flat and true, or to check that tiles are laid without lip. This can be particularly important when working with deliberately out-of-level surfaces such as bathroom floors or shower bases.

Your ears are just as important. Your ears, and the vibrations you otherwise pick up through your body, can tell you if, for example, a tool is cutting correctly, a bearing is failing or that something else is amiss. Sound and vibration can even tell you when mortar in the mixer is ready.

As for hands, one can never say enough. Look after them, clean them, keep them safe from chemicals and protect them from injury—but use them. The more you use them, the more connected you become to the tools and surfaces around you: the more you can feel that a tool is cutting correctly, that those tiles are laid true or that the mortar on your trowel is balanced correctly to butter a brick. You will simply know when and how far to flick that mortar back on the trowel so it is balanced correctly; you won't even have to look. Plasterers learn to hold three tools simultaneously in the one hand: hawk, knife and trowel (See Figure 8.114). Yet when it comes to doing the task, it's the fourth tool that matters.

Enjoy your trade; its gift of skill can take you around the world. No matter the culture, context or task, the most important equipment required will always be available: you.

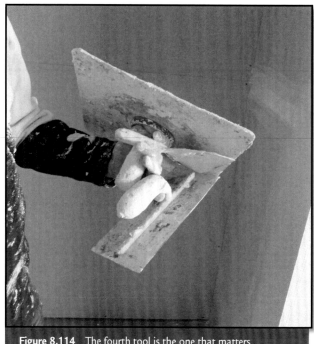

Figure 8.114 The fourth tool is the one that matters

Worksheet 1

Student name: _____

Enrolment year: _____

Class code: _____

Competency name/Number: _____

To be completed by teachers:	
Student competent	☐
Student not yet competent	☐

Task: Read through the sections *Introduction to tools* and *Basic hand tools*, up to and including *Tools for cleaning*, then complete the following.

Q. 1 List three reasons why it is important to use tools with an 'open' hand:

1. _____

2. _____

3. _____

Q. 2 It is important to breathe out, rather than breathing in or holding your breath, when striking an energetic blow because:
- a breathing out relaxes the body at the moment of impact, reducing stress on muscles
- b breathing in uses more of your body muscles and increases body tension, leading to possible damage
- c holding your breath increases tension and can lead to harm being done to your body as the blow is struck
- d all of the above.

Q. 3 What is a key difference between a tradesperson and a handyperson (*bricoleur*)?

_____

_____

_____

Q. 4 Why is it important to use 'the right tool for the right job'?

_____

_____

_____

Q. 5 Which of the abrasive papers below would you choose to remove paint from a surface? State why.

a Open coat: _____

_____

b Closed coat: _____

_____

Q. 6 From the section *Tools for setting out*, and Chapter 7 (*Levelling procedures*), list four common set-out tools for your specific trade:

1. _____
2. _____
3. _____
4. _____

Q. 7 List two natural and two manufactured grits for abrasive paper:

Natural

1. _____
2. _____

Manufactured

1. _____
2. _____

Q. 8 Why is it important to wear a P1 dust mask when sanding?

_____
_____
_____

Q. 9 List three tools that fit into the knife category:

1. _____
2. _____
3. _____

Q. 10 To drive a slotted screw deeper into a material, you would use:
    a a flat screw driver
    b a hammer
    c a nail punch
    d an Allen key.

Worksheet 2

Student name: _____

Enrolment year: _____

Class code: _____

Competency name/Number: _____

To be completed by teachers:	
Student competent	☐
Student not yet competent	☐

Task: Read through the section *Plastering and rendering tools*, then complete the following.

Q. 1 In selecting a trowel, what are three factors you are looking for?

1. _____
2. _____
3. _____

Q. 2 Why must a trowel be clean before doing any work, and kept clean during and after the job?

_____
_____
_____

Q. 3 Why does a taping knife have a curve in the edge of the blade?

_____
_____
_____

Q. 4 A plasterer's adhesive knife has a long blade:
  a for balance
  b to keep their hands clean
  c so it can be used as a cutting blade
  d to get more adhesive on the knife.

Q. 5 In plastering, what's a hawk?

_____
_____
_____

Q. 6 When would you need a 'float'?

_____
_____
_____

PAINTING AND DECORATING, AND MORTAR TRADES

Q. 7 List three advantages of a taping box:

1. _____
2. _____
3. _____

Q. 8 When might you use a plasterer's square?

_____
_____
_____

Q. 9 What role does a utility knife have in plastering?

_____
_____
_____

Q. 10 When might you use a keyhole saw?

_____
_____
_____

Q. 11 List three reasons why jointing boxes are frequently used on today's building sites:

1. _____
2. _____
3. _____

Worksheet 3

Student name: _____

Enrolment year: _____

Class code: _____

Competency name/Number: _____

To be completed by teachers:	
Student competent	☐
Student not yet competent	☐

Task: Read through the section *Bricklaying tools*, then complete the following.

Q. 1 In selecting a trowel, what are three factors you are looking for?

1. _____
2. _____
3. _____

Q. 2 Why is it important to clean your trowel after each work shift?

_____

_____

_____

Q. 3 When might you use a bolster instead of a trowel to cut a brick?

_____

_____

_____

Q. 4 A Dutch pin:
   a only holds string lines in place
   b is used by European tradespeople to keep plans pinned together
   c is a metal dowel or rod driven into the ground when setting out
   d is driven into mortar to hold profiles in place.

Q. 5 Corner blocks are;
   a the bricks or concrete blocks at the corner of a building
   b the concrete blocks (only) at the corner of a building
   c the large corner block-work highlighted on some older-style buildings
   d plastic blocks that hold string lines in place on profiles or brick corners.

Q. 6 When would you need a 'gauge trowel'?

_____

_____

_____

Q. 7 List three places you might use a plugging chisel:

1. _____
2. _____
3. _____

Q. 8 When might you use a 'G' cramp?

_____
_____
_____

Q. 9 How is a brickie's barrow different from wheelbarrow?

_____
_____
_____

Q. 10 List three places you might use a profile:

1. _____
2. _____
3. _____

Q. 11 List three dangers/risks of using a cement mixer:

1. _____
2. _____
3. _____

Worksheet 4

Student name: _____

Enrolment year: _____

Class code: _____

Competency name/Number: _____

To be completed by teachers:	
Student competent	☐
Student not yet competent	☐

Task: Read through the section *Painting tools*, then complete the following.

Q. 1   Name four different styles of brush, then list the identifying characteristics and uses of each:

1.
Name: _____
Characteristics: _____
Use: _____

2.
Name: _____
Characteristics: _____
Use: _____

3.
Name: _____
Characteristics: _____
Use: _____

4.
Name: _____
Characteristics: _____
Use: _____

Q. 2   What is a mahlstick used for?

_____
_____
_____

Q. 3 What are 'drop sheets' used for?

_____
_____
_____

Q. 4 In painting, what is a 'kettle'?

_____
_____
_____

Q. 5 You are at a new job with a new roller. You should:
- **a** wash it thoroughly, then begin painting
- **b** wash it, dry it, rub it down and then begin painting
- **c** rub it down, wash it, dry it, then use it on an unimportant surface first
- **d** always use a new roller fresh from the pack on your most important wall.

Q. 6 What is a 'masking tool' or 'masking machine' used for?

_____
_____
_____

Q. 7 You are required to paint a circle with a radius of 750 mm on a wall with its centre in a specific spot. You would:
- **a** try to find a large lid or disc the right size to draw around
- **b** use a set of trammel heads with pencils
- **c** drive a nail in the wall and tie a string and pencil to it
- **d** make lots of marks on the wall 750 mm around the centre and sketch it in.

Worksheet 5

Student name: _____

Enrolment year: _____

Class code: _____

Competency name/Number: _____

To be completed by teachers:	
Student competent	☐
Student not yet competent	☐

Task: Read through the section *Tile laying tools*, then complete the following.

Q. 1 Why is the number of tools required by tile layers increasing?

_____
_____
_____

Q. 2 The material used on scoring tips of tile cutters is:
   a silicon carbide
   b tungsten carbide
   c aluminium oxide
   d titanium.

Q. 3 What is the difference between quarry tile nippers and other nippers?

_____
_____
_____

Q. 4 What is a gauge trowel used for?

_____
_____
_____

Q. 5 When grouting, why is a grout float believed to be better than a squeegee?

_____
_____
_____

Q. 6 What sort of drill bit is required for drilling porcelain?

_____
_____
_____

Q. 7 Of the many hitting instruments available, which might be useful to you as a tile layer?

_____
_____
_____

Q. 8 Where might you use a combination square?

_____
_____
_____

Q. 9 A grout saw is:
    a used to cut pieces of hard grout to fit in between corner tiles
    b used to remove unwanted grout from between tiles
    c a type of diamond saw for cutting porcelain tiles
    d a small saw for cutting plastic or metal trim.

Worksheet 6

Student name: _____

Enrolment year: _____

Class code: _____

Competency name/Number: _____

To be completed by teachers:
Student competent ☐
Student not yet competent ☐

Task: Read through the section *Plant and equipment*, then complete the following.

Q. 1 List three reasons why builders have moved away from temporary power poles for supply of 240-volt electricity:

1. _____
2. _____
3. _____

Q. 2 List three important factors when positioning a temporary power pole:

1. _____
2. _____
3. _____

Q. 3 You have a 5 kVa portable generator; the number of power tools you can run is:
 a not more than two
 b not more than five
 c as many as you want
 d not more than eight.

Q. 4 The maximum length of a 10A extension lead with a 1.5 mm² conductor area is:
 a 32 m
 b 32 mm
 c 25 m
 d 40 m.

Q. 5 How far off the ground must power leads be supported?

_____

Q. 6 When can you use a double adaptor on-site?

_____

Q. 7 List three types of blades that are generally used in angle grinders:

1. _____
2. _____
3. _____

Q. 8 List three points that can reduce kickback from a power saw:

1. _____
2. _____
3. _____

Q. 9 When changing blades on saws, or belts on sanders, you must always:
_____

Q. 10 What are three critical things that you must never do with compressed air?

1. _____
2. _____
3. _____

Q. 11 A handrail must be supplied to a working platform that is higher than:
_____

Q. 12 The base to height ratio used for positioning a ladder is:
    a 4:1
    b 10:4
    c 1:40
    d 1:4

REFERENCES AND FURTHER READING

Web-based resources

Plastering and rendering
<www.proplaster.com.au> Pro Plaster Products
<www.wallboardtools.com.au> Wallboard Tools
<www.plasterproductsonline.com.au> Plaster Products Online

Bricklaying
<www.brickiestoolshed.com.au> Brickies Tool Shed
<www.diamondway.com.au/bricklaying_tools.html> Diamond Way Australia Pty Ltd
<www.btengpl.com> BT Engineering Australia

Tile laying
<www.tilersonline.com.au> Tilers Online
<www.tilersdirect.com.au> Tilers Direct and Trade
<www.tilers-express.com> Tilers Express. The Tilers Tool Shop

Painting
<www.paintaccess.com.au> Paint Access.com.au
<www.paintersconnected.com.au> Painters Connected
<www.austbrush.com.au> Australian Brushware Corporation

Codes of Practice

The following publications are available online at <www.workcover.nsw.gov.au>:

Work near overhead power lines: Code of Practice

Storage and handling of dangerous goods: Code of Practice

Workplace amenities: Code of Practice

Amenities for construction work: Code of Practice

Australian Standards

AS 1892, Portable ladders

AS/NZS 1576, Scaffolding

AS/NZS 3760:2010, In-service safety inspection and testing of electrical equipment

GLOSSARY

accident An undesirable or unfortunate happening that occurs unintentionally and usually results in harm, injury, damage, or loss.

accident report form A form used to report serious work-related illnesses, injuries or dangerous occurrences to the state authority, as required by law. It gives information about the employer or workplace, the injured or ill person, and the injury, illness or dangerous occurrence.

Act A formal decision or law made by an official body, such as a legislature, court or other governmental authority.

agenda A formal list, plan or outline of things to be done in a specific order, especially a list of things to be discussed at a meeting.

architecture The art and science of designing and constructing buildings, as well as a style or fashion of building, especially one that is typical of a period of history or of a particular place.

area The measurement of surface: the *area* of a flat, or plane, figure is the number of square metres the figure contains. Area can be found by multiplying the length of a figure by its width or breadth.

Art Deco A decorative style characterised by symmetrical and often geometric shapes.

Art Nouveau A decorative style characterised by organic or plant-like forms and by flowing curves and lines.

availability The assurance that the system or product will work instantly when required.

backsight The first sight taken during a survey; it sights 'back' to a previous known or nominated height/position.

barricade A defensive barrier constructed in order to limit or prevent entry, and signify that a hazard or danger exists.

benchmarks Permanent or temporary marks of a known height above the original datum (sometimes used as the datum with a nominated height).

blade 1. A tool, or component of a tool, with a cutting edge (e.g. a knife, or the blade of fibre cement shears) 2. A strapped quantity of bricks in a brick pack. One pack of bricks generally holds five blades.

blob (pad) footings Square, rectangular or round footings placed under piers or posts. May contain reinforcement, depending on the load to be carried.

block plan See *site plan*.

body language Communication by other means than by using words, e.g. through facial expressions, bodily mannerisms, postures and hand gestures that can be interpreted as unconsciously communicating someone's feelings or psychological state. Also described as *non-verbal communication*.

brick veneer construction A building technique consisting of a timber frame that supports the load and forms the real structure, surrounded by a leaf of brickwork to give the appearance of a solid brick structure.

bricoleur A handyman or woman whose work is often viewed with some suspicion because they do not use tools or methods in a 'normal' way.

brick veneer External walls constructed of a timber frame lined internally with plasterboard and with an external skin or veneer of brick.

cavity brick External walls constructed of two skins of brick, separated by a gap (cavity) and held by cavity ties, with a drip groove to prevent moisture from travelling from the outer skin to the inner skin of brick when the outer skin becomes wet.

checklist A list of tasks, items, or points for consideration, verification, action or checking purposes.

Codes of Practice Documents issued by a state or territory OH&S Regulating Authority that provide a set of practical, commonsense, industry-acceptable ways of dealing with the *Occupational Health and Safety Act* and working safely.

committee A person or group of persons elected or appointed to perform some service or function, such as to investigate, report on or act upon a particular matter.

communication The exchange of information between people, e.g. by means of speaking, writing or using a common system of signs or behaviour.

cone A solid shape that has a circular base and a uniformly curved surface which tapers to the apex.

conflict A disagreement or clash between ideas, principles or people.

conventional roofing Roofing for which the timber frame is cut out and assembled on-site.

customer 1. (internal) The person who receives your work next. 2. (external) A person who buys goods and services.

cylinder A solid shape that has ends formed by circles of equal diameter.

dangerous goods Substances that have the potential to cause immediate harm.

datum A reference point, line, or surface from which elevations are measured.

demarcation A dispute that occurs when a job done by one member of a union should in fact belong to a member of another union.

details (on a working drawing) Sectional views drawn to a larger scale than sectional elevations, showing specific requirements that cannot be drawn accurately to scale on sectional elevations.

dimensions Measurement in length, width and depth (thickness).

Earth leakage device A safety switch designed to detect earth leakage and cut power within microseconds. See also *RCD unit*.

efficient Performing or functioning in the best possible manner with the least waste of time and effort.

elevated work platform See *EWP*.

elevation A scale drawing of any side of a building or other structure, in accordance with the compass point direction it faces, that provides information relating to vertical measurements and external finishes.

enterprise agreement A contract between an employer and employees on wages and conditions of work in the employer's workplace.

enterprise bargaining A bargaining process in which the employer negotiates directly with employees with regard to wages, conditions and work practices for that particular workplace.

environment 1. Our environment (everything that constitutes the planet on which we live). 2. The immediate things around us or an incident that has occurred—e.g. noise, heat, wind. Environmental controls include air conditioning and dust suppressant activities.

ergonomics The engineered relationship between tools and or the work environment generally, and humans (more properly, the study of this). How 'user friendly' tools or equipment are – do they make you tired too quickly or actually cause you harm with use.

EWP (elevated work platform) Mechanically or hydraulically raised platforms used to reach high work areas. Can be self-propelled or static.

fall The amount that one reading in a survey is lower than the previous reading.

fitches Small paint brushes for trimming or detail.

flammable materials Any substances that can be easily ignited and that will burn rapidly.

floor plan A horizontal section of the building as viewed from above.

fold out A diagram of the inside of a room used for determining tiling or painting requirements.

footings The lowest part of a building, designed to distribute the load of the building over the foundation.

foresight The last sighting taken in a survey prior to moving the instrument or concluding the survey.

formwork An assembly or temporary construction to support and shape freshly mixed concrete until it sets and hardens.

fuel Any combustible material—that is, any solid, liquid or gas that can burn.

hazard report forms Forms used by workers to report potential hazards to management, via their immediate supervisor, so that the hazard can be removed at the earliest possible time.

hazard Any situation, substance, activity, event or environment that could potentially cause injury or ill health.

horizontal A straight line or surface that is level, or parallel with a water horizon.

incident Something irregular and therefore notable or reportable. Accidents and near misses are incidents, as are specific bullying or harassment actions.

industrial relations The field that looks at the relationship between management and workers, particularly groups of workers represented by a union.

injuries register book A record of all injuries that occur at a workplace, kept on-site.

injury Physical damage to the body or a body part.

instruction A spoken or written statement of what must be done, especially delivered formally, with official authority, or as an order or direction.

intermediate sight In a survey, all the sightings taken from the one instrument position (station) after the first (backsight) and before the last (foresight).

isometric projection A pictorial drawing with lines drawn parallel to the axis at 30°.

legislation The process of writing and passing laws, especially by a governmental assembly or official body.

linear metre A unit of measurement, measuring the distance between two points in a straight line.

line of collimation A straight line that passes through the optical centre of a levelling instrument (or the laser-generating equivalent).

maintainability The assurance that parts and service are readily available and that the system or product can be repaired if necessary.

manual handling An activity requiring a person to use force to lift, lower, push, pull, carry, move or hold any type of object.

material safety data sheet (MSDS) A document prepared by the supplier or manufacturer of a chemical product, clearly stating its hazardous nature, ingredients, precautions to follow, health effects, safe handling/storage information and how to respond effectively in an emergency exposure situation.

material The substance or substances of which a thing is made or composed.

meeting An occasion, in either a formal or informal setting, where people gather together to discuss something.

message A communication in speech, writing, or signals, containing some information, news, advice, request, opinion, fact, emotion, knowledge, warning, or any one of the many things people need to impart to others.

millimetre A unit of length equal to one thousandth of a metre.

minutes An official record of what is said or done during a meeting.

non-toxic waste material All wastes created on a building site that do not produce either a toxic or poisonous health hazard or a toxic threat to the environment. They may, however, cause hazards to workers and the environment in other ways.

occupational health and safety (OH&S) Refers to all factors and conditions that influence, or could influence, health and safety in the workplace.

orthographic projection A basic (single-angle) form of working drawing, consisting of three related views: plan, elevation and section.

parallax error An error that occurs when the mutual point of focus of a telescope's eyepiece and focal lenses are not on the reticle or crosshairs. It is noticeable as the sighted image appears to move up and down behind the crosshairs.

parallelogram A four-sided figure that has parallel opposite sides.

percentage A proportion stated in terms of one-hundredths of an item or quantity.

perpendicular 1. Vertical (as an alternative to 'plumb'). 2. At 90° to another surface or line (as an alternative to 'square').

personal protective equipment (PPE) Safety clothing and equipment, designed to protect a worker's head, eyes/face, hearing, airways/lungs, hands, feet and body when exposed to harmful substances or environments, or specific hazards.

perspective view See *pictorial drawing*.

pictorial drawing A drawing representing the visual appearance of the completed project or construction.

plane A flat, horizontal surface.

plans A term used to represent all drawings, including sections and details, and any supplemental drawings for complete execution of a specific project.

plant The equipment and machinery necessary for carrying on an industrial or engineering activity.

plumb Exactly perpendicular; vertical.

polygon A five-sided figure.

prism A solid shape with two ends formed by straight-sided figures that are identical and parallel to one another.

procedure An established or correct method of doing something.

profile A vertical, square section post placed at the corner or ends of brickwork. Offers a reference plane and course heights.

pyramid A solid shape with a square base, and with triangular sides that terminate at the apex.

quadrilateral A four-sided figure.

quality The level of excellence that goes into a product or service.

quantity An exact or specified amount or measure.

ratio The amount of one (or more) elements in relation to another.

reliability The assurance that a system or product will continue to work for its guaranteed life.

reduced level (RL) The height above or below sea level (or a nominated datum) of a particular geographical location or point.

RCD unit (residual current device) A safety trip switch for 240-volt power supplies. See also *Earth leakage device*.

record A document that shows what kinds of activities are being performed or what kind of results are being achieved. It always documents and provides evidence about the past.

recycle The process by which materials that would otherwise become waste are collected, separated or processed for reuse as raw materials or finished goods.

Regulations Legally enforceable rules that set out the duties of particular groups of people in controlling the risks associated with specific OH&S hazards.

reportable incident Serious work-related illnesses, injuries or dangerous occurrences that, by law, must be reported to the state or territory Regulatory Authority on an accident report form.

resource efficiency A practice in which the primary consideration of material use begins with the concept of 'Reduce-Reuse-Recycle-Repair', stated in descending order of priority.

reticle A frame of lines, generally crosshairs, inside the optical train of a telescope or tube and used for sighting.

rise The amount that one reading in a survey is higher than the previous reading.

risk assessment Considers the effectiveness of existing OH&S controls and then evaluates the probability and the potential severity of specific hazardous events and exposures. On the basis of such an assessment, organisations decide whether or not the risk is acceptable.

roof structure The term given to the roof framing, eaves and roof covering.

safe work method statement (SWMS) A statement that describes how work is to be carried out, identifies the work activities assessed as having risks, and describes the control measures to be applied to the work activities. It includes a description of the equipment used in the work, Standards or Codes to be complied with and the qualifications of the personnel doing the work.

Safety induction training Compulsory training for all workers in the construction industry, designed to familiarise workers with the site risks and hazards and to assist them in developing a general understanding of site safety.

scale drawing The reduction of the dimensions of an object (e.g. a house) to a size suitable for drawing.

section drawing An elevation cut through the building in the position and direction indicated on the floor plan. It shows a cross-section from the bottom of the footings and through the walls, ceilings and roof structure.

sequence The order in which things are arranged, actions are carried out, or events happen.

signage Graphic designs, such as symbols, emblems or words, used for identification or as a means of giving directions or warning.

SI units Metric units of measurement used in the construction industry, taken from the *Système International d'Unités*.

site plan A plan determining the location of the building on the building block. Also referred to as a *block plan*.

slab-on-ground A method of construction combining the floor and the footing into one reinforced, monolithic concrete unit.

specification A precise description of all construction and finishing, including workmanship, not shown on the working drawings of a building.

square 1. A shape with four sides of equal length where the diagonal, corner to corner, lengths are also equal. 2. Two lines or surfaces at right angles (90°) to each other. 3. To be 'in square': when the diagonals of a rectangular figure are of equal length. 4. A tool for marking out at 90° or checking for 'square'.

stadia lines Two small lines above and below the horizontal cross hair in a dumpy level. Used for determining distance.

station 1. In surveying, the location of the survey instrument. 2. In trussed roofing, the position of a specific truss (generally a girder or truncated girder truss).

strip footings A continuous reinforced strip of concrete around the outside of a building to support the external walls.

substrates The materials or surfaces that another material is laid upon or adhered to.

supplier A person giving customers goods and materials with which to work.

survey Determining location and/or elevation of a geographical site, or of particular aspects of the site.

sustainability The development of economic, social and industrial practices that can be maintained into the foreseeable future.

suspended slab floor A reinforced concrete floor suspended above the ground and supported on brick walls.

suspended timber floor A floor that is built off the ground and generally supported or 'suspended' by means of bearers and joists, which in turn are held up by stumps or piers.

symbol A letter, figure or other character or mark (or a combination of these) used to designate something.

teamwork The cooperative or coordinated effort of a group of persons acting together towards a common goal.

tribrach A three-point levelling base.

true An expression indicating that the constructed surface, surfaces, or other work, is as it should be.

trussed roofing Roofing for which the trusses are fabricated off-site, then transported to the site and lifted into position. Trussed roofs are supported on the external walls only.

volume The space occupied by an object or substance, measured in cubic metres (m^3). Volume = length × width × height.

waste Unused or unusable material as a by-product of work.

waste management The processes and activities involved in dealing with waste before and after production. Includes minimisation, handling, processing, storage, recycling, transport and final disposal.

waste minimisation Practices and processes which reduce, as much as possible, the amount of waste generated, or the amount which requires subsequent treatment, storage and disposal.

wind (wynd) A term used to express whether two surfaces, planes or edges are parallel.

working drawings Drawings that consist of three related views—plan, elevation and section—and give a complete understanding of the building. Working drawings show the layout of the building, the setting-out dimensions, and the spaces and parts of the building, and give specific information about the junctions between the parts of the building.

INDEX

A
abbreviations 206
abrasive stones 298–9
abrasives 298–9
accident report form 37, 38–9
accidents 1, 29, 32–3, 35, 38
Act 2
acute hazards 30
acute health effects 22–4
adhesive application 314
adhesive knife 296, 303
agenda 156, 157
air spray guns 324–5
airless spray guns 324, 325
Allen key 299
aluminium folding trestles 329
aluminium recycling 102
American-pattern trowel 305–6
angle grinders 320–1
apprenticeships 80, 82
architect 88
architecture 75
   Art Deco 75–6
   Art Nouveau 75
   early colonial 74–5
   Federation 75–6
area 175
   measurement of 176–9
asbestos cement 77, 328
asbestos cement cutters 298–9
assessment of risks 29, 35
auger bit 299
Australian Builders Labourers Federation 76
Australian Chamber of Commerce and Industry (ACCI) 80
Australian New Apprenticeships System 80
Australian Standards 206, 250
automatic optical levels *see* optical automatic levels
automatic taping box 304
availability (key term) 97
aviation snips 297–8
award 79
award restructuring 81

B
back injuries 8
backsight 261, 262
bar chart 91

barricade 121–2
barricade tapes 122, 150–1
benchmarks 251–2
blades
   for angle grinders 320
   bound brick 326–7
   cleaning metal blades 128
blob footings 215–16
block brushes 310
block plan 202
body language 144, 146–7, 153–4
bolster 297
bolt cutters 298
bonuses 92
brad hammer 292, 293
brick barrows 326–7
brick carrier 308–9
brick hammer 306, 307
brick profile 308
brick saws 327–8
brick trolleys 326–7
brick veneer construction 76, 77, 217
bricklaying hand tools 305–9
bricoleur 289
broad knives 296
brushes 309–11
builder 88
building industry
   career paths 84–5
   developments in construction 76
   developments in technology 77–8
   history 71–5
   industry structure 77
   National Training Reform Agenda 80, 81–2
   roles 88
   sectors 85
   trade categories 77
   women in 78
building inspector 88

C
CAD (computer-aided design) 77–8
career paths 84–5, 87
carry grip 13
caulking gun 316
cavity brick 217–18
cavity walls 76

ceilings 256
cement products 328
CFMEU (Construction, Forestry, Mining and Energy Union) 76
chain block pulley system 12
chalk line 239, 240
checklist 94, 128
chemical hazards 16, 21–5, 31
   dangerous goods 24–5
   disposal 24
   good work practices 25
   storage 25
   toxic chemicals 31
cherry pickers 331
chisels 297, 306, 307
chronic hazards 30
chronic health effects 24
circle cutters 303
circular saws 321–2
civil operations 84, 86
claw hammer 292, 293
cleaning
   cleaning levelling equipment 274
   cleaning up 128
   hand tools for 300–1
   metal blades 128
   personal cleaning procedures 15–16
   personal protective equipment 21
client 88
clothing *see* PPE (personal protective equipment)
Codes of Practice 2, 7
combination square 291
commercial buildings 77
commercial construction 85
committees *see* workplace committees
communication 100–1
   barriers to 141–2
   clear communication 152–5
   feedback 143
   getting the message 145–7
   on-site 143–5
   process of 142–3
   three aspects of 140–2
compasses 291–2
compensation 39–40
competency-based training 85
compressors, portable 323–4
concrete mixers 325–6
concrete slab placement 125
cones 183–4, 185
conflict 101
conflict resolution 79, 158

construction manager 88
construction workers 84
continuing professional development (CPD) 95 6
conventional roofing 219
cordless drills 322–3
corner block 306, 307
corner trowel 302–3
cost savings 96
costing 90, 91–2, 180–3
CPD (continuing professional development) 95–6
cramps 300
cranes 11
creams 19–20
cross pein hammer 292, 293
crowbars 11, 294
curved nipper 314
customer (key term) 97
cylinders 183, 185

D

danger hazard signs 27–8, 149
dangerous goods 24–5
dangerous goods labels 24, 25, 151–2
datum 251, 256
decorating hand tools 309–13
delays
   anticipation of 89–90
   causes of 90
demarcation 79
details 201
diagrams/sketches 144, 152 *see also* working drawings
diamond drill 299
digital level 238, 239
dimensions 172, 203
disposal
   of chemical hazards 24
   of non-toxic wastes 130
dispute resolution 79, 158
domestic construction 85
   alternative methods 218
   environmental controls 220–2
   flooring systems 216–17
   footings 214–16
   residential buildings 77
   roof structures 218–19
   site drainage 220–2
   substrates 220
   wall structures 217–18
doors 209
dot lasers 242–3
draftsperson 88
drawings *see* working drawings

drill bits 299
drills (cordless and mains powered) 322–3
drop sheets 312
dry wall hammer 303
dual lifting 9–10
dumpy levels *see* optical automatic levels
dust hazards 328
dust masks 19
dust suppression 15
Dutch pin 307
duty of care 33

E
ear muffs 18
ear plugs 18
ears as tools 331–2
earth leakage device 318
education *see* training
efficiency 97
electrical equipment
   extension leads 318–19
   power boards 319
   safety signs 28–9
   testing and tagging 318
electrical supply *see* power supplies
elevated work platforms 331
elevation 200–1
emergency information signs 28, 149
emergency procedures 40, 42
   fire hazards 43
   firefighting equipment 42–3
   first aid 40–2
   responsible personnel 40
employer associations 80
ensuite 123–4
enterprise agreement 81
enterprise bargaining 81
environment 78, 106
Environmental Protection Authorities 106, 129
equipment *see* plant and equipment; tools and equipment
ergonomics 289
EWPs (elevated work platforms) 331
extension handles 312
extension ladders 329–30
extension leads 318–19
eye protection 17–18
eyes as tools 331–2

F
face protection 17–18
face shields 17–18
facial expressions 146–7

fall 261
fatigue 8
fibro 77, 328
fibro cutters 298
files 298
fire blankets 46
fire classes 44
fire extinguishers 42
   carbon dioxide (CO_2) 46
   dry chemical powder—AB(E) 45–6
   foam 45
   water 45
fire hazards 43
fire hose reels 47
fire orders 42
fire prevention 43
fire signs 28, 149–50
first aid 40–2
first aid kits 41
fitches 310
fixtures and fittings 208
flammable materials 43
float 304, 315
floats, sanding 304–5
floor plan 200, 207, 209
floor scraper 303
flooring systems 216–17
foam brushes 310
fold out 211–14
footings 214–16
foresight 261, 262
forklifts 12
formwork 14
four-fold rule 173, 240–1
four-peg test 273
fuel 43

G
Gantt chart 91
gauge trowels 306, 315
general construction 84, 86
general foreperson 88
generators, portable 317–18
Geocentric Datum of Australia 251
gloves 19
graphics 207
grievance procedure 79
grout rakes 315
grout saws 315
grout sponge 315
grouters 315
guards for tools and equipment 21

H

hacksaw 295
hammers 292–4, 306, 307
hand creams 19–20
hand signals 146–7, 153–4
hand stones 298–9
hand tools *see also* power tools
  bricklaying 305–9
  choosing quality 288–9
  for cleaning 300–1
  for cutting, scraping and spreading 295–9
  for hitting 292–4, 306, 307
  holding correctly 289–91, 311
  for holding or cramping 300
  for levering 294–5
  painting and decorating 309–13
  plastering and rendering 302–5
  for screwing 299
  for setting out 291–2
  tile laying 313–16
hand trolleys 11, 12
hand trucks 11
hands as tools 331–2
handsaw 295
hawk 302–3
hazard report forms 33, 34
hazard warning signs 148–9
hazards 2 *see also* chemical hazards
  acute hazards 30
  biological hazards 31
  chronic hazards 30
  control procedures 32
  (danger) hazard signs 27–8, 149
  effects on human body 32–3
  fire hazards 43
  groups 30, 31
  hazard warning signs 27
  non-toxic wastes 13
  physical hazards 16, 30–1
  rectification 14
  responsibility and duty of care 33
  safe work method statements (SWMS) 33–5
  safety hazards 30
  signs and labels *see* safety signs and labels
  stress hazards 31
  with vacuum cleaners 328
  workplace hazards 30
  workplace inspections 33
hearing protection 18
heart disease 8–9
heavy engineering 84, 86
HIA (Housing Industry Association) 76

hoardings 121–2
hoists 11
holding hand tools 289–91, 311
hole saw 299
horizontal 236
horizontal sections 209
housekeeping 13–16
  dust suppression 15
  functions 14
  personal cleaning procedures 15–16
  tools and equipment 14–15

I

incentives 92
incidents 35 *see also* accidents; injuries
induction cards 6–7
industrial action 79
industrial construction 85–6
industrial relations 78–9
injuries 1, 8–9, 32–3
  injuries register book 40
  injury management 39
instruction 119–20, 126
intermediate sight 261, 262
isometric projection 199–200

J

jacks 11–12
jemmy bar 294
jobber bit 299
jointer 307–8
jointing box 304, 305
jointing knife 302–3
just in time principle 122–3

K

kettles 312
keyed chuck 322
keyhole saw 296
knives 296–7

L

labels *see* safety signs and labels
ladders 329–31
land surveyor 88
larry hoe 307
laser levels 241–2
laser line generators 242–3, 248–51
  advantages and disadvantages 249
  classes of lasers 249–50
  PPE 250
  safe use of 249–51
  wet trades and 251

laser safety officers 250–1
legislation 2
levelling equipment *see also* laser line generators; optical automatic levels
  cleaning and storage 274
  four-peg test 273
  tool list 238–44
  two-peg test 269–72
levelling procedures
  checking a surface for level 246
  checking a surface for plumb 246–7
  choosing right tool 253–5
  determining relative heights of various points using datum 257
  drawing a level or horizontal line 244–6
  finding difference in height between two points 247, 254–5
  key terms and concepts 235–7, 251–3
  maintaining or plotting a continuous height 255
  maintaining or plotting height relative to a datum 256
  other procedures 251
  plumbing a line 246
  safety 237–8
  sighting terms 261
  transferring a height 248
  using stadia lines 252, 268–9
levels 238, 239, 241–2 *see also* optical automatic levels
levers 11
licensing 95
lifting 9, 327
lifting grips 12
lifting tackles 11–12
lights 144–5
line level 238, 239
line of collimation 252–3
line pin 306–7
linear metres 176
locking pliers 300
London-pattern trowel 305–6
lowering 9

M

mahlsticks 312
maintainability (key term) 97
mallets 292, 293
mandatory (must do) signs 27, 148
manhole cover lifter 11
manual handling 7–13, 31
  clothing 13
  dual lifting 9–10
  lifting 9, 327
  lowering 9
  mechanical aids 10–12

  pushing and pulling 10
  safe work practices 12–13
  shovelling 10, 11
manually levelled lasers 241–2
mash hammers 292, 293
masking tools 312–13
Maslow's hierarchy of needs 71
masonary bit 299
material requirements 122–4
material safety data sheets (MSDS) 22–4
material unit quantities 174
materials, recyclable 104
maulstick 312
MBA (Master Builders Association) 76
measurement 175
  of area 176–9
  calculation of percentages 185–6
  calculation of solid shapes 183–5
  calculation of surface area 184–5
  linear measurement 176
  measuring tools 173–4
  ratios 186
  units of measure 172, 174–5
  of volume 179–80
mechanical aids 10–12
meetings *see* workplace meetings
message 140–2
metres (m) 172
millimetres (mm) 172
minutes 100, 156, 157
mortar hoe 307
mortar pin 306–7
mosaic nipper 314
MSDS (material safety data sheets) 22–4
multi-grips 300
muscle injuries 8

N

nail punch 292, 293
National Occupational Health and Safety Commission 2
National Training Information Service (NTIS) 82
National Training Reform Agenda 80, 81–2
near misses 38
nippers 313–14
non-toxic waste 13
non-toxic waste disposal 130
notched trowels 314, 315

O

occupational health and safety (OH&S) 1
  Codes of Practice 2, 7
  committees 83

occupational health and safety (OH&S) *continued*
   legislation 2, 3
   meetings 155
   offences and penalties 4–5
   reasons for introduction 2–4
   regulating authorities 5
   regulations 2, 3
   rights and responsibilities of employers and employees 4
   safe work practices 7–13
   site induction 5–7, 83
odours 145
off-site 84, 86
OH&S *see* occupational health and safety
optical automatic levels 243
   booking a basic site survey 263–7
   checking the booking 267–8, 269
   parallax error 258–60
   rise and fall method 257, 261
   setting up 257–8
   sighting terms 261
   taking a reading 260
orthographic projection 200
oval brushes 310

P
pad footings 215–16
paint buckets 312
paint kettles 312
paint pads 310
paint pots 312
paint rollers and trays 311–12
paint scrapers 296
painting hand tools 309–13
pallet trucks 12
paper hanger brushes 312
papers 298
parallax error 173, 258–60
   removing 260
parallelograms 177
parrot nipper 314
pendulum automatic level lasers 242
percentage 185–6
permanent marks (PM) 251
perpendicular 235, 236, 237
personal cleaning procedures 15–16
personal development 99
personal protective equipment *see* PPE
personal responsibility 97
perspective view 199
Phillips screwdriver 299
physical hazards 16
pictograms 144
pictorial drawings 199

pictures *see* diagrams/sketches
pinch bar 294
plain trowels 314
plan 142, 198
   site plan 202
plan pegs or marks 251–2
plan reading *see* working drawings
plane 253–4
plank platforms 328–33
planning
   planning methods 89
   purpose of 118–21
plant and equipment
   heavy 323–31
   plant 35
   power supplies 316–19
   powered hand tools 316, 320–3
plasterer's square 305
plastering hand tools 302–5
pliers 300
plugging chisel 306, 307
plumb 236
plumb bob 240
   checking a surface for plumb 246–7
   plumbing a line 246
pointing trowels 306
polling 10
polygons 177, 178, 179
portable compressors 323–4
portable generators 317–18
power boards 319
power supplies 316–19
   portable generators 317–18
   safety precautions 318–19
   temporary power poles 317, 318
power tools
   hand tools 316, 320–3
   mains-powered tools 319–23
PPE (personal protective equipment) 5, 16
   airways/lungs 18–19
   assessing requirements 20–1
   body protection 20
   cleaning 21
   clothing 13, 20
   foot protection 20
   hand protection 19–20
   head protection 16–18
   hearing protection 18
   for laser line generators 250
   with power tools 320
principal certifying authority (PCA) 89
prisms 183, 185
problem-solving 126–8

procedures 234 *see also* levelling procedures
professional development, continuing 95–6
profile 308
profile clamp 308
prohibition (don't do) signs 27, 148
project manager 88
punch 292, 293
pushing 10
putty knife 296, 303
PVC nipper 314
pyramids 184, 185

Q
quadrilaterals 177, 178
quality 97
   quality assurance officer 88
   quality assurance standards 94, 97
   quality control principles 97
   quality tools 288–9
quantities 175
   cost calculations and 180–3
   material unit quantities 174
quantity surveyor 88
quarry nipper 314

R
raker 308
rat-tailed brushes 310
ratios 186
RCD unit 318
readings 253, 257, 260
record 100, 156, 157
recycling 102, 103–4
reduced level (RL) 252
regulating authorities
   for electrical equipment 318
   for environmental protection 106
   for OH&S 5
   for tradespeople 96
reliability (key term) 97
rendering hand tools 302–5
reportable accidents 35
reportable incidents 38
reporting on completed work 128–9
residential buildings *see* domestic construction
residual current device unit 318
resource efficiency 102–5, 106
respirators 19
respiratory cartridges 18–19
respiratory disease 8–9
reticle 259
retractable metal tape 173–4, 241

rise 261
rise and fall method 257, 261
risk assessment 29, 35
RL (reduced level) 252
rollers 11
rollers and trays 311–12
roof structure 218–19
rotary laser levels 241–2, 273
round brushes 310
rubber boots 20
rulers 240–1 *see also* survey staff

S
safe work method statements (SWMS) 33–5, 36
safety *see also* safety signs and labels
   with angle grinders 320–1
   with barrows and trolleys 327
   with brick saws 327–8
   with circular saws 321–2
   with drills (cordless and mains powered) 322–3
   with EWPs 331
   with laser line generators 249–51
   laser safety officers 250–1
   in levelling procedures 237–8
   with portable compressors 324
   power precautions 318–19
   with sanders 323
   with screw guns 323
   site preparation 121–2
   with spray guns 325
   with trestles, planks and ladders 328–33
   with vacuum cleaners 328
   in work practices 7–13
safety boots 20
safety helmets 16–17
safety induction training 5–7, 83
safety joggers 20
safety signs and labels 361–8
   (danger) hazard signs 27–8, 149
   dangerous goods labels 24, 25, 151–2
   for electrical equipment 28–9, 318
   emergency information signs 28, 149
   erection of 150
   fire signs 28, 149–50
   hazard warning signs 27, 148–9
   location of 150
   mandatory (must do) signs 27, 148
   placement 29
   prohibition (don't do) signs 27, 148
   restriction signs 27
   safety signs and tags 25–9
sanders 323

sanding blocks 304–5
sanding floats 305
sandpaper 298
sash cutters 309–10
sawdust 15
saws 295–6
saws, brick 327–8
scaffolding 328–33
scale drawing 202–3
scale rule 173
scissor lift 331
scissors 297–8
scrapers 296, 312
screw guns 323
screwdrivers 299
scriber 315
section drawings 201, 209
sections in-ground 206
self-levelling lasers 242
sequence 93–4, 120–1
sequencing 92–4
services 84, 86
setting out 291–2
shears 297–8
sheet lifter 12
shifter 294
shovelling 10, 11
shovels 302
SI units 175
side cutters 300
signage 95 *see also* safety signs and labels
silica-based dusts 328
site drainage
   problems of poor drainage 220–1
   surface water removal 221–2
site induction 5–7, 83
site manager 88
site plan *see* plan
site security 94
site skips 301
sketches *see* diagrams/sketches
Skills Express 85
slab-on-ground 215, 216
sledge-hammer 292, 293
small tool 304
smells 145
sockets 295
solid shape calculation 183–5
sounds 144
spacers 315
spade bit 299
spanner 294

specification 198, 199, 203–4, 205
specification reading *see* working drawings
spirit level 238
   checking a surface for plumb 246
   checking for error 244–5
   plumbing a line 246
   transferring a height 247
spoken language 143, 145–6, 152–3
spray guns 324–5
square 235, 237
squares
   hand tools 291, 305
   laser 242–3
squeegee 315
stadia lines 252, 268–9
Standards Australia 97
station 252
steel square 291
stencil brushes 310
stepladders 330
Stonehenge 71
storage
   of chemicals 25
   of levelling equipment 274
   of materials 123
straight 235
straight and true 235
straight edge 238, 239, 247
straight nipper 313–14
stress hazards 31
string lines 239
strip footings 215
substrates 220
suction grip 12, 315–16
sun protection 20
sun shades 17
supplier (key term) 97
surface area calculation 184–5
surface water removal *see* site drainage
survey 261
survey staff 243, 244, 260
suspended slab 216–17
suspended timber floor 216
sustainability 102–5
sustainable housing 106–8
SWMS (safe work method statements) 33–5, 36
symbols 148, 204–9

T

tags *see* safety signs and labels
tape measures 240–1
   retractable metal tape 173–4, 241

taping box, automatic 304
taping knife 303
team meetings 155
teamwork 98–9
technological development 77–8
telephone 146
temporary benchmarks (TBM) 251–2
tile cutters 313
tile laying hand tools 313–16
tilting drum mixers 325–6
timber frame 217
time management 89–92
   anticipation of delays 89–90
   causes of delays 90
   costing 90, 91–2
   forecasting 89
   planning methods 89
   time charts 89, 90
   use of bonuses and incentives 92
tin snips 297–8
tools and equipment 14–15, 21 *see also* hand tools; power tools
   checking 125–6
   checklist for 94
   determining requirements 124–5
   hands, eyes and ears as tools 331–2
   for levelling 238–44, 253–5
touch sensations 145
trade categories 77
trade unions 76, 79–80
   meetings 155
   members rights and obligations 80
tradesperson 289
trailers 329
training
   career path and 87
   competency-based 85
   traineeships 80, 82
   training levels 84–5
   training packages 82
trammel heads 291–2
trapezoid 178
traversing 247
trestles 328–33
tribrach 257
trim nipper 314
tripod 240, 241
trolleys, brick 326–7
trowels
   for bricklaying 305–6
   for plastering and rendering 302–3
   for tile laying 314–15

true 235, 236
trussed roofing 219
two-peg test 269–72
two-way radio 146

U

underground warning tapes 151
units of measure 172, 174–5
universal key 298
utility knife 296

V

vacuum cleaners 15, 328
volumes 175, 179–80

W

wall brushes 309
wall structures 217–18
Warrington hammer 292, 293
waste 96, 102
   Environmental Protection Authorities and 129
   resource recovery program 105
   waste management 129
   waste minimisation hierarchy 103–4
water level 243
   transferring a height 247
waterproofing sequence 120–1
wedges 315
wet saws 321–2, 327–8
wetting down 15
wheelbarrows 11, 326
wheelsets 11
win/win solution 158
wind 236, 237
wind-up tape measure 241
windows 207
wire brush 301
WorkCover 39
workers' compensation 39–40
working drawings 172 *see also* diagrams/sketches
   abbreviations 206
   calculating scales 202–3
   details 201
   dimensions 203
   elevation 200–1
   floor plan 200, 207, 209
   function of 200
   key users 198–9
   pictorial representation 199
   plan and document reading 209–11
   scale drawings 202
   sections 201

working drawings *continued*
  sheet sizes 203, 204
  site or block plan 202
  special details 202
  specifications 203–4
  symbols 204–9
  title block 203, 204
workplace committees 33, 82–3, 102, 155
  consultative committees 82
  OH&S committees 83
  works committees 83
workplace competencies 99
workplace maintenance *see* housekeeping
workplace meetings 100–2
  formal meetings 156
  informal meetings 156–7
  participation in 157–8
  types of 155
workplace signage *see* safety signs and labels
workplace structure 86–9
Worksafe Australia 2
written language 143, 146, 153

Important safety and hazard signs

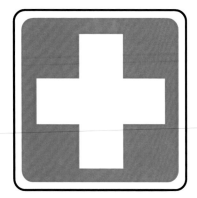  

IMPORTANT SAFETY AND HAZARD SIGNS

FIRST AID

EMERGENCY EXIT

CAUTION LOW HEADROOM

CAUTION
SLIPPERY FLOOR

 DANGER   DANG

IMPORTANT SAFETY AND HAZARD SIGNS

IMPORTANT SAFETY AND HAZARD SIGNS

WATER

FOAM

DRY CHEMICAL

CO_2

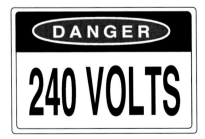